Plant Microbiome Paradigm

Ajit Varma • Swati Tripathi • Ram Prasad
Editors

Plant Microbiome Paradigm

Springer

Editors
Ajit Varma
Amity Institute of Microbial Technology
Amity University
Noida, Uttar Pradesh, India

Swati Tripathi
Amity Institute of Microbial Technology
Amity University
Noida, Uttar Pradesh, India

Ram Prasad
Department of Botany
Mahatma Gandhi Central University
Motihari, Bihar, India

ISBN 978-3-030-50394-9 ISBN 978-3-030-50395-6 (eBook)
https://doi.org/10.1007/978-3-030-50395-6

This Springer imprint is published by the registered company Springer Nature Switzerland AG.
The registered company address is: Gewerbestrasse 11, 6330 Cham, Switzerland

Contents

Chapter 1
Interaction of Epiphyllic Bacteria with Plant Cuticles

Filip Fuchs, Charlotte Petruschke, and Lukas Schreiber

Abstract The phyllosphere is one of the largest ecological niches on our planet. It is formed by the plant cuticle, which is a highly impermeable, hydrophobic biopolymer covering all primary aboveground plant organs protecting them against desiccation. Although living conditions in the phyllosphere are considered harsh, a great variety of microorganisms can live within this habitat. Commensals as well as pathogenic can be found on the plant surface competing for niches and rare nutrient sources. It was found that the physicochemical properties of the cuticle are modified actively by epiphyllic microorganisms. This modification by microorganisms can lead to enhanced wetting of the leaf surface. This is promoted by the secretion of biosurfactants by epiphyllic bacteria promoting and as a consequence leaching of solutes from the apoplast to the plant surface can be enhanced.

1.1 Introduction

With the rise of the land plants approximately 500 million years ago, a new complex and versatile ecological niche appeared, the phyllosphere, consisting of all green aboveground parts of plants. It is estimated that the phyllosphere is approximately as big as the surface of the whole planet which displays it as the greatest biological surface on earth (Kenrick and Crane 1997; Lindow and Brandl 2003). Most of this area is colonized by a different mixture of microbial species, predominantly bacteria, yeast, and filamentous fungi and to some extend protozoa and even algae and mosses (Morris et al. 2002; Singh et al. 2019). The degree of colonization strongly depends on various factors starting with the plant species, the habitat of the host plant (e.g., tropical rain forest, coniferous woodland, or grassland), and the age of the leaf (Kinkel 1997). Additionally to these rather constant factors, the phyllosphere and its inhabitants are affected by rapidly changing climatic conditions such as temperature, humidity, and irradiation. Over the course of a day or with fast changes of

F. Fuchs · C. Petruschke · L. Schreiber (✉)
Institute of Cellular and Molecular Botany (IZMB), University of Bonn, Bonn, Germany
e-mail: s6fifuch@uni-bonn.de; c.petruschke@uni-bonn.de; lukas.schreiber@uni-bonn.de

A. Varma et al. (eds.), *Plant Microbiome Paradigm*,
https://doi.org/10.1007/978-3-030-50395-6_1

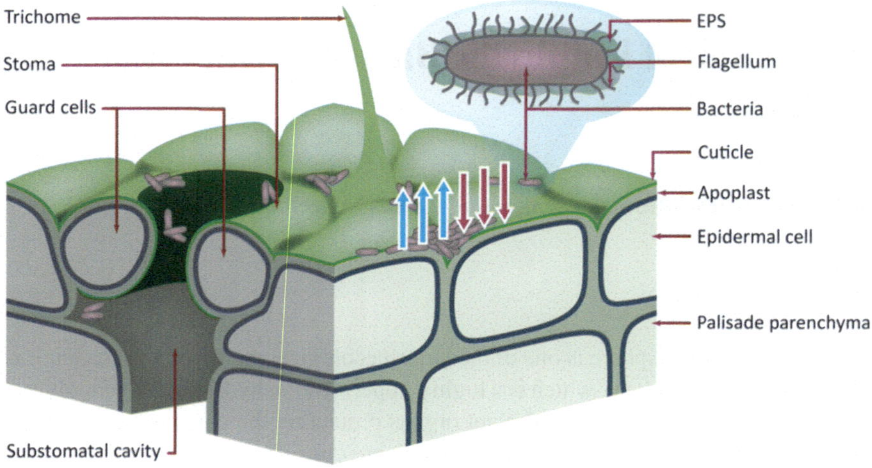

Fig. 1.1 Schematic view on leaf cross section and surface: bacterial conglomerates preferentially settling above cell-cell junctions and at the base of trichomes. Stomata may serve as possible infiltration sites for pathogenic microorganisms. Arrows (blue and red) indicate the potential exchange of molecules (water, irons, sugars, hormones) between bacteria and leaves taking place across the plant cuticle. In the upper right part of the scheme, a magnified bacterial cell produces extra polymeric substances which together with motility are crucial for survival in the phyllosphere

weather conditions, those fluctuations are rapidly followed by changes in the density and number of microorganisms colonizing the leaf surface (Leben 1965). Bacteria are dominating the phyllosphere by far in both number and diversity with up to 10^6–10^7 bacteria per cm^2. It is assumed that 96% of those bacteria live as commensals with no effect on their host plants' health or fitness, whereas 2% are believed to be pathogens, and another 2% can be referred to as plant growth-promoting bacteria (PGPB) (Lindow and Leveau 2002). These PGPBs could contribute to the overall fitness of their host plants by inducing systemic resistance, actively producing plant growth hormones such as auxin, or suppressing pathogens by producing antimicrobial compounds (Brandl et al. 1996; Vorholt 2012; Prasad et al. 2015). The most abundant forms of colonization in the phyllosphere are biofilms or aggregates on hydrophobic leaf surfaces (Lindow and Leveau 2002). The vast majority of bacterial cells on leaves are clustered and embedded in extra polymeric substances (EPS), covering as a thin layer the outer surface of the plant cuticle, preferentially above cell-cell junctions or at the base of trichomes (Fig. 1.1) (Monier and Lindow 2004). The cuticular membrane (CM) is a lipophilic, extracellular biopolymer covering outer epidermal cell walls of leaves and fruits which are exposed to the atmosphere (Schönherr 1982). The cuticle was the evolutionary answer to the biggest problem plants had to face when they conquered land habitats: desiccation. On one side the greatly enlarged surface area of plants results in a more efficient absorption of photosynthetic active radiation (PAR) and promotes a rapid gas exchange of carbon dioxide and oxygen; on the other side, due to an ubiquitous gradient in water potential between atmosphere (low) and leaf (high), a bigger surface is coercively connected with an increase in water loss (Schreiber and Schönherr 2009). The CM serves as an efficient transport barrier for the passive diffusion of water and ultimately protects the plant from rapid desiccation (Schönherr 1982).

In the past the main focus of research in plant microbe interaction was dealing with the hidden half of plants, the so-called rhizosphere where uptake and allocation of water as well as minerals by the plant root system take place (Varma et al. 2019, 2020). A tremendous amount of plant/microbe interactions is taking place in the rhizosphere (Hiltner 1904; Whipps 2001; Prasad et al. 2020). In recent years microbiology of the phyllosphere gained increasing significance, and it is no longer neglected. To describe and understand the underlying mechanisms in water and solute transport within the phyllosphere and the entanglement of plant and microbe physiology, combining classical plant ecophysiology with microbiological approaches represent the main research questions in this field.

1.2 The Hydrophobic Plant Cuticle as Interface Between Epiphyllic Bacteria and Plants

The bulk of terrestrial biomass is produced by plants via photosynthesis (Groombridge and Jenkins 2002). To improve efficiency a common trend in evolution is the enlargement of surface area, which ultimately leads to the rise of spatial two dimensional leafs structures as we know them today. The success of land plants goes along with their ability to protect themselves from desiccation. This is ensured by the plant cuticle which covers all primary aboveground plant organs (Schönherr 1982). The cuticle is an extracellular lipid polymer of hydroxy fatty acids, which are esterified and in addition often linked by ether bonds and direct carbon–carbon bonds between the monomers (Pollard et al. 2008; Villena et al. 1999). Additionally polysaccharides like cellulose and pectins could be detected within the CM, mainly at the inner side of the CM facing the primary epidermal cell wall. There they are emanating from the epidermal cell wall into the cutin polymer and thus contribute to the structure of the CM or its attachment to the cell wall (Guzman et al. 2014; Segado et al. 2016). It has been suggested that these polar polymers form aqueous paths of transport within the lipophilic cutin polymer thus promoting the diffusion of polar and charged solutes and ions across the CM (Schreiber and Schönherr 2009). Together with intracuticular and epicuticular wax, the cutin polymer forms a hydrophobic highly impermeable barrier (Tukey 1970; Schönherr 1982). Cuticular waxes are diverse in their chemical composition (Buschhaus and Jetter 2011) consisting mainly of two fractions namely monomeric linear long-chain aliphatic compounds and pentacyclic triterpenoids. Those waxes are solid and partially crystalline at room temperature (Reynhardt and Riederer 1994). Due to this highly ordered structure of cuticular waxes on the molecular level, they seal the plant cuticle and make it not only the main barrier for passive diffusion of water into the atmosphere but also hinder dissolved organic and inorganic solutes to pass (Schreiber and Schönherr 2009). Whereas waxes are responsible for establishing the diffusion barrier, the cutin polymer serves as a stable matrix for wax deposition (Kolattukudy 1984; Nawrath

2006). Epicuticular waxes on the surface can form different kinds of three-dimensional structures such as scales, platelets, and spikes depending on their diverse chemical compositions (Barthlott et al. 1998). These microscopic structures in the nanometer range significantly increase the leaf surface roughness and consequently impede the attachment of bacteria, fungal spores, and other microscopic invaders. The increased roughness also promotes the self-cleaning "lotus effect," when water droplets cannot attach to a leaf surface and roll off the leaf taking dust particles with them (Barthlott and Neinhuis 1997). However, with increasing leaf age, the surface roughness decreases due to erosion of the epicuticular wax crystals, and roughness becomes less decisive for the attachment of epiphyllic microorganisms (Neinhuis and Barthlot 1998). The prevalent conditions on the leaf surface are aridity, lack of nutrients, and rapidly changing temperatures. Due to these circumstances, the phyllosphere is considered an uninviting and harsh habitat for microorganisms. To cope with these various stresses, bacteria evolved different strategies such as growing as a biofilm or producing biosurfactants. Biofilms are the predominant form of bacterial living on the planet (Flemming and Wingender 2010). A biofilm is a conglomerate of bacterial cells surrounded by EPS which is attached to any inert or living surface. There are biofilms at the interface of a solid phase and the atmosphere or at the interface of a solid and a liquid phase or even between two liquid phases (Jenkinson and Lappin-Scott 2001). In most cases biofilms harbor multiple species, in different niches within the biofilm (Xavier and Foster 2007). There are physiological dependencies between different bacteria but also competition among them making a biofilm a very complex and diverse habitat for microorganisms. The bulk mass of a biofilm consists of extracellular matrix (ECM), actively segregated by their inhabitants. The majority of the ECM consists of EPS. Polysaccharides as well as extracellular DNA (eDNA) form the largest proportion of EPS, followed by proteins and various lipid compounds, mostly phospholipids or lipopolysaccharides (Branda et al. 2005). This highly hydrated periphery forms a slimy matrix in which the bacterial cells are embedded. The biofilm lifestyle on a leaf surface has many advantages for their inhabitants. The hydrated ECM prevents the bacteria from desiccation and can serve as a sink for toxic metabolites. In case of starvation, some components can be used as carbon or energy source (Sutherland 2001). Pigments can accumulate in the ECM protecting the bacteria from strong irradiation, and additionally the biofilm displays a certain protection against grazing protozoa (Flemming and Wingender 2010). Biofilm formation often starts with single motile bacteria propagated by wind and rain splash (Lindow 1996).The initial interaction between a bacterial cell and a surface is defined by the cell surface hydrophobicity, nonspecific van der Waals, and electrostatic forces. This loose contact is reinforced by surface/host-specific adhesins, located on the cell surface or on bacterial appendages such as pili and fimbriae (Romantschuk 1992; Vorholt 2012). This results in the irreversible attachment of the bacterial cell to the surface. Once a bacterial cell has successfully attached to the leaf surface, it starts multiplying and forms small aggregates or microcolonies embedded in EPS. Those aggregates grow and eventually fuse with other cell aggregates to form a mature biofilm (Ramey et al. 2004). The final, yet important stage of biofilm development is the dispersal of pioneer cells. Single "swarming" cells detach from the biofilm and actively move to new uncolonized regions of the surface to establish new microcolonies. For this development, communication within the aggregates and with the surrounding communities is essential. Classical biofilms are highly ordered. EPS production and maintenance of the three-dimensional biofilm structure and population size are strongly connected with quorum sensing and the ability for cell–cell communication (Parsek and Greenberg 2005).

1.3 Changes in Wettability and Permeability of Hydrophobic Leaf Surfaces Due to Colonization with Epiphyllic Bacteria

Due to the hydrophobic epicuticular wax layer and their three-dimensional structures, leaf surface wetting is strongly impaired on young barely colonized leaves. It was shown several times that leaf wettability increases with the age of the leaf (Cape and Percy 1993; Neinhuis and Barthlott 1998). This is normally ascribed to changes in the physicochemical properties of the cuticle. Leaves are constantly exposed to wind, water, irradiation, and microscopic dust particles leading to the erosion of epicuticular waxes (van Gardingen et al. 1991) and thus diminishing the hydrophobic character of the leaf surface. Another, additional, long neglected factor might be the colonization with epiphyllic bacteria. Studies have shown a clear negative connection between the level of colonization (increases) and the contact angle of water (decreases) on hydrophobic leaf surfaces of *Hedera helix* and *Juglans regia* and silanized glass surfaces (Knoll and Schreiber 1998, 2000). Using SEM approaches this effect may be explained by the coverage/coating of the epicuticular waxy microstructure with microbial biofilm. Further analysis by gas chromatography could not correlate changes in wetting property with altered qualitative and quantitative wax composition; thus maybe other factors facilitate the enhanced wetting properties of colonized hydrophobic surfaces. Biosurfactants could be one of those factors. Biosurfactants are surface active agents produced by several species of microorganism (Ron and Rosenberg 2001). Biosurfactants are very versatile in terms of their chemical composition and biological purpose. All surfactants are by definition amphiphilic molecules often composed of a hydrophobic fatty acid tail and a hydrophilic head group. In the case of bacterial surfactants, this includes a wide range of molecules such as glycolipids, lipopeptides, polysaccharides, proteins, and lipoproteins (Muthusamy et al. 2008; Raaijmakers et al. 2010). Initially it was speculated that the sole function of biosurfactants is the promotion of the emulsification of liquid (hydrophobic) hydrocarbons, to elevate the available surface for degradation processes by some bacteria. Recent studies changed that constricted view on the function of biosurfactants (Raaijmakers et al. 2010). Some of those recently discovered functions are the promotion of bacterial swarming on moist surfaces (Berti et al. 2007); biofilm formation, maintenance, and dispersal (Ron and Rosenberg 2001); zoosporicidal or antimicrobial activities (Raaijmakers et al. 2010); and even being a virulence factor themselves (Hildebrand et al. 1998; Burch et al.

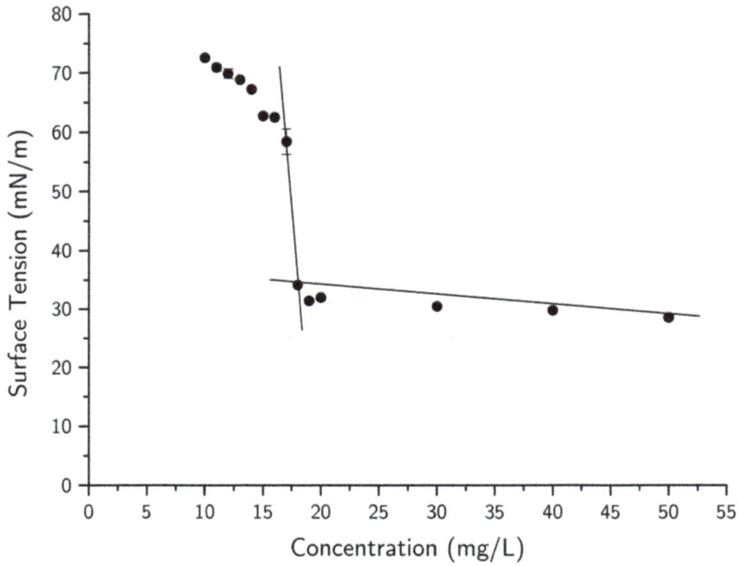

Fig. 1.2 Determination of the critical micelle concentration (CMC) of the biosurfactant produced by *P. fluorescens* wild type. The biosurfactant was extracted with ethyl acetate from the supernatant of mature biofilm emulsions and resuspended in H_2O. The surface tension was measured with the Drop Shape Analyser DSA_{25} (Krüss GmbH). The determined CMC was approximately 18 µg/ml

2014). Biosurfactants also reduce the surface tension of water and thus decrease the contact angle on hydrophobic surfaces, promoting wettability (Rosenberg 1985). The epiphyllic bacterium *Pseudomonas fluorescens* SBW25 produces an amphiphilic cyclic lipopeptide acting as a biosurfactant. The critical micelle concentration (CMC) of the purified biosurfactant was 18 µg/ml (Fig. 1.2). Solutions above the CMC showed a reduced surface tension between 28 and 30 mN/m. Transposon mutants of different epiphyllic *Pseudomonas* species, impaired in their production of biosurfactants, do not promote a lower surface tension (Fig. 1.3). It was shown in studies that biosurfactant production is connected with several kinds of motility in *phyllospheric* species (Lindow and Brandl 2003). Swarming experiments revealed biosurfactant deficient strains are unable to spread over the surface of semisolid agar plates (0.4% agar instead of 1%) (Fig. 1.4). The production of biosurfactants is connected to the epiphillic lifestyle. The supernatant of liquid cultures of *P. fluorescens* is biosurfactant free and does not exhibit reduced surface tension. Contrary, *P. fluorescens* grown as a biofilm on agar plates produces notable amounts of biosurfactant. In their natural habitat, this enhances water availability on the leaf surface and promotes the dispersal of bacterial cells. It was shown that technical surfactants can elevate the permeability of isolated cuticles for water (Riederer and Schönherr 1990) and dissolved compounds by 10- to 100-fold (Schreiber and Schönherr 2009). In comparison, water permeability of isolated cuticles before and after inoculation with *P. fluorescens* for 10–12 days revealed an increase in transpiration by just 40–60%. If this relatively small effect is caused by the biosurfactant or due to the degradation of epicuticular waxes is unknown. Additionally most

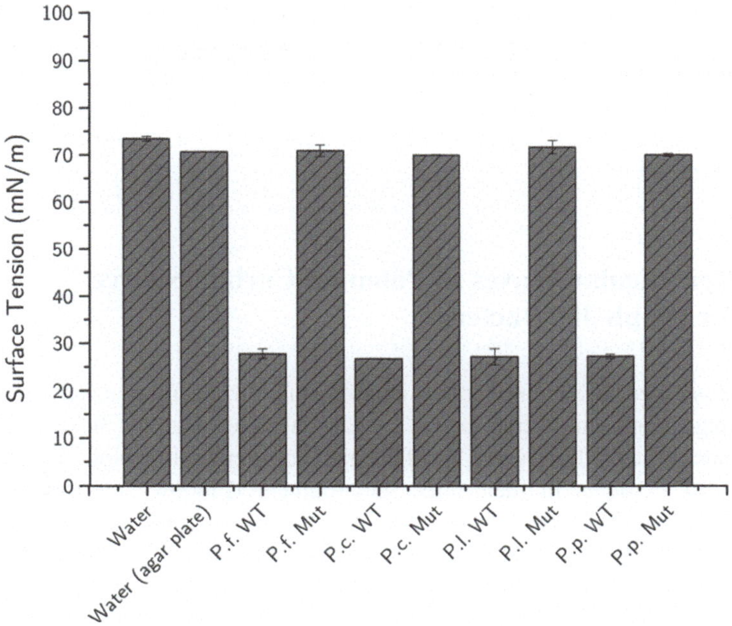

Fig. 1.3 Surface tension of supernatants of different *Pseudomonas* (*fluorescens*, *cedrina*, *lurida*, and *poae*) wild types and their corresponding mutants, deficient in biosurfactant production: Mature biofilms were resuspended in H_2O; the OD_{600} was adjusted to 1, and after centrifugation the surface tensions of the supernatant were measured with the Drop Shape Analyser DSA_{25} (Krüss GmbH)

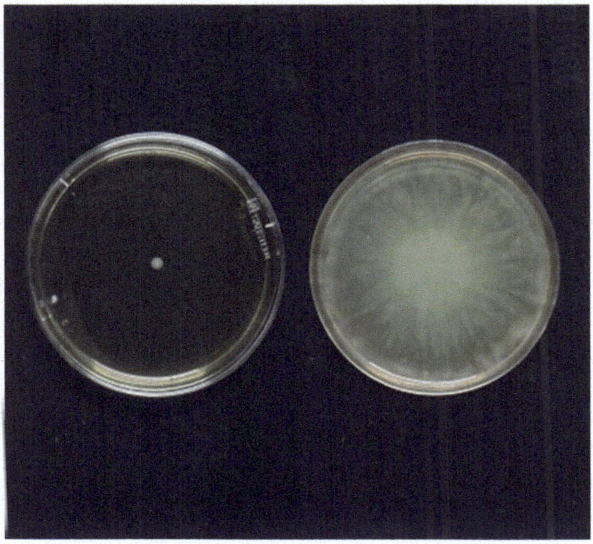

Fig. 1.4 Swarming behavior of *P. fluorescens* on 0.4% agar plates: (left) biosurfactant mutant of *P. fluorescens* after 24 h. (Right) wild type of *P. fluorescens* after 24 h

biosurfactants have a hygroscopic effect, binding water molecules from the atmosphere establishing a thin layer of water on the leaf surface. This should promote the leaching effect, the diffusion of organic molecules from the apoplast through the CM into the water film which serves as sink (Burch et al. 2014). This consequently increases the amount and the availability of nutrients in the phyllosphere and promotes microbial life.

1.4 Epicuticular Waxes as Potential Carbon Source for Epiphyllic Bacteria

The leaf surface is an unfavorable environment for microorganism to survive. Oscillating abiotic and biotic factors demand a high metabolic flexibility of the inhabitants of the phyllosphere. One of the major challenges for epiphyllic bacteria is the lack of nutrients in their direct environment (Lindow and Brandl 2003). Methylotrophs which are a widespread heterogeneous group among epiphyllic microorganisms (Iguchi et al. 2015) use methane or methanol as their sole carbon and energy source. Since methanol is considered to be abundant on leaves (MacDonald and Fall 1993), methylotrophs often account for the primal microorganisms colonizing a leaf. But most heterotrophic microorganisms rely on more complex organic compounds as carbon and energy source. A common strategy is the consumption of leached leaf sugars (Mercier and Lindow 2000). Bacterial microcolonies and biofilms are commonly found along leaf veins, at the bases of trichomes or at depressions at the junctions of epidermal cells (Lindow and Brandl 2003; Monier and Lindow 2004). These specific areas of the leaf surface are known to provide increased amounts of nutrients (Remus-Emsermann et al. 2012). As previously mentioned, many leaf colonizers are able to produce biosurfactants to increase the wettability of the hydrophobic cuticle (Bunster et al. 1989), which in turn increases their mobility on the leaf surface to favorable growth sides (Mercier and Lindow 2000). The increased wettability is also linked to an increased nutrient leaching. Due to the prolonged presence of a water film on the hydrophobic leaf surface, solutes diffuse across the cuticle in a higher amount (Remus-Emsermann et al. 2011). Next to the leaching effect (Tukey 1970), which mainly promotes the availability of leaf sugars, it is suggested that components of the cuticle itself could serve as carbon source for epiphytic bacteria. In this part the focus is on the possible degradation of wax components of the cuticle by epiphyllic bacteria. Within the cutin polymer, intracuticular waxes are deposited building the main diffusion barrier (Zeisler and Schreiber 2016). Cuticular waxes are diverse in their chemical composition (Buschhaus and Jetter 2011) consisting mainly of two fractions, namely, monomeric linear long-chain aliphatic compounds such as alcohols, aldehydes, and alkanes with chain lengths varying between C_{20} and C_{30} and pentacyclic triterpenoids, usually C_{30} molecules, deriving from the terpenoid pathway. The aliphatic compounds are made of C_{16} and C_{18} fatty acids, which are elongated and further modified with different functional groups (Kunst and Samuels 2003). Esters formed between long-chain fatty acids and alcohols are characterized by exceptionally long-chain lengths between C_{40} and C_{60}. Aside from these two main fractions, other substance classes such as ketones, secondary alcohols, and dioles have been reported and characterized to a minor extend as constitutes of the CM. Depending on species and its taxonomic group, the proportion of these compounds could vary strongly (Jetter et al. 2006). It is known that among others, bacteria of the genus *Rhodococcus* and *Pseudomonas* are capable of catabolizing a wide range of hydrocarbons (Wentzel et al. 2007). Usually strains capable of hydrocarbon degradation are discovered in oil-contaminated soil or aquatic systems contaminated with insoluble hydrocarbons (Nilanjana and Preethy 2011). It is further speculated that some phyllospheric strain may use aliphatic compounds of cuticular waxes as carbon and/or energy source. Degradation experiments were conducted with isolated *Fagus sylvatica* cuticular wax incubated for 28 days with the epiphyllic bacterium *Rhodococcus fascians*, in M9 mineral media without another carbon or energy source. Subsequently the chemical analysis of the wax compounds with gas chromatography and mass spectroscopy revealed a noticeable decrease of the primary alcohols and a pronounced decrease of the total wax amount (Fig. 1.5). Besides bacteria there is also convincing microscopical evidence that epiphyllic fungi can alter the three-dimensional epicuticular wax structure (Zeisler-Diehl et al. 2018) although it is still unclear how this effect is obtained. Nevertheless epiphyllic bacteria have an impact on the wax structure and potentially composition and hence may alter the transport barrier leading to an increased leaching of nutrients.

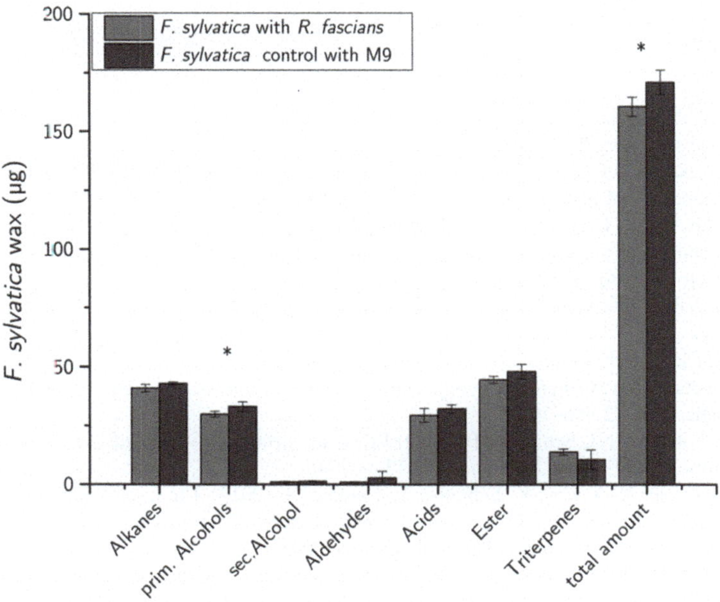

Fig. 1.5 Wax degradation of *Fagus sylvatica* wax inoculated with *Rhodococcus fascians* bacteria. *Fagus sylvatica* wax was inoculated with bacteria for 28 days. Significances tested with student *t*-test. $p < 0.05$ = significant difference between *F. sylvatica* wax inoculated with bacteria and control wax inoculated with m9 media

This metabolic activity together with the production of biosurfactants could contribute to survival of epiphytic bacteria in the nutrient low environment on the leaf surface.

1.5 Conclusion

Although living conditions for microorganisms on hydrophobic leaf surfaces are suboptimal due to oscillating biotic and abiotic factors, the phyllosphere displays the greatest colonized biological surface on earth. Adaptations to this ecological niche, like growing in mixed multispecies biofilms or altering the physicochemical properties of the plant cuticle, ensure the success of epiphyllic microorganisms. Specific interactions between epiphytes and the plant cuticle can promote leaf surface wetting and thus elevate the level of accessible nutrients. Enhanced wettability also promotes the dispersal of motile bacteria on the leaf surface and thus the colonization of new habitats. The biofilm lifestyle facilitates the possibility of multispecies consortia in which complex interactions like metabolic entanglements ensure advantages for the inhabitants. Plant growth-promoting bacteria can actively suppress pathogen species or promote plant health by induced systemic resistance. This could be of great interest for agricultural applications like reduced pesticide usage. Further plant physiological analytical and molecular biological experiments should improve our knowledge of the microbial ecology of the phyllosphere.

References

Barthlott W, Neinhuis C (1997) The purity of sacred lotus or escape from contamination in biological surfaces. Planta 202:1–8

Barthlott W, Neinhuis C, Cutler D, Ditsch F, Meusel I, Theisen I, Wilhelmi H (1998) Classification and terminology of plant epicuticular waxes. Bot J Linn Soc 126:237–260

Berti AD, Greve NJ, Christensen QH, Thomas MG (2007) Identification of a biosynthetic gene cluster and the six associated lipopeptides involved in swarming motility of *Pseudomonas syringae* pv. tomato DC3000. J Bacteriol 189:6312–6323

Branda SS, Vik S, Friedman L, Kolter R (2005) Biofilms: the matrix revisited. Trends Microbiol 13:20–26

Brandl M, Clark EM, Lindow SE (1996) Characterization of the indole-3 acetic acid (IAA) biosynthetic pathway in an epiphytic strain of *Erwinia herbicola* and IAA production in vitro. Can J Microbiol 42:586–592

Bunster L, Fokkema NJ, Schippers B (1989) Effect of surface-active *Pseudomonas spp.* on leaf wettability. Appl Environ Microbiol 855:1340–1345

Burch AY, Zeisler V, Yokota K, Schreiber L, Lindow SE (2014) The hygroscopic biosurfactant syringafactin produced by *Pseudomonas syringae* enhances fitness on leaf surfaces during fluctuating humidity. Environ Microbiol 16:2086–2098

Buschhaus C, Jetter R (2011) Composition differences between epicuticular and intracuticular wax substructures: how do plants seal their epidermal surfaces? J Exp Bot 62:841–853

Cape JN, Percy KE (1993) Environmental influences on the development on of spruce needle cuticles. New Phytol 125:787–799

Flemming HC, Wingender J (2010) The biofilm matrix. Nat Rev Microbiol 8:623–633

Groombridge B, Jenkins M (2002) World atlas of biodiversity: Earth's living resources in the 21st century. University of California Press, Berkeley

Guzman P, Fernandez V, Graca J, Cabral V, Kayali N, Khayet M, Gil L (2014) Chemical and structural analysis of *Eucalyptus globulus* and *E. camaldulensis* leaf cuticles: a lipidized cell wall region. Front Plant Sci 5:481. https://doi.org/10.3389/fpls.2014.00481

Hildebrand PD, Braun PG, McRae KB, Lu X (1998) Role of the biosurfactant viscosin in broccoli head rot caused by a pectolytic strain of *Pseudomonas fluorescens*. Can J Plant Pathol 20:296–303

Hiltner L (1904) Über neuere Erfahrungen und Probleme auf dem Gebiete der Bodenbakteriologie unter besonderer Berücksichtigung der Gründüngung und Brache. Zentralblatt Bakteriol 2:14–25

Iguchi H, Yurimoto H, Sakai Y (2015) Interactions of methylotrophs with plants and other heterotrophic bacteria. Microorganisms 3:137–151

Jenkinson HF, Lappin-Scott HM (2001) Biofilms adhere to stay. Trends Microbiol 9:9–10

Jetter R, Kunst L, Samuels A (2006) Composition of plant cuticular waxes. In: Riederer M, Müller C (eds) Biology of the plant cuticle. Blackwell, Oxford, pp 145–175

Kenrick P, Crane PR (1997) The origin and early evolution of plants on land. Nature 389:33–39

Kinkel LL (1997) Microbial population dynamics on leaves. Annu Rev Phytopathol 35(327):347

Knoll D, Schreiber L (1998) Influence of epiphytic micro-organisms on leaf wettability: wetting of the upper leaf surface of *Juglans regia* L. and of model surfaces in relation to colonisation by micro-organisms. New Phytol 140:271–282

Knoll D, Schreiber L (2000) Plant-microbe interactions: wetting of ivy (*Hedera helix* L.) leaf surfaces in relation to colonization by epiphytic microorganisms. Microb Ecol 40:33–42

Kolattukudy PE (1984) Biochemistry and function of cutin and suberin. Can J Bot 62:2918–2933

Kunst L, Samuels AL (2003) Biosynthesis and secretion of plant cuticular wax. Prog Lipid Res 42:51–80

Leben C (1965) Epiphytic microorganisms in relation to plant disease. Annu Rev Phytopathol 3:209–230

Lindow SE (1996) Role of immigration and other processes in determining epiphytic bacterial populations. In: Morris CE, Nicot PC, Nguyen-The C (eds) Aerial plant surface microbiology. Plenum, New York, pp 155–168

Lindow SE, Brandl MT (2003) Microbiology of the phyllosphere. Appl Environ Microbiol 69:1875–1883

Lindow SE, Leveau JHJ (2002) Phyllosphere microbiology. Curr Opin Biotechnol 13:238–243

MacDonald RC, Fall R (1993) Detection of substantial emissions of methanol from plants to the atmosphere. Atmos Environ A Gen Top 27:1709–1713

Mercier J, Lindow SE (2000) Role of leaf surface sugars in colonization of plants by bacterial epiphytes. Appl Environ Microbiol 66(1):369–374

Monier JM, Lindow SE (2004) Frequency, size, and localization of bacterial aggregates on bean leaf surfaces. Appl Environ Microbiol 70:346–355

Morris CE, Barnes MB, McLean RCJ (2002) Biofilms on leaf surfaces: implications for the biology, ecology and management of populations of epiphytic bacteria. In: Lindow SE, Hecht-Poinar EI, Elliott VJ (eds) Phyllosphere microbiology. APS, St Paul, pp 139–155

Muthusamy K, Gopalakrishnan S, Ravi T, Sivachidambaram P (2008) Biosurfactants: properties, commercial production and application. Curr Sci 94:736–747

Nawrath C (2006) Unraveling the complex network of cuticular structure and function. Curr Opin Plant Biol 9:281–287

Neinhuis C, Barthlott W (1998) Seasonal changes of leaf surface contamination in beech, oak and ginko in relation to leaf micromorphology and wettability. New Phytol 138:91–98

Nilanjana D, Preethy C (2011) Microbial degradation of petroleum hydrocarbon contaminants: an overview. Biotechnol Res Int 2011:941810, 13 p. https://doi.org/10.4061/2011/941810

Parsek MR, Greenberg E (2005) Sociomicrobiology: the connections between quorum sensing and biofilms. Trends Microbiol 13:27–33

Pollard M, Beisson F, Li Y, Ohlrogge JB (2008) Building lipid barriers: biosynthesis of cutin and suberin. Trends Plant Sci 13:236–246

Prasad R, Kumar M, Varma A (2015) Role of PGPR in soil fertility and plant health. In: Egamberdieva D, Shrivastava S, Varma A (eds) Plant growth-promoting rhizobacteria (PGPR) and medicinal plants. Springer International Publishing, Switzerland, pp 247–260

Prasad R, Chhabra S, Gill SS, Singh PK, Tuteja N (2020) The microbial symbionts: potential for the crop improvement in changing environments. In: Tuteja N, Tuteja R, Passricha N, Saifi SK (eds) Advancement in crop improvement techniques. Elsevier, Amsterdam, Netherlands, pp 233–240

Raaijmakers JM, de Bruijn I, Nybroe O, Ongena M (2010) Natural functions of lipopeptides from *Bacillus* and *Pseudomonas*: more than surfactants and antibiotics. FEMS Microbiol Rev 34:1037–1062

Ramey BE, Koutsoudis M, von Bodman SB, Fuqua C (2004) Biofilm formation in plant–microbe associations. Curr Opin Microbiol 7:602–609

Remus-Emsermann MNP, de Oliveira S, Schreiber L, Leveau JHJ (2011) Quantification of lateral heterogeneity in carbohydrate permeability of isolated plant leaf cuticles. Front Microbiol 2:1–7. https://doi.org/10.3389/fmicb.2011.00197

Remus-Emsermann MNP, Tecon R, Kowalchuk GA, Leveau JHJ (2012) Variation in local carrying capacity and the individual fate of bacterial colonizers in the phyllosphere. ISME J 6:756–765

Reynhardt EC, Riederer M (1994) Structures and molecular dynamics of plant waxes. II Cuticular waxes from leaves of *Fagus sylvatica* L and *Hordeum vulgare* L. Eur Biophys J 23:59–70

Riederer M, Schönherr J (1990) Effects of surfactants on water permeability of isolated plant cuticles and on the composition of their cuticular waxes. Pestic Sci 29:85–94

Romantschuk M (1992) Attachment of plant pathogenic bacteria to plant surfaces. Annu Rev Phytopathol 30(1):225–243

Ron EZ, Rosenberg E (2001) Natural roles of biosurfactants. Environ Microbiol 3:229–236

Rosenberg E (1985) Microbial surfactants. Crit Rev Biotechnol 3:109–132

Schönherr J (1982) Resistance of plant surfaces to water loss properties of cutin, suberin and associated lipids. In: Lange OL et al (eds) Encyclopedia of plant physiology. Springer, Berlin, pp 153–179

Schreiber L, Schönherr J (2009) Water and solute permeability of plant cuticles. Measurement and data analysis. Springer, Heidelberg

Segado P, Dominguez E, Heredia A (2016) Ultrastructure of the epidermal cell wall and cuticle of tomato fruit (*Solanum lycopersicum* L.) during development. Plant Physiol 170:935–946

Singh D, Raina TK, Kumar A, Singh J, Prasad R (2019) Plant microbiome: a reservoir of novel genes and metabolites. Plant Gene. https://doi.org/10.1016/j.plgene.2019.100177

Sutherland IW (2001) The biofilm matrix: an immobilized but dynamic microbial environment. Trends Microbiol 9:222–227

Tukey HB (1970) The leaching of substances from plants. Annu Rev Plant Physiol 21:305–324

Van Gardingen PR, Grace J, Jeffree CE (1991) Abrasive damage by the wind to the needle surface of *Picea sitchensis* (Bong) Carr and *Pinus sylvestris* L. plant. Cell Environ 14:185–193

Varma A, Swati T, Prasad R (2019) Plant microbe interface. Springer International Publishing, Switzerland. ISBN 978-3-030-19831-2. https://www.springer.com/gp/book/9783030198305

Varma A, Swati T, Prasad R (2020) Plant microbe symbiosis. Springer International Publishing, Switzerland. ISBN 978-3-030-36247-8. https://www.springer.com/gp/book/9783030362478

Villena JF, Domínguez E, Stewart D, Heredia A (1999) Characterization and biosynthesis of non-degradable polymers in plant cuticles. Planta 208:181–187

Vorholt JA (2012) Microbial life in the phyllosphere. Nat Rev Microbiol 10:828–840

Wentzel A, Ellingsen TE, Kotlar HK (2007) Bacterial metabolism of long-chain n-alkanes. Appl Microbiol Biotechnol 76:1209–1221

Whipps JM (2001) Microbial interactions and biocontrol in the rhizosphere. J Exp Bot 52:487–511

Xavier JB, Foster KR (2007) Cooperation and conflict in microbial biofilms. Proc Natl Acad Sci 104:876–881

Zeisler V, Schreiber L (2016) Epicuticular wax on cherry laurel (*Prunus laurocerasus*) leaves does not constitute the cuticular transpiration barrier. Planta 243:65–81

Zeisler-Diehl VV, Barthlott W, Schreiber L (2018) Plant cuticular waxes: composition, function, and interactions with microorganisms. In: Wilkes H (ed) Hydrocarbons, oils and lipids: diversity, origin, chemistry and fate. Handbook of hydrocarbon and lipid microbiology. Springer, Heidelberg. https://doi.org/10.1007/978-3-319-54529-5_7-1

Chapter 2
Plant Microbiome and Its Important in Stressful Agriculture

Bahman Khoshru, Sajjad Moharramnejad, Nahid Hosseinzadeh Gharajeh, Behnam Asgari Lajayer, and Mansour Ghorbanpour

Abstract The benefits of the green revolution in current agriculture are well-obvious. Use and application of soil and plant microbiome's potential can overcome part of agricultural limits. The issue of crop production in current world is complicated by decreasing proper farmland, biotic and abiotic stresses and high equipment and labor costs. Therefore, sustainable and appropriate agriculture would depend on the utilization of microorganisms and new methods, which will aid as a marginal approach for more crop yield in the aftertime. The abiotic and biotic stresses are key constraints for food quality, crop yield, and global food security. Effective microbes have a substantial role in biotic and abiotic stresses management, reduce chemical fertilizers and increase the yield of plant cultivar's by affecting elemental cycling. Additionally, high solicitation for food and crowd everyday increasing, increment necessity of how to use the microbiome for more crop yields and decrease losses affected by environmental stresses. In this chapter the role of the plant microbiome as a new strategy was investigated that may be responsible for increase in crop productivity and eventually by effectively answering biotic and abiotic stresses leads to food security.

B. Khoshru · B. Asgari Lajayer
Department of Soil Science, Faculty of Agriculture, University of Tabriz, Tabriz, Iran

S. Moharramnejad
Crop and Horticultural Science Research Department, Ardabil Agricultural and Natural Resources Research and Education Center, AREEO, Moghan, Iran

N. H. Gharajeh
Faculty of Agriculture, Department of Plant Breeding and Biotechnology, University of Tabriz, Tabriz, Iran

M. Ghorbanpour (✉)
Faculty of Agriculture and Natural Resources, Department of Medicinal Plants, Arak University, Arak, Iran
e-mail: m-ghorbanpour@araku.ac.ir

A. Varma et al. (eds.), *Plant Microbiome Paradigm*,
https://doi.org/10.1007/978-3-030-50395-6_2

2.1 Introduction

The increasing trend of global warming may lead to global climate change in the coming years. Accordingly, food security is being threatened by the consequent environmental stresses (Battisti and Naylor 2009; Delangiz et al. 2019). This is deteriorated by the fact that by 2050, the world population is estimated to reach 8.9 billion (Singh et al. 2011). Beside climate variation, the growing population, vast agricultural practices and thus soil health reduction for crop cultivation are considered as vital factors affecting agricultural sustainability (Wassmann et al. 2009). With the elevating population rate and unsustainable traditional agricultural system (Masciarelli et al. 2014), producing sufficient amount of food which meets the world demand is a serious issue for farmers as well as policymakers. Moreover, the excessive consumption of fertilizers, pesticides, and weedicides with chemical origin in agriculture leads to fundamental loss in diversity of beneficial soil microbes. The constant effect of abiotic and biotic stresses on our agroecosystem directly changes health and fertility features of soil and crop productivity. Plant growth and productivity are negatively affected by numerous stress factors which are mainly classified as abiotic and biotic types. They occur either naturally or due to human induction. Abiotic and biotic stresses account for respectively 50% and 30% loss of worldwide agricultural productivity. Abiotic stress includes drought, heavy metal pollution in addition to suboptimal salinity and temperature. Stress condition affects morphology, physiology, biochemistry, and even on its molecular characteristics of a plant. Some of the main stress factors resulted from climate change include intense salinity, temperature, water limitation, and heavy metal pollutants. The biotic stress is itself influenced by abiotic stress factors which consequently lead to reduction in crop productivity, soil microbial diversity, soil fertility, and severe competition for nutrient resources (Chodak et al. 2015; Singh et al. 2019). The use of beneficial plant-associated microbial community promoting plant growth/development under harsh conditions is of great attention. The microorganisms such as mycorrhizal fungi, such as mycorrhizal fungi and plant growth promoting bacteria (PGPB) including plant growth promoting rhizobacteria (PGPR) would contribute to environmental stability and shifting toward sustainable agriculture (Prasad et al. 2005, 2020). Based on their effect, three groups of (a) beneficial, (b) deleterious, and (c) neuter types of plant-associated microbes exist. The members of *Azospirillum, Azotobacter, Pseudomonas, Bacillus, Burkholderia, Enterobacter*, *Klebsiella, Serratia,* and *Variovorax* genera are regarded as PGPR which maintain plant health under both optimal and suboptimal conditions promote plant growth and development. However, the majority of plant growth promoting microbes (PGPM) and arbuscular mycorrhizae (AM) are sensitive to drought, heavy metal, and salinity stress (Vimal et al. 2017). Therefore, it is challenging to establish biofertilizers compatible with the stress condition. Nevertheless, some of these microorganisms have the ability to simultaneously overcome the adverse conditions and consolidate plants through developing efficient mechanisms. In this regard, an efficient approach is needed to investigate interaction of plant-microbe in maintaining plant growth and conferring disease resistance upon sustainable agriculture (Finkel et al. 2017). In this chapter, we focus on beneficial effect of tolerant beneficial plant-associated microbes and their status in agriculture and the situation of operations performed to enhance sustainable agricultural production.

2.2 Environmental Stress and Biochemical Changes in Plants

Several abiotic factors influence development of plant development and limit production of crop, variant rates of acidic situations unfavorably effect on nutrients of soil that cause a nutrient deficiency in plant and disrupt normal physiological ability for growth and development of plant (Emamverdian et al. 2015). Prolonged disposal to stress of salinity resulted toxicity within the cell along with interruption of osmotic equivalency (Saghafi et al. 2019a). Effect of ionic followed with osmotic stresses leads to altered plant growth and development (Munns and Tester 2008). Forbearance to salinity stress needed to regulate ionic and osmotic balance in the cells. For resistance toward salinity, plants protect delicate plant tissues from vast salinity area or by emanating ions out of roots or keeping ions away from the cytoplasm (Silva et al. 2010; Saghafi et al. 2019b). During freezing conditions, some plants developed a mechanism to cope up with cold temperatures by elevating their defense response by the process of cold acclimation (Thomashow 2010). The plants after sensing the stress, show a quick and compelling reaction to initiate an intricate stress-specific signaling by synthesizing plant hormone and cumulation of phenolic acids and flavonoids (Qin et al. 2011). Abiotic stresses are primary cause for the generation of reactive oxygen species (ROS). The ROS are including oxygen and organized as a natural corollary of the metabolism of oxygen and have key roles in homeostasis and cell signaling. The generation and elimination of ROS are at balance under natural conditions, whereas under biochemical changes and stress of environments, it fazes this equilibrium by increasing the production of ROS (Nath et al. 2017; Kapoor et al. 2019; Kundu et al. 2020). The ROS is very dangerous and toxic for the cell structures of organism as they unfavorably affect the function and structure of the biomolecules. The ROS is produced in plants in mitochondria, peroxisomes, and chloroplasts (Asgari Lajayer et al. 2017; Ghassemi et al. 2018). Hydrogen peroxide (H_2O_2) and oxygen radicals are produced in mitochondria due to the overreduction of the electron transport chain. Chloroplasts are the main source of the production of H_2O_2 and O_2 (Davletova et al. 2005), due to higher oxygen pressure and reduced molecular oxygen than in other organelles in the electron transport chain within PSI (Dat et al. 2000). These superoxides are changed to H_2O_2 either spontaneously or by the operation of the enzyme superoxide dismutase (SOD). Hydroxyl radical's production is also carried out by H_2O_2. It has been reported that peroxisomes are a major producer of H_2O_2 and responsible for the formation of superoxides (O^{2-}).

They damage the biomolecules such as proteins, lipids, carbohydrates, and DNA, which leads to cell death (Foyer and Noctor 2005).

2.2.1 Interactions Between Plant and Microbes in Stress Conditions

One major question that has struck the minds of many researchers is whether the interplay between microbes and plants can help the plant to increase its resistance to stress. To answer this question, it should be noted that the PGPM helping plant to stress tolerance growth and survival under adverse (Nadeem et al. 2014). Some direct and indirect mechanisms were used ubiquitous via microbes to improve development and plant growth during stress conditions. Different molecular and biochemical mechanisms are used by microbes to stimulate development and growth. For example, inoculation plants with PGPM stimulate plant growth by regulating nutritional balance and hormonal, generating plant growth regulator and inducing persistence versus phytopathogens (Spence and Bais 2015). The results of Sabeti Amirhendeh et al. indicated that inoculated tobacco roots with *Azotobacter chroococcum* increased N uptake, qualitative characteristics and yield and application of biological fertilizers is a suitable strategy to move toward sustainable agriculture. The PGPM also produce defined metabolites which decreased pathogen crowd around plant adjacent. For instance, Złoch et al. (2016) reported that siderophore produced by these microbes in rhizosphere reduced plant growth, because it reduced the iron availability to certain pathogens. Moreover, they can also simplify plant growth by producing plant hormones and solubilize phosphate fixing atmospheric nitrogen (Ahmad et al. 2011). Vardharajula et al. (2011) also reported certain other mechanisms consist nutrient mobilization, generation of exopolysaccharide, rhizobitoxine, etc. that accompany the plant to overcome the critical environment. Rhizobitoxine ameliorate growth and development of plant under stress situation by preventing ethylene generation (Kumar et al. 2009). In addition, microbes can progress the plant growth and development by important enzymes like 1-aminocyclopropane-1-carboxylic acid (ACC) deaminase, glucanase, and chitinase in stress situation (Farooq et al. 2009). Further, sigma factor in some bacteria is responsible for modification of gene expression under harsh condition to defeat undesirable effects (Gupta et al. 2013). The interaction present between the roots of higher plants with fungi is another strategy of growth/development promotion. AMs are the most common mycorrhizae present in agricultural lands which offer advantageous role in cycling and absorption of nutrients and their translocation to the plant. This way they neutralize negative effects of stress condition and maintain the actual growth/development of the plant. PGPMs are expected to be an appropriate alternative to chemical and inorganic fertilizers to establish sustainable agriculture and respond food security concerns through plant-microbe interactions. The PGPB promote plant growth/development via three possible ways

(Govindasamy et al. 2010): (a) improving nutrition and growth of the host plant; (b) antagonistic attitude toward pathogens and stimulation of defense mechanism; and (c) contribution to build up sustainable agriculture. The latter may lead to increase in the yield and nutritional quality of food grains in addition to saving 20–25% of the costs related to chemical fertilizers or pesticides under fluctuating climate. Taking the advantages of sustainable agriculture, financial income will further be improved because of organic food production.

2.2.2 *Abiotic Stress*

Plants are affected by various unpredicted disturbances, and among them, abiotic stress is the prime cause of limiting the crop production in worldwide (Wang et al. 2003; Saghafi et al. 2018; Asgari Lajayer et al. 2019; Khoshmanzar et al. 2019). Abiotic stresses are the undesirable impacts of nonliving factors on the active organisms in a certain environment. These stresses are constituting an important limitation to sustainable agriculture. The effect of abiotic factor on the plant growth and development depends on its power or quantity. Any alteration from such optimal environmental conditions, which is deficit in the chemical or physical environment, is considered as abiotic stress and critically impacts on plant growth, development, and productivity (Bray 2000). They are chronic features of nearly all the world's climatic regions since several critical environmental risks and these threats are mobilized by global climate change and population growth (Gleick 1994). Anomaly environmental conditions generate abiotic stresses that are the primary restrictive factors for limiting crop production (Grayson 2013). Abiotic stresses comprise of heat, cold, drought, alkaline conditions and salinity, waterlogging, light intensity, and nutrient deficiency. Drought has affected 64% of the worldwide land area, salinity 6%, anoxia 13%, soil alkalinity 15%, mineral starvation 9%, and cold 57% (Bailey-Serres and Voesenek 2008). Of the world's 5.2 billion ha of dryland agriculture, 3.6 billion ha is influenced by the issues of soil erosion, degradation, and salt stress (Riadh et al. 2010). Plants adapt with the rapid alteration and affliction of ecological conditions as a result of their natural metabolic mechanisms (Simontacchi et al. 2015). Deviations in the external environment conditions could put the plant metabolism out of homeostasis (Foyer and Noctor 2005) and make the need for the plant to harbor some metabolic and genetic mechanisms in the cell (Gill and Tuteja 2010). Plants retain a variety of defense mechanism to combat abiotic stress conditions (Yolcu et al. 2016). These mechanisms involve in the metabolic reprogramming in cellular system to enable biophysicochemical processes of the external conditions (Massad et al. 2012). Several time, plants reduced the burden of abiotic stresses with the help of the inhabitant microbiome (Turner et al. 2013). Microbes are the integral part of ecological system and important for crop production. Microorganisms are important inhabitants of seeds also and proliferate as germinate in the soils to form mutual associations at the surface or endophytic associations inside the roots, stems, or leaves. Plant microbiome gives principal support to the plants in securing supplements, opposing against infections, and enduring abiotic stresses (Turner et al. 2013). Microbial inherent metabolic and hereditary abilities make them reasonable organisms to cope up with environmental challenges (Singh 2014). Their communications with the plants incited a few fundamental responses that improved the metabolic mechanism of the plants for defense against abiotic stress conditions (Nguyen et al. 2016). Several studies reported the imperative characteristics of the microbial communications with plants that propose mechanisms based on plant-microorganism associations that accentuated the biochemical, molecular, and cellular mechanisms of plant defense against stresses (Onaga and Wydra 2016). Studies on plant microbiome at molecular, physiological, and biochemical levels observed that plant-microbe's associations communicate plant responses against stress conditions (Farrar et al. 2014). Developments of technologies also simplified comprehension of gene redaction systems, RNAi-mediated gene suppressing, technology of mutant, proteomic analysis, and metabolite profiling are going to disclose massive molecular data that assisted in raising our conception of microbe interactions.

2.2.2.1 Drought and Its Impact on Crop Productivity

Drought is known as deleterious environmental stress which is of great attention for environmentalists as well as agricultural scientists. It is a critical agriculture issue throughout the world which limits plant growth, development, and productivity. Drought stress affects the majority of agricultural lands in the world and consequently human life and economy are being influenced (Disante et al. 2011). Drought has a wide range of implications on growth parameters as well as stress responsive genes during the adverse condition. Low water content decreases cell size, turgor pressure, and membrane integrity and leads to production of reactive oxygen species and leaf senescence which altogether reduce productivity of crop plants. In spite of that, limited water condition causes a series of morphological, physiological, and molecular alterations in plants, such as height reduction, elevated ethylene production, lipid peroxidation, changes in chlorophyll content, membrane function and protein conformation due to accumulation of free radicals, photosynthesis apparatus damage, photosynthesis inhibition, and cell death (Lata and Prasad 2011; Tiwari et al. 2016). It should be highlighted that in the near future the impact of climate change will lead to more frequent and intense occurrences of drought.

The microbiome associated to plants can affect many plant properties including development and biotic and abiotic stress tolerance (Rolli et al. 2015; Panke-Buisse et al. 2015). One of the important tensions for the world's farmers is drought stress. The relief of the microbiome to plant conformity to drought stress is trivially understood. Rolli et al. (2015) assayed in vivo eight isolates, over 510 strains, for their potency to patronage grapevine and *Arabidopsis* growth under drought stress; they indicate that promoting of plant growth activity is dependent to stress and not a per se strain features. Likely, inoculated pepper plant with chosen strains under watered and drought situations represented a pattern of stress-dependent plant growth-promoting by enhancing shoot and leaf biomass and shoot length and increasing photosynthesis in drought-challenged grapevine, with a deep positive efficacy on sensitive drought rootstock. Totally, these eventuates show that the examined bacteria considerable contributed to plant conformity to drought by stress-induced promotion of plant growth. *Achromobacter piechaudii* ARV8 raised drought stress tolerance in pepper and tomato by 1-aminocyclopro pane-1-carboxylate (ACC) deaminase. The workmanships which present drought stress tolerance in plants stand mainly contemplative. However, it is thinkable that the separation of ACC in plant by bacterial will prevent ethylene generation which finally decrease stress of plant and provide normal plant growth (Duan et al. 2009; Yang et al. 2009). Alternative investigation highlighted the positive effect of bacterial priming on seedlings of wheat in drought stress condition (Timmusk et al. 2017) that incremented plant biomass by 78% and meliorated photosynthesis five times at intensive drought. Inoculation of wheat with *Burkholderia phytofirmans* PsJN had an incremented photosynthesis, higher water use efficiency, and high chlorophyll add-up and grain efficiency than the control under water shortage in the field situations (Naveed et al. 2014). Likely, inoculated maize with both *Enterobacter* sp. FD17 and *B. phytofirmans* had better performance compared to controls (Naveed et al. 2014). The bacterial *Pseudomonas putida*, *Pseudomonas* sp., and *Bacillus megaterium*, isolated from extremely water-stressed soil, promoted plant growth under drought situations (Marulanda et al. 2009). Inoculation of sunflower seedlings with *Pseudomonas* sp. Strain GAP-45 increased durability and plant biomass under stress of drought (Sandhya et al. 2009). It is conceivable that inoculation of bacteria can efficiently root colonies resulting in constant soil aggregates and finally increased stress tolerance. In other study, inoculation of maize plants with *Pseudomonas* strain GAP-45 caused to increased antioxidant and compatible solutes under condition of water deficit. The microbes isolated from roots of plants growing under extreme dry conditions, in tomato, grapevine, olive, and pepper plants, meliorated the growth of alternative host species under similar growth situations (Marasco et al. 2013). This solution of stress-persistence strategy has the potential to save costs, time, and effort. Inoculation of lettuce with *Pseudomonas mendocina* and arbuscular mycorrhizal fungi (*Glomus intraradices*) that resulted in antioxidant catalase activity at drastic drought situations implying to practicable use of microbes in relief of oxidative stress (Kohler et al. 2008). The accumulation of protein along with glutathione and ascorbate has important impress in preserve plant metabolic functions and admitting protection at drought conditions. Lavender plants inoculated with *Glomus intraradices* accumulated these compounds and presented great drought tolerance by improving water contents, root biomass, and N and P contents (Marulanda et al. 2007).

2.2.2.2 Mechanism of Drought Stress Tolerance

Water deficiency and drought are the main cause of plant destruction and crop losses worldwide. Based on the official statistics, drought and related stresses are the cause of about 67% of crop losses in USA in the last 50 years (Comas et al. 2013). Drought-tolerant microbes improve plant growth and development under water deficiency. In order to maintain low water potential, microbes have adopted different evolutionary, adaptation, and tolerance mechanisms. They include formation of thick wall, entering dormant stage, accumulation of osmolytes, and production of exopolysaccharides (EPS). To face negative effects of drought in plants and soil, the plant-associated microbes develop various mechanisms. Apart from water content supply, they provide nutritional and optimum environmental requirements for constant plant growth. The beneficial rhizosphere colonized microbes perform their plant growth/development promotion by either direct or indirect mechanisms. The possible mechanisms are via (1) phytohormones including indole-3-acetic acid (IAA), cytokinins, and abscisic acid (ABA); (2) bacterial exopolysaccharides; (3) ACC deaminase; and (4) induced systemic tolerance. To maintain their growth and development, plants produce phytohormones (Farooq et al. 2009; Porcel et al. 2014). PGPRs are also capable of synthesizing plant hormones which, during stress condition, act as stimulus for plant growth and division. For instance, under drought stress IAA (the most active auxin) produced by PGPR regulates cell division, shoot growth, differentiation of vascular tissue and adventitious and lateral root (Goswami et al. 2015). Inoculation of seeds or plants with PGPR results in high concentration of ABA, the important growth regulator, to modify plant physiological and molecular features by regulating root hydraulic conductivity and drought-related transcription levels so that it can tolerate water deficient condition (Jiang et al. 2013). This has been demonstrated in *Arabidopsis thaliana* on which *Azospirillum brasilense* has improved drought tolerance effect mainly through elevation of ABA production (Cohen et al. 2015). ACC deaminase produced by bacteria during drought stress hydrolyzes 1-Aminocyclopropane-1-carboxylate (ACC), immediate precursor of ethylene, into ammonia and alpha-ketobutyrate (Bal et al. 2013). Microbes and PGPRs of drought-tolerant type in maize are shown to maintain the growth, water potential, and water loss during stress condition. Production of free amino acids such as proline and sugar in plants are increased, and the antioxidant activity is decreased by the impact of beneficial microbial inoculants (Vardharajula et al. 2011). During drought stress, soybean is demonstrated to have low chlorophyll content and correspondingly reduced photosynthesis. Inoculation with *Pseudomonas putida* H-2-3 counteracts drought stress by compensating the actual chlorophyll content and biomass development. Combinational inoculants of endophytic and rhizospheric PGPR elevate the plant ability to tolerate stress conditions. The microbial exopolysaccharide may improve plant tolerance against drought in some plants. Inoculation of three drought-tolerant bacterial strains of *Alcaligenes faecalis* (AF3), *Proteus penneri* (Pp1), and *Pseudomonas aeruginosa* (Pa2), through proline, protein, and sugar production, showed potential increase in relative water content of maize (Naseem and Bano 2014). A variety of biochemical, physiological, and molecular mechanisms have been developed in bacteria to help them survive drought conditions such as spore formation, synthesis of compatible solutes and EPS (Chithrashree et al. 2011). The latter fortifies plants against water limitation (Sandhya et al. 2009). In drought condition, accumulation of compatible solutes including betaine, glycine, proline, and trehalose contribute bacteria to maintain their functional protein content, membrane integrity, and permeability. Plant growth, biomass, nutrient uptake, relative water content, and briefly plant survival under drought stress are promoted by inoculation with certain combination of mycorrhizal specific bacteria and drought-tolerant microbial community. The combination of *Bacillus thuringiensis* and *P. putida,* due to proline accumulation of in shoot and root, leads to reduction in electrolyte leakage and stromal conductance (Ortiz et al. 2015).

2.2.2.3 Salinity Stress: The Main Restricting Factor of Agriculture

Salinity stress, as the prevalent abiotic stress, affects the majority of agricultural lands and modern agriculture worldwide. High level of ions and osmotic stress in saline condition has toxic effect on microbes and limit their activity leading to poor plant growth/development. Osmosis stress occurs due to the reduced water potential in soil from which plant have difficulty to uptake water and nutrients. Soil salinity is the result of presence of cations such as Ca_2^+ (calcium), K^+ (potassium), Na^+ (sodium), and anions like Cl^- (chloride) and NO_3^- (nitrate). Inadequate rainfall and weak weathering of soil are the main reasons for accumulation of salts as electrically charge ions in the soil (Shrivastava and Kumar 2015). Various features of plants including seed germination, nodulation process, agricultural productivity, water and nutrient uptake, crop yield, ecological, and physicochemical balance along with nitrogen fixation are fundamentally affected by salt stress (Shrivastava and Kumar 2015). Nitrogenase, the enzyme responsible for nitrogen fixation, is significantly reduced in osmotic stress. Soil salinity limits the water uptake by roots, and since abnormal salty water within cells are toxic, plant growth is suppressed. Just like plant growth, microorganism functioning is affected by salinity mainly due to osmotic effect and ion toxicity. Compared to bacteria, fungi are more sensitive and vulnerable to osmotic stress. Osmotic potential of lower and higher than normal makes water uptake from soil difficult for plants and microorganisms. The studies declare that inoculation with PGP and endophytic microbe community alleviates deleterious effects of salt on plants. PGP microbes develop direct and indirect mechanisms to promote plant growth during salinity stress. In addition, negative effects of saline condition are effectively ameliorated by PGPB through biofilm formation (Kasim et al. 2016). Compared to the control samples, *Azospirillum*-inoculated lettuce seeds showed improved germination and vegetative growth under salinity stress (Barassi et al. 2006). A separate study demonstrated that co-inoculation with *Pseudomonas stutzeri* lowering destructive effects of soil salinity in salt-tolerant and salt-susceptible chili peppers (Bacilio et al. 2016). While, some microbial species are demonstrated to mitigate salinity stress activity of biofilm formation in barely grains (Kasim et al. 2016). Salt-tolerant AM fungi and bacteria efficiently elevate salinity tolerance in certain plants. During salinity stress, co-inoculation of *R. intraradices* and *Massilia* sp. RK4 contributes root colonization of Arbuscular mycorrhizal fungi (AMF) as well as nutrient accumulation in maize.

As a matter of fact, plant association with fungi and microbes efficiently assist maize plant to tolerate excessive salinity (Krishnamoorthy et al. 2016).

The microbes of halotolerant survival under soil stress salinity and exhibit properties for assistance plants to growth at high-salinity condition. From 130 rhizobacterial strains isolated from wheat plants cultured under saline situations, 24 isolates tolerated high rates (8%) of NaCl stress. The compilers imputed this tolerance to several genes, hormones, and proteins such as nifH, IAA, siderophores, and gibberellin (Upadhyay et al. 2009). It has been reported that halotolerant bacterial strains isolated from Korea increased plant development under salinity conditions via decreasing ethylene production (Siddikee et al. 2010). New halotolerant diazotrophic bacteria isolated from roots of *Salicornia brachiate* (extreme halophyte), with properties such as IAA production, phosphorus solubilization, and ACC deaminase activity, represents other potential candidates (Jha et al. 2012). According to Arora et al. (2014), 17 of 20 bacteria isolated from salt-tolerant plant species easily grew in culture at 7.5% NaCl and 2 of 17 grew in 10% NaCl. Giri and Mukerji (2004) reported that the inoculation of arbuscular mycorrhizal fungi (AMF) has enhanced tolerance of salinity stress in host plant. The enhanced growth in saline soils can because of incremented phosphate and diminished Na^+ shoot concentration compared to uninoculated treatment. However, investigation on the bacterial and AM species potency to impel protective proteins and osmoprotectants is demanded. The mentioned reports offer that plants under stress conditions can easily recruit various bacterial strains with wide utilities for plants grown under salt stress. Collectively, this phenomenon has been denominated induced systemic tolerance (Yang et al. 2009).

2.2.2.4 Mechanism of Salinity Stress Tolerance

Salts are a natural component in water and soils. The process of increasing the salt content is known as salinization that is an important stress for most plants. It is uninterruptedly increasing owing to climate change. Soil salinity stresses in plants have two ways: (1) high concentrations of salts within the plant can be toxic, and (2) high concentrations of salts in the soil make it harder for roots to extract water and soluble nutrients. Therefore, diversity of salinity stress-tolerant microbes is involved in promotion of growth under stress condition. Plants have different mechanisms to deal with salinity stress. The mechanisms of salinity tolerance divided into three main categories: (1) tolerance to osmotic stress, (2) Na^+ exclusion from leaf blades, and (3) tissue tolerance. The direct mechanisms of plants include phytohormones production (e.g., cytokinin, auxin, gibberellins, and ethylene), nutrient mobilization, siderophore production, and nitrogen fixation (Hayat et al. 2010). These mechanisms lead to increase surface area, root length and root volume, and number by nutrient uptake. An important indirect mechanism is to reduce the affluence of disease agents and plant pathogens. ACC deaminase produced by root colonizing rhizobacteria converts ACC to ammonia and alpha-ketobutyrate and thus lowers ethylene. The enzyme rhizobitoxine inhibits production of ethylene and increases nodulation under stress condition (Shahzad et al. 2017). PGPB, by accumulating osmolytes in the plant cell cytoplasm and maintaining their cell turgor, ensure plant growth under osmotic stress. EPS produced by microbes, through binding with cations and removing it from the plant access, interacts with salinity (Vardharajula et al. 2011). Co-inoculation of plants with PGPR strains of Rhizobium and Pseudomonas can help the plant grow in saline soil and ameliorate destructive effects of salinity (Bano and Fatima 2009). Two saline soil isolated rhizospheric bacteria of *Bacillus pumilus* and *Bacillus subtilis* demonstrated PGPR features of IAA production, phosphate solubilization, ammonia and hydrogen cyanide (HCN) production, and tolerance against salt stress condition (Damodaran et al. 2013). Bano and Fatima (2009) reported that PGPRs of *Rhizobium* and *Pseudomonas* ameliorate salt-affected maize. The induced tolerance is due to elevated proline production and selective uptake of K ions in addition to diminished electrolyte leakage and osmotic potential. Co-inoculation with *Bacillus pumilus* and *Pseudomonas pseudoalcaligenes* leads to high glycine betaine content and improves salinity tolerance in rice. During salt stress, *Acinetobacter* spp. and *Pseudomonas* sp. accumulate IAA and ACC deaminase to ensure plant growth in barley and oats. During salt stress, *Bacillus pumilus* and *P. pseudoalcaligenes* are reported by Jha and Subramanian (2014) to decrease lipid peroxidation and superoxide dismutase activity in salt sensitive rice GJ-17. During salinity stress, PGPRs stimulate physical and chemical changes which lead to induced systemic tolerance (IST) and enhanced growth. PGPRs induce desirable root/shoot growth along with decreased disease susceptibility of cotton to fungi such as *Fusarium solani* and tolerance against red rot disease (Egamberdieva et al. 2015). In salinity affected wheat crop, plant growth, grain weight, and total dry weight are improved by salt-tolerant *Azospirillum* strains (Nia et al. 2012). In maize, ABA hormone, by acidification of apoplast, plays its primary role of tolerating salinity stress. Under stress condition, enhanced storage life, growth, and product quality are detected in the lettuce seeds inoculated with *Azospirillum* (Fasciglione et al. 2015). *Hartmannibacter diazotrophicus* E19, the PGPR isolated from *Plantago winteri*, helps barley (*Hordeum vulgare* L.) maintain its growth in osmotic stress (Suarez et al. 2015). The PGPR isolates of *Bacillus polymyxa* BcP26, *Pseudomonas alcaligenes* PsA15, and *Mycobacterium phlei* MbP18 are capable of surviving in the saline lands such as calcisol soil (Egamberdiyeva 2007). PGPB co-inoculation of *Zea mays* with *Rhizobium* and *Pseudomonas* induces high proline production in addition to decreased electrolyte leakage, selective uptake of K ions, and maintenance of relative water content leading to plant salt tolerance (Bano and Fatima 2009). The PGPR strains of *P. aeruginosa*, *P. fluorescens*, and *P. stutzeri* isolated from tomato rhizosphere were found to contain high sodium chloride concentration and stimulate accumulation of ACC deaminase and phytohormones in tomato and promote its salinity tolerance (Bal et al. 2013; Tank and Saraf 2010). Tricarboxylic acid (TCA) cycle is involved in tolerance behavior of tomato plant in salinity condition (de la Torre-González et al. 2017). Improved growth and elevated salt tolerance is reported in the rice inoculated *Bacillus amyloliquefaciens* NBRISN13 (SN13) through up- or downregulation of different genes (Nautiyal et al. 2013). Phytohormone produced by endophytic bacteria also helps plants maintain their growth in salinity stress. *Bacillus amyloliquefaciens* RWL-1 produces ABA and auxin which stimulates *Oryza sativa* to tolerate against high saline condition (Shahzad et al. 2017). Apart from endogenous plant hormones, exogenous jasmonic acid (JA) and plant growth promoting endophytic bacteria in *Solanum pimpinellifolium* are proven to neutralize negative salinity impacts on plant (Khan et al. 2017). Oxidative and osmotic stresses are considered as consensus secondary stress resulted from salinity and drought. These stresses are defeated by PGPB through induced systemic resistance (ISR). Defense mechanism includes the main steps of phytohormone and antioxidant production as well as osmotic adjustment. Against salinity and drought stress, microbial strains and plants accumulate compatible osmolytes to maintain the growth. Microbe-induced plants show high proline concentration under salinity and drought stress.

2.2.2.5 Stress of Heavy Metals (HMs)

Industrialization, anthropogenic activities, and overuse of chemical fertilizers resulted to pollution of heavy metals in soils, and these metals have intense effect on plants and ultimately, it endangers human and animals' health. Therefore, heavy metal pollution and contaminated soils are a serious global environmental problem. Generally, HMs are distinctive as metals with densities more than 4 g/cm^3, nondegradable, atomic weights, or atomic numbers (Ma et al. 2016a). Despite the heavy metals are extant in many aspects of novel life but have comparatively scarce amount in the earth's crust . They are also poisonous at low concentration. To protect the toxic effects and heavy metal contamination, it is a completely necessary principle to eliminate these heavy metals from environment. There are many techniques for heavy metals remediation, but they have many limitations such as the destruction of the soil structure and cost problems (Glick 2010). Using bacterial mixtures and other microorganisms for bioremediation of heavy metals is a new approach for heavy metals elimination. Phytoremediation is an appearing method that develops utilization of plants and plant microbiomes to refine heavy metals from contaminated soils. Moreover, it is an economical and sustainable method to remove heavy metals (Chirakkara et al. 2016). In this method, application of microbial communities boosted the phytoremediation performance. The microbes are more sensitive compare to other living organism therefore can be a nice index for heavy metal stress identity (Chen et al. 2014). In the recent decade, bioremediation and application of microbial diversity for heavy metals remedy were successful. The reason for these successes is summarized by economic aspects, aesthetic approach environment friendly and also applying in various environmental conditions. Some plant growth-promoting materials, for example, plant hormones (gibberellins, cytokinins, and IAA), siderophores, and ACC deaminase, are produced via microbes of plant to play critical roles in plant health in contaminated soils (Santoyo et al. 2016). Excessive contamination of soil can decrease root developments mainly owing to oxidative stress, which restricts the extent of phytoremediation (Hu et al. 2016). Also, the nutrients deficiency and low density of microbes could be restrict phytoremediation (Gerhardt et al. 2009). Cadmium, lead, manganese, chromium, copper, zinc, and aluminum are common heavy metals. Metalloids are chemical element that exhibits some properties of metals, and some of those similar to arsenic (As) and antimony (Sb) also show toxicity (Pandey 2012). Bacteria of rhizosphere be worthy special attention between the microbes participating in phytoremediation as these able directly meliorate the process proficiency via changing oxidation/ reduction reactions and soil pH (Rajkumar et al. 2012). *Microbacterium* sp. G16 and *Pseudomonas fluorescens* G10 remarkably enhanced the Pb solubility in *Brassica napus* compared with uninoculated controls and were mostly ascribed to phosphate solubilization, ACC deaminase, siderophores, and IAA. Many endophyte genera such as *Enterobacter, Serratia, Bacillus*, *Burkholderia*, and *Agrobacterium* lead to enhance the phytoremediation extent and production of biomass in contaminated soils to heavy metal (Zheng et al. 2016; Feng et al. 2017). Furthermore, the mycorrhizal fungi have important performance in phytoremediation because of tolerance to heavy metals and hyper accumulation of them.

2.2.2.6 Microbial-Assisted Remediation of Heavy Metals

During metal stress condition, the microorganisms tolerant to heavy metals such as firmicutes, mycorrhiza, and rhizobacteria maintain plant growth and removal of heavy metals. They do this through different mechanisms of efflux, impermeability to metals, volatilization, EPS sequestration, metal complexation, enzymatic detoxification in addition to nitrogen fixation, nutrient mobilization, siderophores, and phosphate solubilization (Verma et al. 2013; Ahmad et al. 2011). Moreover, the production of ethylene, IAA, and ACC deaminase is reduced, and ultimately disease is repressed by the effect of heavy metal-tolerant microbes (Glick 2010). Different living and nonliving forms of microbial biomass are being applied to remove heavy metals. In this regard, the bacterial and fungal cell wall properties and the corresponding functional group are delicately investigated (Vijayaraghavan and Yun 2008). Bioaccumulation of heavy metals by microorganisms and their removal from the soil is a fine solution for plants in metal polluted soil. Proteobacteria, firmicutes, and actinobacteria are able to remove excessive concentration of As, Mn, and Pb from heavy metal contaminated soil (Zhang et al. 2015). The copper (Cu) concentration > 1 mM inhibits the actual growth of *Vicia faba* (Fatnassi et al. 2015). By co-inoculation with PGPR and rhizobia, detrimental effects of metal pollution are decreased. AM fungi, through decreasing H_2O_2 and malonaldehyde (MDA) ameliorates negative effects of cadmium stress (Hashem et al. 2016). Jing et al. (2014) demonstrated that metal-tolerant *Enterobacter* sp. and *Klebsiella* sp. offer their beneficial association through plant growth factor production. Further, in contaminated soil the cadmium-resistant PGPB of *Klebsiella* sp. BAM1 and *Micrococcus* sp. MU1 induce cadmium mobilization, root elongation, and improve plant growth (Prapagdee et al. 2013). Plant growth and nutrition acquisition was improved by siderophore production and phosphate solubilization, induced by arsenic-resistant bacteria (ARB) which were isolated from *Pteris vittata*

(Ghosh et al. 2015). Two bacterial species of *Azospirillum brasilense* Az39 and *Bradyrhizobium japonicum* E109 efficiently propagate in arsenic (As) contaminated soil and improve plant growth through As accumulation in cell biomass (Armendariz et al. 2015). In total, PGPR have promotion effect on plant development under contamination of heavy metals or restrain their inclusion in the tissues of plant (Li et al. 2007).

2.2.2.7 Mechanism of Heavy Metals Removal Assisted by PGPM

Phytoremediation is regarded as an advanced approach rather than conventional methods whose efficiency is achieved by the use of PGPM as heavy metal removing agents in contaminated soil (Glick 2010). It is an efficient novel and healthy method for removing heavy metals. PGPB make the metals available for accumulation and remove them through chemical and physical process (Ullah et al. 2015). Microbes neutralize heavy metals by reducing their high toxic concentration through extracellular/intracellular accumulation, sequestration, and biotransformation (Babu et al. 2013; Qian et al. 2012). Another group of microbes removes heavy metals by their complete degradation. *Pseudomonas* sp. MBR, for instance, demonstrates the ability in biotransformation of single and complex form of ions (Qian et al. 2012). Phytoremediation by PGP microbes are developed via either direct or indirect mechanisms.

2.2.2.8 Direct Mechanism of Phytoremediation

Main direct processes involved in phytoremediation by PGPM include accumulation, bioavailability of heavy metals, and their solubilization (Vymazal and Březinová 2016). Plant-associated microbes activate numerous mechanisms in contaminated soils for removing of heavy metals. Siderophore, produced by plant-associated microbes, is an organic compound with low molecular weight which chelates heavy metals and makes them available in rhizosphere. Siderophore performs its role by ferric iron chelation, high affinity to metals and formation of complexes and their transportation into cytosol (Das et al. 2007; Saha et al. 2016; Złoch et al. 2016). Among heavy metals, formation of siderophore iron complex and their transfer to cytosol is the most common chelation process (Złoch et al. 2016). Some microbes, other than rhizobacteria, which settle inside plant tissues, are capable of phytoremediation. Endophytic bacteria through providing mineral nutrient, enzymes, and growth regulators demonstrate metal resistance features and hence promote plant growth during metals stress. Rhizosphere bacteria are the main producers of siderophore (Ma et al. 2016a, b; Złoch et al. 2016). During metal stress and poor nitrogen condition, endophytic microbes are able to synthesize nitrogenase enzyme and provide appropriate nitrogen concentrations to the related plants. Gupta et al. (2013) have isolated stem endophytic genera of *Acinetobacter*, *Burkholderia*, *Rahnella*, and *Sphingomonas* from *Populus trichocarpa* and *Salix sitchensis* capable of synthesizing nitrogenase enzyme and atmospheric nitrogen fixation. During long-term nitrogen deficiency, endophytic bacteria increase the rate of nitrogen fixation (Gupta et al. 2013). The phytohormones (especially auxin) produced by endophytic bacteria improve nutrient uptake and root growth. Further, PGPMs produce some organic acids with low molecular weight which are significant elements of phytoremediation. Citric, gluconic, and oxalic acids are considered as the most efficient agents of mobilizing heavy metals (Ullah et al. 2015; Janoušková et al. 2006). Additionally, mobility of heavy metals such as Hg, As, Se, and Cr is extensively affected by reduction or oxidation reactions. Some metals show less solubility in high oxidation state rather than low oxidation state. Metalloids retain their solubility in both oxidation state and ionic form (Bolan et al. 2014). Bio-methylation is another method for mobilizing heavy metals which includes the transfer of methyl group by bacterial activity. A large number of bacteria mediate methylation of Pb, Hg, Se, As, Tn, and Sn (Bolan et al. 2014). Phytochelatins (PCs) are cysteine-rich peptides/enzymes which bind to metals and are basically synthesized from glutathione by some fungi and plants (Gadd 2010).

2.2.2.9 Indirect Mechanism of Phytoremediation

Indirect mechanism related to phytoremediation includes inhibiting pathogen infection, increasing the accumulation of heavy metals and improving plant growth and development. High concentration of heavy metals in rhizosphere interferes with nutrient uptake and plant growth is limited. The microbes of plant growth promoting type provide nutrient during harsh conditions. During metal stress condition, PGPB by symbiotic association, perform atmospheric nitrogen fixation and supply the plant nutrition (Nonnoi et al. 2012). Large amount of phosphorus, as the important necessary element of plant growth, exists in the soil. However, it exists in complex form which is unavailable to plants. Phosphorus mostly exists as insoluble form in the soil (Lavakush et al. 2014). Microbes synthesize organic acids which through acidification solubilize the insoluble materials and hence make them accessible to the plant. During metal stress, entophytic bacteria through controlling the pathogens or ISR enhance plant growth (Ma et al. 2016a, b). In conclusion, a variety of microbes have potent efficiency in phytoremediation and heavy metals removal which ultimately promotes plant growth and contribute plants to survive under heavy metal stress.

2.2.2.10 Temperature Stress

One of the other consequences of climate change is frequent occurrence and intense degree of temperature stress. Alike cold condition, heat stress (HS) is concerned as the world crucial abiotic stresses threatening food security and crop productivity. Temperature stress causes significant changes in water content (transpiration), plasma membrane disrupture, enzyme malfunctioning, impaired photosynthesis activity, and restrained cell division and plant growth. The tropical and subtropical regions, such as India, are mainly touched by global climate change (Rodell et al. 2009; Alam et al. 2017). Several cell and cell membrane components are affected by temperature, for example, heat increments fluidity of the cell while cold rigidifies it. Heat stress is caused by a number of physiological and biological resources. Heat stress is a critical abiotic stress causing substantial changes in plant hormone concentration and the related responses. During high temperature, JA concentration is increased to several folds. Complex regulatory mechanisms are involved in plants to induce tolerance to the crop plant. A myriad of plant species has developed adaptation approaches to suboptimal temperature. The fluctuations in environmental factors stimulate several physiological changes enabling plant species to acclimatize and survive in heat stress and cold shock. Production and accumulation of enzymes along with osmolytes are the mechanisms that plants use to defend against heat stress. Major functional proteins of temperature stress include ROS-scavenging enzymes (ascorbate peroxidase and catalase) and heat shock proteins (HSP20, HSP 60, HSP70, HSP 90, and HSP100) (Qu et al. 2013; Kotak et al. 2007). However, many of the crop plants cannot tolerate extreme temperatures. Hence, there is an imperative necessity to investigate tolerance mechanisms in response to extreme temperature.

2.2.2.11 Tolerance Mechanisms of Heat Stress

The climate change of international board (IPCC 2007) presented that universal temperatures are foretell to increment by 1.8–3.6 °C by the terminal of twenty-first century because of changes in climate. High temperatures are a main impediment in crop production also microbial colonization, that these results in significant cellular damage such as degradation of protein and assembly. The organism's response to high temperature is the generation of a particular class of polypeptides as heat shock proteins (HSPs). Adaptation to stress in microorganisms demonstrates a process of complex multilevel regulatory that can involve many genes (Srivastava et al. 2008), such that microbes expand diverse conformity strategies to action the stress. Some microbe's action transcend at high temperatures that can be momentous for plants at high temperature. The bacteria *P. putida* strain NBR10987 represented thermos-toleration in the stress of drought in chickpea rhizosphere and was ascribed to the overexpression of stress sigma factor (δs) and formation of thick biofilm (Srivastava et al. 2008). Specified bacterial strains struggle with stress conditions by generating exopolysaccharides (EPS), which contain incomparable cementing and water holding specifications, and play essential roles in stress toleration by water maintenance and formation of biofilm. The seedlings of sorghum inoculated with *Pseudomonas* AKM-P6 strain had meliorated tolerance to stress of heat via boosted physiological and metabolic proficiency furthermore indicating a unique interplay of inducible proteins in toleration of heat by microbes usage (Ali et al. 2009). The stress of low temperature is a main limiting agent to production of crops since it has damaging effects on growth and development of plants. Inoculated grapevines with

B. phytofirmans PsJN increases tolerance to low temperatures (nonfreezing). Inoculation of this bacteria promote faster accumulation of metabolites and stress-relevant proteins, which resulted high affective insistence to low temperature, that indicate a positive priming effect on plants (Theocharis et al. 2012). Grapevine roots inoculated with *B. phytofirmans* PsJN lead to increased root growth, higher plant biomass, and incremented physiological acting at low temperature. Subsequent analysis disclosed that inoculated plants meaningfully incremented proline, starch, and phenolic rates compared with uninoculated control plants, which boosted low temperature toleration in grapevine plants (Barka et al. 2006). Usually soybean symbiotic activities (nodule infection and nitrogen fixation) are inhibited by low temperature; however, soybean inoculated with both *Bradyrhizobium japonicum* lead to faster growth at 15 °C (Zhang et al. 1996). Inoculation of wheat seedlings with *Pseudomonas* sp. strain PPERs23 increased lengths of root and shoot and content of amino acid, total phenolics and chlorophyll. Also, inoculated seedlings of wheat had increased physiologically usable iron, proline, protein, anthocyanins, relative water contents and decreased ratio of Na^+/K^+ and leakage of electrolyte, lead to raised toleration of cold (Mishra et al. 2009). The many bacterial strains have been reported for increased cold-tolerance stress in plants (Selvakumar et al. 2008). *Pseudomonas cedrina*, *Arthrobacter nicotianae*, and *Brevundimonas terrace* adapted for low temperature exhibition multifunction plant growth promoting potency (Yadav et al. 2014). The bacteria isolated from pea plant root nodules of low temperature growing have effective biofertilizer competency in low temperature (Meena et al. 2015).

2.2.2.12 Stress of Nutrient Deficiency

The mankind can be applicate of useful microbes to increase the sustainability of running systems of agriculture. The microbiome of rhizosphere has significant roles in handling of plant nutrients (Adhya et al. 2015). The famous examples consist of PGP rhizobacteria and mycorrhizal fungi involved in phosphorus uptake (Miransari 2011). The plants generally obtain nutrients from the phyllosphere and rhizosphere (Turner et al. 2013). Management of the plant nutrient entails optimal usage of atmosphere, water, soil factors, and NPK fertilizers (Miao et al. 2011) along with advantageous microbiome to assistance ameliorate the efficiency of nutrient usage. An abound of investigation is existent on the advantage of symbionts such as mycorrhizal fungi for transferring essential macro- and microelements such as phosphorus and other nutrients from soil to growing plants (Salvioli et al. 2016). Except of *Rhizobium* and *Bradyrhizobium*, various other bacteria of endophytes have been presented to provide symbiosis with plants for nitrogen fixation bioavailability in unspecialized host tissues even in the inexistence of nodules (Santi et al. 2013). For instance, *Cyanobacteria* coexist with a range of plants from different clads (such as gunnera, cycads, and lichens) and create heterocysts appropriate for biological nitrogen fixation with association of nitrogenase enzyme (Santi et al. 2013). In a study disclosed that 77 of 102 isolated bacteria from root of sugarcane fix nitrogen and solubilize phosphorus, respectively, without any problem (Leite et al. 2014). Similarly, analysis of the cowpea plant rhizosphere performing sequencing of 16S rRNA presented that *Burkholderia* and *Achromobacter* species with association *Rhizobium* and *Bradyrhizobium* can nodulate cowpea roots and carry out BNF (Guimarães et al. 2012). Several reports have represented that algal genera such as *Anabaena*, *Phormidium*, and *Aphanocapsa* able to fix atmospheric nitrogen in paddy fields (Hasan 2013). To consider the significance of necessary plant nutrients, it would be rational to find out bacterial species that influence uptake of macro- and micronutrient in plants at variant defective and toxic situations (Mapelli et al. 2012). Also, plant microbiomes can boost the several trace elements uptakes such as Fe and Ca (Lee et al. 2010). Generally, the microbiome plant in rhizosphere act vital roles in degrading organic compositions that are needed not only for their durability but also for plant development in nutrient deficiency soils (Bhattacharyya et al. 2015).

2.2.3 Biotic Stress

Biotic stress occurs as the consequence of damage in an organism caused by other living organisms. Plant's root exudates in addition to other chemicals produced by plants attract a variety of microbes. The main damage causing pathogens or infectious organisms of the crop plants include fungi, bacteria, pests, harmful insects, weeds, and viruses (Ramegowda and Senthil-Kumar 2015). They have undesirable and deleterious effects of nutrient imbalance, unstable hormonal regulation, and physiological disorder. Plenty of plants modify their gene expression and confront biotic stress through acclimatization and adaptation processes. Nonetheless, some nonpathogenic microbes are able to suppress a series of pathogen-caused diseases. Therefore, to replace pesticides and chemical fertilizers, the use of beneficial microbes and PGPM as biological control has been regarded as an appropriate and stable alternative approach. Under abiotic stress condition, the biocontrol agents have beneficial effect against soilborne diseases in terms of crop yield and plant growth/development. The plant-associated bacteria and fungi assist in colonization of root hair and promotion of plant growth, health, and development. PGPMs are the cost-effective and eco-friendly tools for suppressing diseases. They defend against pathogens by activation of cellular components including cell wall reinforcement, cellular disruption, and production of secondary metabolites. JA, ethylene, and salicylic acid (SA) play an important role in signal transduction and defense procedure (Verhage et al. 2010; Bari and Jones 2009). Co-inoculation with PGPR and mycorrhizae alleviates destructive effects of biotic stress. Through improving growth properties and decreasing plant susceptibility to disease, PGPRs and mycorrhizae protect plant against pathogens (Dohroo and Sharma 2012).

2.2.3.1 Mechanism of Biotic Stress Tolerance

The naturally occurring plant-microbe interactions, through providing nutrient mobilization and protection against pathogens, are vital for the plant to reach its desired actual growth (Shoebitz et al. 2009). Through microbe-plant association various elicitors are released which induce plant physiological, biochemical, and molecular changes in plants ensuring disease resistance for several months. In biotic stresses, ROS production and oxidative disruption are considered as the important tolerance mechanisms (Miller et al. 2010). Microbes trigger defensive mechanisms via systemic acquired resistance (SAR) or induced systematic resistance (ISR) pathways. The latter may be reinforced by plant growth promoting microbes of nonpathogenic root associated types, while SAR is correlated with pathogenesis-related (PR) proteins and involves modifications of gene expression. Gene stimulation and expression in ISR and SAR follow different patterns depending on the elicitor and the induced regulatory pathways (Nawrocka and Małolepsza 2013). During biotic stress, PGPMs stimulate SAR pathway which induces accumulation of PR proteins and SA, while ISR relies on jasmonate- and ethylene-regulated pathways (Bari and Jones 2009; Salas-Marina et al. 2011). Reactive nitrogen oxygen species (NOS) and oxygen species extensively affect JA, ET, and SA production and alleviate the negative effects of pathogens by complex network (Bari and Jones 2009; Choudhary and Johri 2009). Regulatory factors like ethylene have significant role in triggering PR gene expression. Nonpathogenic microbes release elicitors which stimulate resistance in plants.

2.2.3.2 Induced Systemic Resistance of Plants

Induced systemic resistance (ISR) is a resistance mechanism in plants induced by infection. In this regard, infections made by fungal, bacterial, and virus microbes induce resistance in plants against possible future attacks (ISR) (Heil 2001). The ISR infused by phytopathogens, provides plant immunity against wide spectra of pathogens. PGPM-related ISR is activated through allopathic compound production and competition for ecotype and nutrient. Siderophores and antibiotics are the allelochemicals which contribute plant to overcome the pathogens (Choudhary and Johri 2009; Jain et al. 2013). The defense mechanisms induced by PGPM were first reported in carnation and cucumber in response to *Fusarium* sp. (wilt disease pathogen) and foliar disease pathogen (*Colletotrichum orbiculare*) (Compant et al. 2005). Systemic resistance induction by *Bacillus amyloliquefaciens* was reported by Lee et al. to efficiently activate resistance in *Panax ginseng* against *P. cactorum*. Furthermore, *Pseudomonas* strains are able to limit crop disease through ISR. *Paenibacillus* potentially develops ISR and provides effective biological control agent (BCA) for cabbage against black rot disease (*Xanthomonas campestris*) and (Ghazalibiglar et al. 2016). ISR developed by *Bacillus* strains is reported by Chithrashree et al. (2011) in rice against bacterial leaf blight (*Xanthomonas oryzae*).

2.2.3.3 Systemic Acquired Resistance

Systemic acquired resistance (SAR) is developed in response to infections as the active defense mechanisms in plants. SAR is important for plant to resist against disease and to recover from disease. The nature of pathogen is recognized by host plant based on the generated molecular pattern and detoxification processes are adopted by modification of plant gene expression and production of hormones and metabolites (Sunkar et al. 2012). *Arthrobacter* sp. and *Bacillus* sp. isolated from the tomato rhizosphere show great potential as plant growth promoters due to biocontrol properties, IAA production and phosphate solubilization. A number of PGPB such as *P. aeruginosa* PS1 counteract efficiently against fungicides which are naturally produced by fungi. They can be used in soil against fungicides. Ahemad and Khan (2012) demonstrated that siderophores, phytohormones, ammonia, and hydrogen cyanide were accumulated under stress condition.

2.3 Cross Talks Between Plants and Microbes During Stress Conditions

There are several cross talks between microbes and plants during their interaction by different signaling molecules. Various microbes are hurtful to plants that limit growth and development. Plants have mechanism to recognize specified compositions abandon from microbes and enhance defense responses. The plant signaling hormones, namely, SA, JA, and ethylene, are used to activate defense mechanism during the interaction between microbes and plants in response to stress conditions (Yi et al. 2014; Singh et al. 2019). Plants identify pathogens by detecting extracellular molecules that are called pathogen-associated molecular patterns (PAMPs) or microbe-associated molecular patterns, namely, Ef-TU proteins, bacterial flagella, lipopolysaccharides, and peptidoglycans (Boller and Felix 2009), and/or intracellular effector proteins or tissue damage using pattern recognition receptor (PRR) proteins located on the cell surface or within the cell (Rivas and Thomas 2005). The plant immune system comprises of four levels. In level 1, PAMPs of microbes are recognized and bind to specific PRRs located on the cell surface that triggers the plant immune system and leads to enhanced immunity (PTI), which prevents colonization and proliferation (Bakker et al. 2012; Newman et al. 2013). In level 2, several pathogens induced effectors that enhance virulence. The effectors hinder with PTI and lead to effect or triggered susceptibility. In level 3, nucleotide-binding leucine-rich repeat receptor proteins recognize the effector, which activated the effector-triggered immunity (ETI) that leads to disease resistance. In level 4, natural selection has motivated pathogens to conquer ETI by emerging effectors promoting virulence till plants have developed new receptors.

2.3.1 *Plant Microbiome and Physiological Mechanism of Against Stress*

It is clear that relationships between phytomicrobiome and plants are ancient and long time. Several studies have enhanced our understanding on physiological methods associated with roots, chemical molecules produced by roots, signaling between microbes and root, and possible defense mechanisms (Pangesti et al. 2013). Researchers have given special care to microbes associated with root in soil among other symbiotic associations between many plants and microorganisms. Mycorrhiza is eminent by fungal colonization inside or outside the cell that helps in nutrient assumption (Field et al. 2015; Kumar et al. 2017). Rhizobacteria form root nodules of leguminous plants, involve in nitrogen fixation, and deliver it to the plants (Gourion et al. 2015). These affiliations have given data about mutualistic relationship since plants have created constitutive and inducible defense mechanism to keep away from destructive communications.

2.4 PGPB Impacts on Plants

PGPB (plant growth promoting bacteria) are improving plant growth and tolerance against environmental stress (Prasad et al. 2015). Plants are exposed to different abiotic stress conditions, and phytohormones play a vital role in signaling such as JA, ABA, ET, and SA that react to stress defending plants from different environmental challenges (Fujita et al. 2006). Further, studies reported that ACC deaminase activity of PGPB could regulate the stresses in plants (Glick 2014). The PGPB not only help in combating to abiotic stresses but also enhance crops productivity including rice, corn, and soybean (Suarez et al. 2015). Improved root colonizing capability of *Pseudomonas* sp. laterally with its ability to synthesize exopolysaccharides prompts improved resistance in respond to salinity stress in rice during germination (Sen and Chandrasekhar 2014). Similarly, it has been demonstrated that inoculation of *Bacillus pumilus* enhances rice development in reaction to stresses of heavy metal and salinity (Khan et al. 2016).

2.4.1 *Phytohormones*

The phytohormones are chemical syntaxes and act a major function in plants defense mechanism. In this state, plants react and adapt to abiotic stresses by matching the phytohormone levels. Phytohormones are produced in one part and show their effects on other parts. Many abiotic stresses, for example, drought, cold, salinity, and physical damages (wounding), are mostly controlled by the phytohormones and stress conditions activate phytohormones signaling ways. A few reports have shown that PGPB fortify plant development by direct or indirect systems. In the direct mechanism, microbes accumulate phytohormones, for example, IAA, gibberellins, cytokinins, and ET, that invigorate plant development as well as regulate the hormone level in plants that can likewise antagonize to phytopathogens (Rajkumar et al. 2013). In indirect mechanism, the microbes actuate plant resistance by producing chemicals that can regulate the hormone level. PGPB can likewise animate plant development by communicating the compound ACC deaminase that severs ACC to α-ketobutyrate and alkali, diminishing the ET level in plants (Sun et al. 2009). Usually, plants synthesize low ET that is valuable for plant development and improvement. Further, amid stress responses in plants, the expanded ET biosynthesis is alluded to as "stress ET" (Stearns and Glick 2003) that is a response to biotic and abiotic stress conditions (Lim and Kim 2013).

2.4.2 Colonization of Root

Rhizobacteria are colonizing plant roots amid various phases of plant development, and they can proliferate on roots to assemble a mutual association among plants and microorganisms, where these communications give advantages to both the partners (Reiter and Sessitsch 2006). The mechanism of the microbial group to metabolize and vie for carbon sources in the rhizosphere is reliant on the synthesis of plant root exudates (Farrell et al. 2014). Once the microscopic organisms colonize the root, they can habit on the surface of the roots (epiphytic) or can enter into the root and spread into the ethereal parts of the plant and vascular tissue cortex (endophytic) (James 2000). Many researchers observed that Gram-negative and Gram-positive microbes enter into the root through the primary root, horizontal roots, and root hair (Huang et al. 2011). Furthermore, it has been reported that *Curvularia protuberata* microorganisms colonize with the root and defense *Dichanthelium lanuginosum* and *Solanum lycopersicum* plants from drought and heat stress conditions (de Zelicourt et al. 2013).

2.4.3 Mechanism of Quorum Sensing

Quorum sensing is the process of communication between cells in bacteria by persuading different chemical. This encourages the microbial groups to react rapidly, hinder contending organisms, enhance supplement uptake, and adjust to changing ecological conditions. Quorum sensing depend on the synthesis of diffusible molecules such as *N*-acyl-homoserine lactones (AHLs) (Hartmann and Schikora 2012). Likewise, it controls bacterial size and populace status. AHLs, autoinducer-2, and 2-heptyl-3-hydroxy-4-quinoline are utilized as a part of cell–cell communication inside the bacterial group to synchronize a few activities and influence them to work more like a solitary unit. These signaling particles are exceptional among the microbial species. AHLs in proteobacteria, gamma-butyrolactones in streptomyces, *cis*-11-methyl-2-dodecanoic acid in *Xanthomonas*, and oligopeptides in Gram-positive microorganisms are act as signaling molecules (Danhorn and Fuqua 2007).

2.4.4 Promotion Stress Resistance

The plant microbiome particularly the root microbiome is engaged in the protection against biotic stresses, by going about as a defensive shield against soilborne pathogens (Weller et al. 2002). The components are incorporating different direct communications with plant pathogens and backhanded associations through the plant by incitement of the immune system of plants (Lugtenberg and Kamilova 2009). In the recent research, it has shown that the microbiome is not only involved in coping with biotic stress, it is also involved in protection against abiotic stress (Bragina et al. 2013). For instance, the plant microbiome has been appeared to be associated with defense against drought as well as high salinities stresses (Rolli et al. 2015). Studies reported that the plant microbiome is likewise associated with cold acclimation, an essential factor constraining the development and yield of crops.

2.4.5 Growth and Development of Plant

The plant microbiome also affects the plant secondary metabolites that result in the development of different metabolism in plant. It has been accounted for the flavor of strawberries and the fabrication of bioactive compounds in medicinal plants (Zabetakis et al. 1999). In an examination on *A. thaliana*, the rhizosphere microbiomes are engaged in insect-feeding characteristics, which were most likely an aftereffect of microbiome-driven changes in the metabolites of leaf (Badri et al. 2013). It has been studied that the expulsion of the flower microbiome of *Sambucus nigra* leads to a decreased terpene emanation in flower, which pivotally involved in fertilization and thus in fruit and seed production (Penuelas et al. 2014).

2.4.6 Phenology of Plant

The past studies on plant microbiome uncovered the abrupt effect of the root microbiome on plant phenology. Some reports stated that soil microbes affect the blooming time of a *Boechera stricta* (Wagner et al. 2014). Essentially, regarding fruitful transplantation of rhizosphere microbiomes from *A. thaliana* to *Brassica rapa* affected their blossoming times, bringing about comparative moves in blooming phenology (Partida-Martinez and Heil 2011). Co-advancement of plants and related microbial groups has been estimated in the light of culture-subordinate outcomes got for the rhizosphere of wheat cultivars (Germida and Siciliano 2001), maize, sugar beet, and lettuce by the use of profound sequencing techniques (Cardinale et al. 2015).

2.5 Practical Plant-Associated Microbiome Providing in Agronomy

It is substantial to perceive interactions of plant-microbe to develop an advantageous soil microbiome. It is uncertain whether such useful microbial associations would be consistent in agricultural soils or not. At normal situations, usually, plant roots and soil type characterize the combination and community of microbial societies with roots of plant. The effect of plant roots and soil type on the plant microbiome has been studied widely (Lareen et al. 2016). The soil properties (physiochemical) have a straight effect on particular microbes and root exudates of plant (Dumbrell et al. 2010). Furthermore, type of soil mostly characterizes the soil biome while plant root exudates of plant tend to appoint a desirable rhizobium. Ultimately, type of soil and plant species are momentous roles which define the combination of rhizosphere and recruit different microbial societies for the establishment of a desirable rhizobiome to enhancement crop productivity and decrease biotic and abiotic stresses (Yeoh et al. 2016). These agents considerably chip in to the elective richness of advantageous microbes in the rhizobiome that can assist to recognize heritable properties to ameliorate plant productivity.

2.6 Plant Microbiome and Prospects for Future Studies

The developing investigation related to the microbiome of plant focus to importance of microbiome for plant growth, health and productivity (Prasad et al. 2018, 2020). While most investigates detections are elementary, centralized researches are essential to solve the elegancies of this extremely complex phenomenon to comprehend dynamics and communication of microbes to use this vastly untapped resource (Celebi et al. 2010). Opportunities for exploiting the plant microbiome for raising crops are numerous and diverse. Plant microbiome has a considerable performance in management of plants stress and presented models for comprehension mechanisms of stress tolerance. Create transgenic plants containing advantageous microbe genes would be another strategy, for example, transgenic plants containing ACC deaminase gene obtained from bacteria. With considering the time and other topics related to the transgenic plants' development, it would be high affordable and eco-friendly to develop microbial inoculants to diminish abiotic stresses (Mengual et al. 2014). Even though investigations have indicated considerable improvements to stress tolerance exploiting PGPM to crops under field situations (Mengual et al.

2014), others have showed negative or opposite results (Nadeem et al. 2014). An efficient strategy for a resistant advantageous crop is microbial consortium utilizations in the agriculture to provide the rhizosphere microbiome to response to stresses (biotic and abiotic) without harming to productivity of plant growth (Trabelsi and Mhamdi 2013). Hence, the stress tolerance mechanisms that microbes give to their hosts require more research to extend appropriate microbial consortia for ready-utilization under various biotic and abiotic stresses. This will entail parallel works at interdisciplinary levels from different fields (microbiologists, molecular biologists, soil scientists, plant breeders, plant physiologists, and agronomists).

2.7 Conclusion

The various kinds of biotic and abiotic stresses including temperature, salinity, flooding, drought, ultraviolet radiation, and heavy metals are effecting characteristics of plant growth, qualities, survivability, and their productivities. These stresses cause crop losses worldwide. Overexpression of salinity, drought, cold, heat, and alkalinity tolerant proteins as a result of modified physiological and biological properties helps crop plants to survive in stress conditions. Stress conditions primarily lead to hormonal imbalance, vulnerability to ion toxicity, nutrient mobilization, reducing crop yield, food quality, and security in addition to suppressing plant growth and development. The proper solution to the stress conditions in plants is to exploit triple interaction of plant-microbe soil in microbe-mediated tools and techniques. During abiotic and biotic stress, fortifying plants through their consolidation with stress-tolerant PGPMs, and mycorrhizae fungi promote plant growth. The microbes perform their promoting effect by providing nutritional requirements, regulating plant hormones and siderophore production, and improving the antioxidant system. Under multiple stresses, defensive mechanisms further involve induced ASR and ISR during. The provision of nutrient and water is augmented by AM during stress condition which gives plant the feature of stress tolerance. Utilization of microbes offers a great solution to secure future food issues while maintaining soil health. Generally, this chapter describes microbes as ecological engineers to neutralize stress conditions. Based on what has been reviewed, we advise scientific societies and policymakers to present an organized future plan for adverse biotic and abiotic conditions and their corresponding destructive effect on global economy as well as food security. Considering the current challenges, there is demanding necessity for future research to characterize possible stress-tolerant PGPMs. For this purpose, a range of thorough studies are required to assess diversity of microbial community to formulate effective microbial spectra to defeat the detrimental effects of global environmental change.

References

Adhya TK, Kumar N, Reddy G, Podile AR, Bee H, Samantaray B (2015) Microbial mobilization of soil phosphorus and sustainable P management in agricultural soils. Curr Sci 108(7):1280–1287

Ahemad M, Khan MS (2012) Alleviation of fungicide-induced phytotoxicity in greengram (*Vigna radiata* L.) [Wilczek] using fungicide-tolerant and plant growth promoting Pseudomonas strain. Saudi J Biol Sci 19(4):451–459

Ahmad M, Zahir ZA, Asghar HN, Asghar M (2011) Inducing salt tolerance in mung bean through coinoculation with rhizobia and plant-growth-promoting rhizobacteria containing 1-aminocyclopropane-1-carboxylate deaminase. Can J Microbiol 57(7):578–589

Alam MA, Seetharam K, Zaidi PH, Dinesh A, Vinayan MT, Nath UK (2017) Dissecting heat stress tolerance in tropical maize (*Zea mays* L.). Field Crop Res 204:110–119

Ali SZ, Sandhya V, Grover M, Kishore N, Rao LV, Venkateswarlu B (2009) Pseudomonas sp. strain AKM-P6 enhances tolerance of sorghum seedlings to elevated temperatures. Biol Fertil Soils 46(1):45–55

Armendariz AL, Talano MA, Oller AL, Medina MI, Agostini E (2015) Effect of arsenic on tolerance mechanisms of two plant growth-promoting bacteria used as biological inoculants. J Environ Sci 33:203–210

Arora S, Patel PN, Vanza MJ, Rao GG (2014) Isolation and characterization of endophytic bacteria colonizing halophyte and other salt tolerant plant species from coastal Gujarat. Afr J Microbiol Res 8(17):1779–1788

Asgari Lajayer B, Ghorbanpour M, Nikabadi S (2017) Heavy metals in contaminated environment: destiny of secondary metabolite biosynthesis, oxidative status and phytoextraction in medicinal plants. Ecotoxicol Environ Saf 145:377–390

Asgari Lajayer B, Khadem Moghadam N, Maghsoodi MR, Ghorbanpour M, Kariman K (2019) Phytoextraction of heavy metals from contaminated soil, water and atmosphere using ornamental plants: mechanisms and efficiency improvement strategies. Environ Sci Pollut Res 26 (9):8468–8484

Babu AG, Kim JD, Oh BT (2013) Enhancement of heavy metal phytoremediation by *Alnus firma* with endophytic *Bacillus thuringiensis* GDB-1. J Hazard Mater 250:477–483

Bacilio M, Moreno M, Bashan Y (2016) Mitigation of negative effects of progressive soil salinity gradients by application of humic acids and inoculation with *Pseudomonas stutzeri* in a salt-tolerant and a salt-susceptible pepper. Appl Soil Ecol 107:394–404

Badri DV, Zolla G, Bakker MG, Manter DK, Vivanco JM (2013) Potential impact of soil microbiomes on the leaf metabolome and on herbivore feeding behavior. New Phytol 198 (1):264–273

Bailey-Serres J, Voesenek LA (2008) Flooding stress: acclimations and genetic diversity. Annu Rev Plant Biol 59:313–339

Bakker MG, Manter DK, Sheflin AM, Weir TL, Vivanco JM (2012) Harnessing the rhizosphere microbiome through plant breeding and agricultural management. Plant Soil 360(1–2):1–3

Bal HB, Nayak L, Das S, Adhya TK (2013) Solation of ACC deaminase producing PGPR from rice rhizosphere and evaluating their plant growth promoting activity under salt stress. Plant Soil 366 (1-2):93–105

Bano A, Fatima M (2009) Salt tolerance in *Zea mays* L. following inoculation with Rhizobium and Pseudomonas. Biol Fertil Soils 45(4):405–413

Barassi CA, Ayrault G, Creus CM, Sueldo RJ, Sobrero MT (2006) Seed inoculation with Azospirillum mitigates NaCl effects on lettuce. Sci Hortic 109(1):8–14

Bari R, Jones JD (2009) Role of plant hormones in plant defence responses. Plant Mol Biol 69:473–488

Barka EA, Nowak J, Clement C (2006) Enhancement of chilling resistance of inoculated grapevine plantlets with a plant growth-promoting rhizobacterium, *Burkholderia phytofirmans* strain PsJN. Appl Environ Microbiol 72(11):7246–7252

Battisti DS, Naylor RL (2009) Historical warnings of future food insecurity with unprecedented seasonal heat. Science 323(5911):240–244

Bhattacharyya D, Yu SM, Lee YH (2015) Volatile compounds from *Alcaligenes faecalis* JBCS1294 confer salt tolerance in *Arabidopsis thaliana* through the auxin and gibberellin pathways and differential modulation of gene expression in root and shoot tissues. Plant Growth Regul 75(1):297–306

Bolan N, Kunhikrishnan A, Thangarajan R, Kumpiene J, Park J, Makino T, Kirkham MB, Scheckel K (2014) Remediation of heavy metal (loid) s contaminated soils–to mobilize or to immobilize? J Hazard Mater 266:141–166

Boller T, Felix G (2009) A renaissance of elicitors: perception of microbe-associated molecular patterns and danger signals by pattern-recognition receptors. Annu Rev Plant Biol 60:379–406

Bragina A, Cardinale M, Berg C, Berg G (2013) Vertical transmission explains the specific Burkholderia pattern in Sphagnum mosses at multi-geographic scale. Front Microbiol 4:394

Bray EA (2000) Response to abiotic stress. In: Gruissem W, Jones R (eds) Biochemistry and molecular biology of plants. American Society of Plant Physiologists, Rockville, pp 1158–1163

Cardinale M, Grube M, Erlacher A, Quehenberger J, Berg G (2015) Bacterial networks and co-occurrence relationships in the lettuce root microbiota. Environ Microbiol 17(1):239–252

Celebi SZ, Demir S, Celebi R, Durak ED, Yilmaz IH (2010) The effect of arbuscular mycorrhizal fungi (AMF) applications on the silage maize (*Zea mays* L.) yield in different irrigation regimes. Eur J Soil Biol 46(5):302–305

Chen P, Jeannotte R, Weimer BC (2014) Exploring bacterial epigenomics in the next-generation sequencing era: a new approach for an emerging frontier. Trends Microbiol 22(5):292–300

Chirakkara RA, Cameselle C, Reddy KR (2016) Assessing the applicability of phytoremediation of soils with mixed organic and heavy metal contaminants. Rev Environ Sci Biotechnol 15 (2):299–326

Chithrashree AC, Udayashankar S, Nayaka SC, Reddy MS, Srinivas C (2011) Plant growth-promoting rhizobacteria mediate induced systemic resistance in rice against bacterial leaf blight caused by *Xanthomonas oryzae* pv. oryzae. Biol Control 59(2):114–122

Chodak M, Gołębiewski M, Morawska-Płoskonka J, Kuduk K, Niklińska M (2015) Soil chemical properties affect the reaction of forest soil bacteria to drought and rewetting stress. Ann Microbiol 65(3):1627–1637

Choudhary DK, Johri BN (2009) Interactions of Bacillus spp. and plants–with special reference to induced systemic resistance (ISR). Microbiol Res 164(5):493–513

Cohen AC, Bottini R, Pontin M, Berli FJ, Moreno D, Boccanlandro H, Travaglia CN, Piccoli PN (2015) *Azospirillum brasilense* ameliorates the response of *Arabidopsis thaliana* to drought mainly via enhancement of ABA levels. Physiol Plant 153(1):79–90

Comas L, Becker S, Cruz VMV, Byrne PF, Dierig DA (2013) Root traits contributing to plant productivity under drought. Front Plant Sci 4:442

Compant S, Duffy B, Nowak J, Clément C, Barka EA (2005) Use of plant growth-promoting bacteria for biocontrol of plant diseases: principles, mechanisms of action, and future prospects. Appl Environ Microbiol 71(9):4951–4959

Damodaran T, Sah V, Rai RB, Sharma DK, Mishra VK, Jha SK, Kannan R (2013) Isolation of salt tolerant endophytic and rhizospheric bacteria by natural selection and screening for promising plant growth-promoting rhizobacteria (PGPR) and growth vigour in tomato under sodic environment. Afr J Microbiol Res 7(44):5082–5089

Danhorn T, Fuqua C (2007) Biofilm formation by plant-associated bacteria. Annu Rev Microbiol 61:401–422

Das A, Prasad R, Srivastava A, Giang PH, Bhatnagar K, Varma A (2007) Fungal siderophores: structure, functions and regulations. In: Varma A, Chincholkar SB (eds) Microbial siderophores, vol 12. Springer, Berlin, pp 1–42

Dat J, Vandenabeele S, Vranová E, Van Montagu M, Inzé D, Van Breusegem F (2000) Dual action of the active oxygen species during plant stress responses. Cell Mol Life Sci 57(5):779–795

Davletova S, Rizhsky L, Liang H, Shengqiang Z, Oliver DJ, Coutu J, Shulaev V, Schlauch K, Mittler R (2005) Cytosolic ascorbate peroxidase 1 is a central component of the reactive oxygen gene network of Arabidopsis. Plant Cell 17(1):268–281

de la Torre-González A, Navarro-León E, Albacete A, Blasco B, Ruiz JM (2017) Study of phytohormone profile and oxidative metabolism as key process to identification of salinity response in tomato commercial genotypes. J Plant Physiol 216:164–173

de Zelicourt A, Al-Yousif M, Hirt H (2013) Rhizosphere microbes as essential partners for plant stress tolerance. Mol Plant 6(2):242–245

Delangiz N, Behrouzi Varjovi M, Asgari Lajayer B, Ghorbanpour M (2019) The potential of biotechnology for mitigation of greenhouse gasses effects: solutions, challenges, and future perspectives. Arab J Geosci 12(5):174

Disante KB, Fuentes D, Cortina J (2011) Response to drought of Zn-stressed *Quercus suber* L. seedlings. Environ Exp Bot 70(2-3):96–103

Dohroo A, Sharma DR (2012) Role of plant growth promoting rhizobacteria, arbuscular mycorrhizal fungi and their helper bacteria on growth parameters and root rot of apple. World J Sci Technol 2(12):35–38

Duan J, Müller KM, Charles TC, Vesely S, Glick BR (2009) 1-aminocyclopropane-1-carboxylate (ACC) deaminase genes in rhizobia from southern Saskatchewan. Microb Ecol 57(3):423–436

Dumbrell AJ, Nelson M, Helgason T, Dytham C, Fitter AH (2010) Relative roles of niche and neutral processes in structuring a soil microbial community. ISME J 4(3):337

Egamberdieva D, Jabborova D, Hashem A (2015) Pseudomonas induces salinity tolerance in cotton (*Gossypium hirsutum*) and resistance to Fusarium root rot through the modulation of indole-3-acetic acid. Saudi J Biol Sci 22(6):773–779

Egamberdiyeva D (2007) The effect of plant growth promoting bacteria on growth and nutrient uptake of maize in two different soils. Appl Soil Ecol 36(2–3):184–189

Emamverdian A, Ding Y, Mokhberdoran F, Xie Y (2015) Heavy metal stress and some mechanisms of plant defense response. Sci World J 2015:756120

Farooq M, Wahid A, Kobayashi N, Fujita D, Basra SM (2009) Plant drought stress: effects, mechanisms and management. In: Sustainable agriculture. Springer, Dordrecht

Farrar K, Bryant D, Cope-Selby N (2014) Understanding and engineering beneficial plant–microbe interactions: plant growth promotion in energy crops. Plant Biotechnol J 12(9):1193–1206

Farrell M, Prendergast-Miller M, Jones DL, Hill PW, Condron LM (2014) Soil microbial organic nitrogen uptake is regulated by carbon availability. Soil Biol Biochem 77:261–267

Fasciglione G, Casanovas EM, Quillehauquy V, Yommi AK, Goñi MG, Roura SI, Barassi CA (2015) Azospirillum inoculation effects on growth, product quality and storage life of lettuce plants grown under salt stress. Sci Hortic 195:154–162

Fatnassi IC, Chiboub M, Saadani O, Jebara M, Jebara SH (2015) Impact of dual inoculation with rhizobium and PGPR on growth and antioxidant status of *Vicia faba* L. under copper stress. C R Biol 338(4):241–254

Feng NX, Yu J, Zhao HM, Cheng YT, Mo CH, Cai QY, Li YW, Li H, Wng MH (2017) Efficient phytoremediation of organic contaminants in soils using plant-endophyte partnerships. Sci Total Environ 583:352–368

Field KJ, Pressel S, Duckett JG, Rimington WR, Bidartondo MI (2015) Symbiotic options for the conquest of land. Trends Ecol Evol 30(8):477–486

Finkel OM, Castrillo G, Paredes SH, González IS, Dangl JL (2017) Understanding and exploiting plant beneficial microbes. Curr Opin Plant Biol 38:155–163

Foyer CH, Noctor G (2005) Oxidant and antioxidant signalling in plants: a re-evaluation of the concept of oxidative stress in a physiological context. Plant Cell Environ 28(8):1056–1071

Fujita M, Fujita Y, Noutoshi Y, Takahashi F, Narusaka Y, Yamaguchi-Shinozaki K, Shinozaki K (2006) Crosstalk between abiotic and biotic stress responses: a current view from the points of convergence in the stress signaling networks. Curr Opin Plant Biol 9(4):436–442

Gadd GM (2010) Metals, minerals and microbes: geomicrobiology and bioremediation. Microbiology 156(3):609–643

Gerhardt KE, Huang XD, Glick BR, Greenberg BM (2009) Phytoremediation and rhizoremediation of organic soil contaminants: potential and challenges. Plant Sci 176:20–30

Germida J, Siciliano S (2001) Taxonomic diversity of bacteria associated with the roots of modern, recent and ancient wheat cultivars. Biol Fertil Soils 33(5):410–415

Ghassemi S, Farhangi-Abriz S, Faegi-Analou R, Ghorbanpour M, Asgari LB (2018) Monitoring cell energy, physiological functions and grain yield in field-grown mung bean exposed to exogenously applied polyamines under drought stress. J Soil Sci Plant Nutr 18(4):1108–1125

Ghazalibiglar H, Hampton JG, de Jong EV, Holyoake A (2016) Is induced systemic resistance the mechanism for control of black rot in *Brassica oleracea* by a Paenibacillus sp.? Biol Control 92:195–201

Ghosh P, Rathinasabapathi B, Ma LQ (2015) Phosphorus solubilization and plant growth enhancement by arsenic-resistant bacteria. Chemosphere 134:1–6

Gill SS, Tuteja N (2010) Reactive oxygen species and antioxidant machinery in abiotic stress tolerance in crop plants. Plant Physiol Biochem 48(12):909–930

Giri B, Mukerji KG (2004) Mycorrhizal inoculant alleviates salt stress in *Sesbania aegyptiaca* and *Sesbania grandiflora* under field conditions: evidence for reduced sodium and improved magnesium uptake. Mycorrhiza 14(5):307–312

Gleick PH (1994) Water, war & peace in the Middle East. Environ Sci Policy Sustain Dev 36 (3):6–42

Glick BR (2010) Using soil bacteria to facilitate phytoremediation. Biotechnol Adv 28(3):367–374

Glick BR (2014) Bacteria with ACC deaminase can promote plant growth and help to feed the world. Microbiol Res 169(1):30–39

Goswami D, Thakker JN, Dhandhukia PC (2015) Simultaneous detection and quantification of indole-3-acetic acid (IAA) and indole-3-butyric acid (IBA) produced by rhizobacteria from l-tryptophan (Trp) using HPTLC. J Microbiol Methods 110:7–14

Gourion B, Berrabah F, Ratet P, Stacey G (2015) Rhizobium–legume symbioses: the crucial role of plant immunity. Trends Plant Sci 20(3):186–194

Govindasamy V, Senthilkumar M, Magheshwaran V, Kumar U, Bose P, Sharma V, Annapurna K (2010) Bacillus and Paenibacillus spp.: potential PGPR for sustainable agriculture. In: Plant growth and health promoting bacteria. Springer, Berlin

Grayson M (2013) Agriculture and drought. Nature 501(7468):S1–S1

Guimarães AA, Jaramillo PM, Nóbrega RS, Florentino LA, Silva KB, de Souza Moreira FM (2012) Genetic and symbiotic diversity of nitrogen-fixing bacteria isolated from soils under agriculture use in the Western Amazon using cowpea as the trap plant. Appl Environ Microbiol AEM-01303

Gupta G, Panwar J, Jha PN (2013) Natural occurrence of *Pseudomonas aeruginosa*, a dominant cultivable diazotrophic endophytic bacterium colonizing *Pennisetum glaucum* (L.) R. Br. Appl Soil Ecol 64:252–261

Hartmann A, Schikora A (2012) Quorum sensing of bacteria and trans-kingdom interactions of N-acyl homoserine lactones with eukaryotes. J Chem Ecol 38(6):704–713

Hasan MA (2013) Investigation on the nitrogen fixing cyanobacteria (BGA) in Rice fields of north-west region of Bangladesh. III: filamentous (Heterocystous). J Environ Sci Nat Resour 6 (1):253–259

Hashem A, Abd_Allah EF, Alqarawi AA, Al Huqail AA, Egamberdieva D, Wirth S (2016) Alleviation of cadmium stress in *Solanum lycopersicum* L. by arbuscular mycorrhizal fungi via induction of acquired systemic tolerance. Saudi J Biol Sci 23(2):272–281

Hayat R, Ali S, Amara U, Khalid R, Ahmed I (2010) Soil beneficial bacteria and their role in plant growth promotion: a review. Ann Microbiol 60(4):579–598

Heil M (2001) Induced systemic resistance (ISR) against pathogens–a promising field for ecological research. Perspect Plant Ecol Evol Syst 4(2):65–79

Hu S, Gu H, Cui C, Ji R (2016) Toxicity of combined chromium (VI) and phenanthrene pollution on the seed germination, stem lengths, and fresh weights of higher plants. Environ Sci Pollut Res 23(15):15227–15235

Huang B, Lv C, Zhuang P, Zhang H, Fan L (2011) Endophytic colonisation of *Bacillus subtilis* in the roots of *Robinia pseudoacacia* L. Plant Biol 13(6):925–931

IPCC (2007) Climate change. The physical science basis. Contribution of working group I to the fourth assessment report of the intergovernmental panel on climate change. Cambridge University Press, Cambridge

Jain S, Vaishnav A, Kasotia A, Kumari S, Gaur RK, Choudhary DK (2013) Bacteria-induced systemic resistance and growth promotion in *Glycine max* L. Merrill upon challenge inoculation with *Fusarium oxysporum*. Proc Natl Acad Sci India Sect B: Biol Sci 83(4):561–567

James EK (2000) Nitrogen fixation in endophytic and associative symbiosis. Field Crop Res 65 (2–3):197–209

Janoušková M, Pavlíková D, Vosátka M (2006) Potential contribution of arbuscular mycorrhiza to cadmium immobilisation in soil. Chemosphere 65(11):1959–1965

Jha Y, Subramanian RB (2014) PGPR regulate caspase-like activity, programmed cell death, and antioxidant enzyme activity in paddy under salinity. Physiol Mol Biol Plants 20(2):201–207

Jha B, Gontia I, Hartmann A (2012) The roots of the halophyte *Salicornia brachiata* are a source of new halotolerant diazotrophic bacteria with plant growth-promoting potential. Plant Soil 356 (1–2):265–277

Jiang S, Zhang D, Wang L, Pan J, Liu Y, Kong X, Zhou Y, Li D (2013) A maize calcium-dependent protein kinase gene, ZmCPK4, positively regulated abscisic acid signaling and enhanced drought stress tolerance in transgenic Arabidopsis. Plant Physiol Biochem 71:112–120

Jing YX, Yan JL, He HD, Yang DJ, Xiao L, Zhong T, Yuan M, Cai XD, Li SB (2014) Characterization of bacteria in the rhizosphere soils of *Polygonum pubescens* and their potential in promoting growth and Cd, Pb, Zn uptake by *Brassica napus*. Int J Phytoremediation 16 (4):321–333

Kapoor D, Singh S, Kumar V, Romero R, Prasad R, Singh J (2019) Antioxidant enzymes regulation in plants in reference to reactive oxygen species (ROS) and reactive nitrogen species (RNS). Plant Gene 19:100182. https://doi.org/10.1016/j.plgene.2019.100182

Kasim WA, Gaafar RM, Abou-Ali RM, Omar MN, Hewait HM (2016) Effect of biofilm forming plant growth promoting rhizobacteria on salinity tolerance in barley. Ann Agric Sci 61 (2):217–227

Khan MA, Gemenet DC, Villordon A (2016) Root system architecture and abiotic stress tolerance: current knowledge in root and tuber crops. Front Plant Sci 7:1584

Khan AL, Waqas M, Asaf S, Kamran M, Shahzad R, Bilal S, Khan MA, Kang SM, Kim YH, Yun BW, Al-Rawahi A (2017) Plant growth-promoting endophyte Sphingomonas sp. LK11 alleviates salinity stress in *Solanum pimpinellifolium*. Environ Exp Bot 133:58–69

Khoshmanzar E, Aliasgharzad N, Neyshabouri MR, Khoshru B, Arzanlou M, Asgari Lajayer B (2019) Effects of Trichoderma isolates on tomato growth and inducing its tolerance to water-deficit stress. Int J Environ Sci Technol. https://doi.org/10.1007/s13762-019-02405-4

Kohler J, Hernández JA, Caravaca F, Roldán A (2008) Plant-growth-promoting rhizobacteria and arbuscular mycorrhizal fungi modify alleviation biochemical mechanisms in water-stressed plants. Funct Plant Biol 35(2):141–151

Kotak S, Larkindale J, Lee U, von Koskull-Döring P, Vierling E, Scharf KD (2007) Complexity of the heat stress response in plants. Curr Opin Plant Biol 10(3):310–316

Krishnamoorthy R, Kim K, Subramanian P, Senthilkumar M, Anandham R, Sa T (2016) Arbuscular mycorrhizal fungi and associated bacteria isolated from salt-affected soil enhances the tolerance of maize to salinity in coastal reclamation soil. Agric Ecosyst Environ 231:233–239

Kumar KV, Srivastava S, Singh N, Behl HM (2009) Role of metal resistant plant growth promoting bacteria in ameliorating fly ash to the growth of *Brassica juncea*. J Hazard Mater 170(1):51–57

Kumar M, Prasad R, Kumar V, Tuteja N, Varma A (2017) Mycorrhizal fungi under biotic and abiotic stress. In: Varma A, Prasad R, Tuteja N (eds) Mycorrhiza. Springer International, Cham, pp 57–70

Kundu P, Gill R, Nehra A, Sharma KK, Hasanuzzaman M, Prasad R, Tuteja N, Singh Gil SS (2020) Reactive oxygen species (ROS) management in engineered plants for abiotic stress tolerance. In: Tuteja N, Tuteja R, Passricha N, Saifi SK (eds) Advancement in crop improvement techniques. Elsevier, Amsterdam, Netherlands, pp 241–262

Lareen A, Burton F, Schäfer P (2016) Plant root-microbe communication in shaping root microbiomes. Plant Mol Biol 90(6):575–587

Lata C, Prasad M (2011) Role of DREBs in regulation of abiotic stress responses in plants. J Exp Bot 62(14):4731–4748

Lavakush YJ, Verma JP, Jaiswal DK, Kumar A (2014) Evaluation of PGPR and different concentration of phosphorus level on plant growth, yield and nutrient content of rice (*Oryza sativa*). Ecol Eng 62:123–128

Lee SW, Ahn IP, Sim SY, Lee SY, Seo MW, Kim S, Park SY, Lee YH, Kang S (2010) Pseudomonas sp. LSW25R, antagonistic to plant pathogens, promoted plant growth, and reduced blossom-end rot of tomato fruits in a hydroponic system. Eur J Plant Pathol 126(1):1–1

Leite MC, de Farias AR, Freire FJ, Andreote FD, Kuklinsky-Sobral J, Freire MB (2014) Isolation, bioprospecting and diversity of salt-tolerant bacteria associated with sugarcane in soils of Pernambuco. Brazil Rev Brasileira Engenharia Agrícola Ambiental 18:73–79

Li WC, Ye ZH, Wong MH (2007) Effects of bacteria on enhanced metal uptake of the Cd/Zn-hyperaccumulating plant, *Sedum alfredii*. J Exp Bot 58(15-16):4173–4182

Lim JH, Kim SD (2013) Induction of drought stress resistance by multi-functional PGPR *Bacillus licheniformis* K11 in pepper. Plant Pathol J 29(2):201

Lugtenberg B, Kamilova F (2009) Plant-growth-promoting rhizobacteria. Annu Rev Microbiol 63:541–556

Ma Y, Rajkumar M, Zhang C, Freitas H (2016a) Inoculation of *Brassica oxyrrhina* with plant growth promoting bacteria for the improvement of heavy metal phytoremediation under drought conditions. J Hazard Mater 320:36–44

Ma Y, Rajkumar M, Zhang C, Freitas H (2016b) Beneficial role of bacterial endophytes in heavy metal phytoremediation. J Environ Manag 174:14–25

Mapelli F, Marasco R, Balloi A, Rolli E, Cappitelli F, Daffonchio D, Borin S (2012) Mineral–microbe interactions: biotechnological potential of bioweathering. J Biotechnol 157(4):473–481

Marasco R, Rolli E, Vigani G, Borin S, Sorlini C, Ouzari H, Zocchi G, Daffonchio D (2013) Are drought-resistance promoting bacteria cross-compatible with different plant models? Plant Signal Behav 8:e26741

Marulanda A, Porcel R, Barea JM, Azcón R (2007) Drought tolerance and antioxidant activities in lavender plants colonized by native drought-tolerant or drought-sensitive Glomus species. Microb Ecol 54(3):543

Marulanda A, Barea JM, Azcón R (2009) Stimulation of plant growth and drought tolerance by native microorganisms (AM fungi and bacteria) from dry environments: mechanisms related to bacterial effectiveness. J Plant Growth Regul 28(2):115–124

Masciarelli O, Llanes A, Luna V (2014) A new PGPR co-inoculated with *Bradyrhizobium japonicum* enhances soybean nodulation. Microbiol Res 169(7-8):609–615

Massad TJ, Dyer LA, Vega G (2012) Costs of defense and a test of the carbon-nutrient balance and growth-differentiation balance hypotheses for two co-occurring classes of plant defense. PLoS One 7(10):e47554

Meena RK, Singh RK, Singh NP, Meena SK, Meena VS (2015) Isolation of low temperature surviving plant growth–promoting rhizobacteria (PGPR) from pea (*Pisum sativum* L.) and documentation of their plant growth promoting traits. Biocatal Agric Biotechnol 4(4):806–811

Mengual C, Schoebitz M, Azcón R, Roldán A (2014) Microbial inoculants and organic amendment improves plant establishment and soil rehabilitation under semiarid conditions. J Environ Manag 134:1–7

Miao Y, Stewart BA, Zhang F (2011) Long-term experiments for sustainable nutrient management in China. A review. Agron Sustain Dev 31(2):397–414

Miller GA, Suzuki N, Ciftci-Yilmaz SU, Mittler RO (2010) Reactive oxygen species homeostasis and signalling during drought and salinity stresses. Plant Cell Environ 33(4):453–467

Miransari M (2011) Hyperaccumulators, arbuscular mycorrhizal fungi and stress of heavy metals. Biotechnol Adv 29(6):645–653

Mishra PK, Mishra S, Selvakumar G, Kundu S, Shankar GH (2009) Enhanced soybean (*Glycine max* L.) plant growth and nodulation by *Bradyrhizobium japonicum*-SB1 in presence of *Bacillus thuringiensis*-KR1. Acta Agric Scand Sect B–Soil Plant Sci 59(2):189–196

Munns R, Tester M (2008) Mechanisms of salinity tolerance. Annu Rev Plant Biol 59:651–681

Nadeem SM, Ahmad M, Zahir ZA, Javaid A, Ashraf M (2014) The role of mycorrhizae and plant growth promoting rhizobacteria (PGPR) in improving crop productivity under stressful environments. Biotechnol Adv 32(2):429–448

Naseem H, Bano A (2014) Role of plant growth-promoting rhizobacteria and their exopolysaccharide in drought tolerance of maize. J Plant Interact 9(1):689–701

Nath M, Bhatt D, Prasad R, Tuteja N (2017) Reactive oxygen species (ROS) metabolism and signaling in plant-mycorrhizal association under biotic and abiotic stress conditions. In: Varma A, Prasad R, Tuteja N (eds) Mycorrhiza. Springer International, Cham, pp 223–232

Nautiyal CS, Srivastava S, Chauhan PS, Seem K, Mishra A, Sopory SK (2013) Plant growth-promoting bacteria *Bacillus amyloliquefaciens* NBRISN13 modulates gene expression profile of leaf and rhizosphere community in rice during salt stress. Plant Physiol Biochem 66:1–9

Naveed M, Hussain MB, Zahir ZA, Mitter B, Sessitsch A (2014) Drought stress amelioration in wheat through inoculation with *Burkholderia phytofirmans* strain PsJN. Plant Growth Regul 73 (2):121–131

Nawrocka J, Małolepsza U (2013) Diversity in plant systemic resistance induced by Trichoderma. Biol Control 67(2):149–156

Newman MA, Sundelin T, Nielsen JT, Erbs G (2013) MAMP (microbe-associated molecular pattern) triggered immunity in plants. Front Plant Sci 4:139

Nguyen D, Rieu I, Mariani C, van Dam NM (2016) How plants handle multiple stresses: hormonal interactions underlying responses to abiotic stress and insect herbivory. Plant Mol Biol 91 (6):727–740

Nia SH, Zarea MJ, Rejali F, Varma A (2012) Yield and yield components of wheat as affected by salinity and inoculation with Azospirillum strains from saline or non-saline soil. J Saudi Soc Agric Sci 11(2):113–121

Nonnoi F, Chinnaswamy A, de la Torre VS, de la Pena TC, Lucas MM, Pueyo JJ (2012) Metal tolerance of rhizobial strains isolated from nodules of herbaceous legumes (*Medicago* spp. and *Trifolium* spp.) growing in mercury-contaminated soils. Appl Soil Ecol 61:49–59

Onaga G, Wydra K (2016) Advances in plant tolerance to abiotic stresses. In: Plant genomics. In Tech

Ortiz N, Armada E, Duque E, Roldán A, Azcón R (2015) Contribution of arbuscular mycorrhizal fungi and/or bacteria to enhancing plant drought tolerance under natural soil conditions: effectiveness of autochthonous or allochthonous strains. J Plant Physiol 174:87–96

Pandey VC (2012) Phytoremediation of heavy metals from fly ash pond by *Azolla caroliniana*. Ecotoxicol Environ Saf 82:8–12

Pangesti N, Pineda A, Pieterse CM, Dicke M, Van Loon JJ (2013) Two-way plant mediated interactions between root-associated microbes and insects: from ecology to mechanisms. Front Plant Sci 4:414

Panke-Buisse K, Poole AC, Goodrich JK, Ley RE, Kao-Kniffin J (2015) Selection on soil microbiomes reveals reproducible impacts on plant function. ISME J 9(4):980

Partida-Martinez LP, Heil M (2011) The microbe-free plant: fact or artifact? Front Plant Sci 2:100

Penuelas J, Farré-Armengol G, Llusia J, Gargallo-Garriga A, Rico L, Sardans J, Terradas J, Filella I (2014) Removal of floral microbiota reduces floral terpene emissions. Sci Rep 4:6727

Porcel R, Zamarreño ÁM, García-Mina JM, Aroca R (2014) Involvement of plant endogenous ABA in *Bacillus megaterium* PGPR activity in tomato plants. BMC Plant Biol 14(1):36

Prapagdee B, Chanprasert M, Mongkolsuk S (2013) Bioaugmentation with cadmium-resistant plant growth-promoting rhizobacteria to assist cadmium phytoextraction by *Helianthus annuus*. Chemosphere 92(6):659–666

Prasad R, Garg AP, Varma A (2005) Interaction of medicinal plants with plant growth promoting rhizobacteria and symbiotic fungi. In: Podila GK, Varma A (eds) Basic research and applications of Mycorrhizae, vol 1. IK International, Delhi, pp 363–407

Prasad R, Kumar M, Varma A (2015) Role of PGPR in soil fertility and plant health. In: Egamberdieva D, Shrivastava S, Varma A (eds) Plant growth-promoting Rhizobacteria (PGPR) and medicinal plants. Springer International, Cham, pp 247–260

Prasad R, Gill SS, Tuteja N (2018) Crop improvement through microbial biotechnology. Elsevier, Amsterdam, Netherlands. ISBN 9780444639882. https://www.elsevier.com/books/crop-improvement-through-microbialbiotechnology/prasad/978-0-444-63987-5

Prasad R, Chhabra S, Gill SS, Singh PK, Tuteja N (2020) The microbial symbionts: potential for the crop improvement in changing environments. In: Tuteja N, Tuteja R, Passricha N, Saifi SK (eds) Advancement in crop improvement techniques. Elsevier, Amsterdam, Netherlands, pp 233–240

Qian J, Li D, Zhan G, Zhang L, Su W, Gao P (2012) Simultaneous biodegradation of Ni–citrate complexes and removal of nickel from solutions by *Pseudomonas alcaliphila*. Bioresour Technol 116:66–73

Qin F, Shinozaki K, Yamaguchi-Shinozaki K (2011) Achievements and challenges in understanding plant abiotic stress responses and tolerance. Plant Cell Physiol 52(9):1569–1582

Qu AL, Ding YF, Jiang Q, Zhu C (2013) Molecular mechanisms of the plant heat stress response. Biochem Biophys Res Commun 432(2):203–207

Rajkumar M, Sandhya S, Prasad MN, Freitas H (2012) Perspectives of plant-associated microbes in heavy metal phytoremediation. Biotechnol Adv 30(6):1562–1574

Rajkumar M, Ma Y, Freitas H (2013) Improvement of Ni phytostabilization by inoculation of Ni resistant *Bacillus megaterium* SR28C. J Environ Manag 128:973–980

Ramegowda V, Senthil-Kumar M (2015) The interactive effects of simultaneous biotic and abiotic stresses on plants: mechanistic understanding from drought and pathogen combination. J Plant Physiol 176:47–54

Reiter B, Sessitsch A (2006) Bacterial endophytes of the wildflower *Crocus albiflorus* analyzed by characterization of isolates and by a cultivation-independent approach. Can J Microbiol 52 (2):140–149

Riadh K, Wided M, Hans-Werner K, Chedly A (2010) Responses of halophytes to environmental stresses with special emphasis to salinity. In: Kader J-C, Delseny M (eds) Advances in botanical research, vol 53. Academic, New York, pp 117–145

Rivas S, Thomas CM (2005) Molecular interactions between tomato and the leaf mold pathogen *Cladosporium fulvum*. Annu Rev Phytopathol 43:395–436

Rodell M, Velicogna I, Famiglietti JS (2009) Satellite-based estimates of ground water depletion in India. Nature 460(7258):999

Rolli E, Marasco R, Vigani G, Ettoumi B, Mapelli F, Deangelis ML, Gandolfi C, Casati E, Previtali F, Gerbino R, Pierotti CF (2015) Improved plant resistance to drought is promoted by the root-associated microbiome as a water stress-dependent trait. Environ Microbiol 17 (2):316–331

Saghafi D, Ghorbanpour M, Asgari LB (2018) Efficiency of rhizobium strains as plant growth promoting rhizobacteria on morpho-physiological properties of *Brassica napus* L. under salinity stress. J Soil Sci Plant Nutr 18(1):253–268

Saghafi D, Ghorbanpour M, Shirafkan Ajirloo H, Asgari LB (2019a) Enhancement of growth and salt tolerance in *Brassica napus* L. seedlings by halotolerant rhizobium strains containing ACC deaminase activity. Plant Physiol Rep 24(2):225–235

Saghafi D, Delangiz N, Asgari Lajayer B, Ghorbanpour M (2019b) An overview on improvement of crop productivity in saline soils by halotolerant and halophilic PGPRs. 3 Biotech 9:261

Saha M, Sarkar S, Sarkar B, Sharma BK, Bhattacharjee S, Tribedi P (2016) Microbial siderophores and their potential applications: a review. Environ Sci Pollut Res 23(5):3984–3999

Salas-Marina MA, Silva-Flores MA, Uresti-Rivera EE, Castro-Longoria E, Herrera-Estrella A, Casas-Flores S (2011) Colonization of Arabidopsis roots by *Trichoderma atroviride* promotes growth and enhances systemic disease resistance through jasmonic acid/ethylene and salicylic acid pathways. Eur J Plant Pathol 131(1):15–26

Salvioli A, Ghignone S, Novero M, Navazio L, Venice F, Bagnaresi P, Bonfante P (2016) Symbiosis with an endobacterium increases the fitness of a mycorrhizal fungus, raising its bioenergetic potential. ISME J 10(1):130

Sandhya VZ, Grover M, Reddy G, Venkateswarlu B (2009) Alleviation of drought stress effects in sunflower seedlings by the exopolysaccharides producing *Pseudomonas putida* strain GAP-P45. Biol Fertil Soils 46(1):17–26

Santi C, Bogusz D, Franche C (2013) Biological nitrogen fixation in non-legume plants. Ann Bot 111(5):743–767

Santoyo G, Moreno-Hagelsieb G, Orozco-Mosqueda MC, Glick BR (2016) Plant growth promoting bacterial endophytes. Microbiol Res 183:92–99

Selvakumar G, Kundu S, Joshi P, Nazim S, Gupta AD, Mishra PK, Gupta HS (2008) Characterization of a cold-tolerant plant growth-promoting bacterium *Pantoea dispersa* 1A isolated from a sub-alpine soil in the North Western Indian Himalayas. World J Microbiol Biotechnol 24 (7):955–960

Sen S, Chandrasekhar CN (2014) Effect of PGPR on growth promotion of rice (*Oryza sativa* L.) under salt stress. Asian J Plant Sci Res 4:62–67

Shahzad R, Khan AL, Bilal S, Waqas M, Kang SM, Lee IJ (2017) Inoculation of abscisic acid-producing endophytic bacteria enhances salinity stress tolerance in *Oryza sativa*. Environ Exp Bot 136:68–77

Shoebitz M, Ribaudo CM, Pardo MA, Cantore ML, Ciampi L, Curá JA (2009) Plant growth promoting properties of a strain of *Enterobacter ludwigii* isolated from *Lolium perenne* rhizosphere. Soil Biol Biochem 41(9):1768–1774

Shrivastava P, Kumar R (2015) Soil salinity: a serious environmental issue and plant growth promoting bacteria as one of the tools for its alleviation. Saudi J Biol Sci 22(2):123–131

Siddikee MA, Chauhan PS, Anandham R, Han GH, Sa T (2010) Isolation, characterization, and use for plant growth promotion under salt stress, of ACC deaminase-producing halotolerant bacteria derived from coastal soil. J Microbiol Biotechnol 20(11):1577–1584

Silva ED, Ribeiro RV, Ferreira-Silva SL, Viégas RA, Silveira JA (2010) Comparative effects of salinity and water stress on photosynthesis, water relations and growth of *Jatropha curcas* plants. J Arid Environ 74(10):1130–1137

Simontacchi M, Galatro A, Ramos-Artuso F, Santa-María GE (2015) Plant survival in a changing environment: the role of nitric oxide in plant responses to abiotic stress. Front Plant Sci 6:977

Singh R (2014) Microorganism as a tool of bioremediation technology for cleaning environment: a review. Proc Int Acad Ecol Environ Sci 4(1):1–6

Singh JS, Pandey VC, Singh DP (2011) Efficient soil microorganisms: a new dimension for sustainable agriculture and environmental development. Agric Ecosyst Environ 140 (3–4):339–353

Singh D, Raina TK, Kumar A, Singh J, Prasad R (2019) Plant microbiome: a reservoir of novel genes and metabolites. Plant Gene. https://doi.org/10.1016/j.plgene.2019.100177

Spence C, Bais H (2015) Role of plant growth regulators as chemical signals in plant–microbe interactions: a double edged sword. Curr Opin Plant Biol 27:52–58

Srivastava S, Yadav A, Seem K, Mishra S, Chaudhary V, Nautiyal CS (2008) Effect of high temperature on *Pseudomonas putida* NBRI0987 biofilm formation and expression of stress sigma factor RpoS. Curr Microbiol 56(5):453–457

Stearns JC, Glick BR (2003) Transgenic plants with altered ethylene biosynthesis or perception. Biotechnol Adv 21(3):193–210

Suarez C, Cardinale M, Ratering S, Steffens D, Jung S, Montoya AM, Geissler-Plaum R, Schnell S (2015) Plant growth-promoting effects of *Hartmannibacter diazotrophicus* on summer barley (*Hordeum vulgare* L.) under salt stress. Appl Soil Ecol 95:23–30

Sun Y, Cheng Z, Glick BR (2009) The presence of a 1-aminocyclopropane-1-carboxylate (ACC) deaminase deletion mutation alters the physiology of the endophytic plant growth-promoting bacterium *Burkholderia phytofirmans* PsJN. FEMS Microbiol Lett 296(1):131–136

Sunkar R, Li YF, Jagadeeswaran G (2012) Functions of microRNAs in plant stress responses. Trends Plant Sci 17(4):196–203

Tank N, Saraf M (2010) Salinity-resistant plant growth promoting rhizobacteria ameliorates sodium chloride stress on tomato plants. J Plant Interact 5(1):51–58

Theocharis A, Bordiec S, Fernandez O, Paquis S, Dhondt-Cordelier S, Baillieul F, Clément C, Barka EA (2012) *Burkholderia phytofirmans* PsJN primes *Vitis vinifera* L. and confers a better tolerance to low nonfreezing temperatures. Mol Plant-Microbe Interact 25(2):241–249

Thomashow MF (2010) Molecular basis of plant cold acclimation: insights gained from studying the CBF cold response pathway. Plant Physiol 154(2):571–577

Timmusk S, Behers L, Muthoni J, Muraya A, Aronsson AC (2017) Perspectives and challenges of microbial application for crop improvement. Front Plant Sci 8:49–55

Tiwari S, Lata C, Chauhan PS, Nautiyal CS (2016) *Pseudomonas putida* attunes morphophysiological, biochemical and molecular responses in *Cicer arietinum* L. during drought stress and recovery. Plant Physiol Biochem 99:108–117

Trabelsi D, Mhamdi R (2013) Microbial inoculants and their impact on soil microbial communities: a review. BioMed Res Int 1–11

Turner TR, James EK, Poole PS (2013) The plant microbiome. Genome Biol 14(6):209

Ullah A, Heng S, Munis MF, Fahad S, Yang X (2015) Phytoremediation of heavy metals assisted by plant growth promoting (PGP) bacteria: a review. Environ Exp Bot 117:28–40

Upadhyay SK, Singh DP, Saikia R (2009) Genetic diversity of plant growth promoting rhizobacteria isolated from rhizospheric soil of wheat under saline condition. Curr Microbiol 59(5):489–496

Vardharajula S, Zulfikar Ali S, Grover M, Reddy G, Bandi V (2011) Drought-tolerant plant growth promoting Bacillus spp.: effect on growth, osmolytes, and antioxidant status of maize under drought stress. J Plant Interact 6(1):1–4

Verhage A, van Wees SC, Pieterse CM (2010) Plant immunity: it's the hormones talking, but what do they say? Plant Physiol 154(2):536–540

Verma JP, Yadav J, Tiwari KN, Kumar A (2013) Effect of indigenous *Mesorhizobium* spp. and plant growth promoting rhizobacteria on yields and nutrients uptake of chickpea (*Cicer arietinum* L.) under sustainable agriculture. Ecol Eng 51:282–286

Vijayaraghavan K, Yun YS (2008) Bacterial biosorbents and biosorption. Biotechnol Adv 26 (3):266–291

Vimal SR, Singh JS, Arora NK, Singh S (2017) Soil-plant-microbe interactions in stressed agriculture management: a review. Pedosphere 27(2):177–192

Vymazal J, Březinová T (2016) Accumulation of heavy metals in aboveground biomass of *Phragmites australis* in horizontal flow constructed wetlands for wastewater treatment: a review. Chem Eng J 290:232–242

Wagner MR, Lundberg DS, Coleman-Derr D, Tringe SG, Dangl JL, Mitchell-Olds T (2014) Natural soil microbes alter flowering phenology and the intensity of selection on flowering time in a wild Arabidopsis relative. Ecol Lett 17(6):717–726

Wang W, Vinocur B, Altman A (2003) Plant responses to drought, salinity and extreme temperatures: towards genetic engineering for stress tolerance. Planta 218(1):1–4

Wassmann R, Jagadish SV, Sumfleth K, Pathak H, Howell G, Ismail A, Serraj R, Redona E, Singh RK, Heuer S (2009) Regional vulnerability of climate change impacts on Asian rice production and scope for adaptation. Adv Agron 102:91–133

Weller DM, Raaijmakers JM, Gardener BB, Thomashow LS (2002) Microbial populations responsible for specific soil suppressiveness to plant pathogens. Annu Rev Phytopathol 40(1):309–348

Yadav J, Verma JP, Jaiswal DK, Kumar A (2014) Evaluation of PGPR and different concentration of phosphorus level on plant growth, yield and nutrient content of rice (*Oryza sativa* L.). Ecol Eng 62:123–128

Yang J, Kloepper JW, Ryu CM (2009) Rhizosphere bacteria help plants tolerate abiotic stress. Trends Plant Sci 14(1):1–4

Yeoh YK, Paungfoo-Lonhienne C, Dennis PG, Robinson N, Ragan MA, Schmidt S, Hugenholtz P (2016) The core root microbiome of sugarcanes cultivated under varying nitrogen fertilizer application. Environ Microbiol 18(5):1338–1351

Yi SY, Shirasu K, Moon JS, Lee SG, Kwon SY (2014) The activated SA and JA signaling pathways have an influence on flg22-triggered oxidative burst and callose deposition. PLoS One 9(2): e88951

Yolcu S, Ozdemir F, Güler A, Bor M (2016) Histone acetylation influences the transcriptional activation of POX in *Beta vulgaris* L. and *Beta maritima* L. under salt stress. Plant Physiol Biochem 100:37–46

Zabetakis I, Moutevelis-Minakakis P, Gramshaw JW (1999) The role of 2-hydroxypropanal in the biosynthesis of 2, 5-dimethyl-4-hydroxy-2H-furan-3-one in strawberry (*Fragaria ananassa*, cv. Elsanta) callus cultures. Food Chem 64(3):311–314

Zhang F, Dashti N, Hynes RK, Smith DL (1996) Plant growth promoting rhizobacteria and soybean [*Glycine max* (L.) Merr.] nodulation and nitrogen fixation at suboptimal root zone temperatures. Ann Bot 77(5):453–460

Zhang J, Wang LH, Yang JC, Liu H, Dai JL (2015) Health risk to residents and stimulation to inherent bacteria of various heavy metals in soil. Sci Total Environ 508:29–36

Zheng YK, Qiao XG, Miao CP, Liu K, Chen YW, Xu LH, Zhao LX (2016) Diversity, distribution and biotechnological potential of endophytic fungi. Ann Microbiol 66(2):529–542

Złoch M, Thiem D, Gadzała-Kopciuch R, Hrynkiewicz K (2016) Synthesis of siderophores by plant-associated metallotolerant bacteria under exposure to Cd^{2+}. Chemosphere 156:312–325

Chapter 3
Plant-Microbe Interactions: Applications for Plant-Growth Promotion and In Situ Agri-waste Management

Anu Kalia and Jayesh Singh

Abstract The microbial genera, both cultivable and uncultivable, associated with the host plant inhabiting in or on different tissues or organ and the environment surrounding the plant body are unique and comprise the plant-associated microbiome (PAM). The PAM profile of the host plant gets altered in response to certain biotic and abiotic stresses and the anthropogenic interventions. Therefore, identification of key groups or genera of the microbes among the whole PAM is critical as these microbes, besides themselves, can modulate several other microbial communities, leading to considerable positive or negative changes in the community structure, abundance, and the overall microbial diversity. This manuscript explores the alterations in PAM particularly in response to the agri-management practices and its role in enhancing growth, vigor, and yield in host crop plants. Further, the precise role of a specific group of PAM, the soil organic matter decomposers, has been discussed with special reference to agri-waste generated by cereal crop cultivation. The practical applications of the PAM studies for improvement in the agricultural sustainability through optimization of the plant microbiome have also been discussed.

3.1 Introduction

The soil microbial communities and their role in ecosystem regulation including waste decomposition, nutrient cycling, transformation, and availability are crucial for crop production (Hartman et al. 2018). These soil-dwelling organisms including the microbes and higher organisms do not exist in isolation rather are interwoven in

A. Kalia
Department of Soil Science, Electron Microscopy and Nanoscience Laboratory, Punjab Agricultural University, Ludhiana, Punjab, India

J. Singh (✉)
Department of Plant Breeding and Genetics, Punjab Agricultural University, Ludhiana, Punjab, India
e-mail: jayesh_singh@pau.edu

A. Varma et al. (eds.), *Plant Microbiome Paradigm*,
https://doi.org/10.1007/978-3-030-50395-6_3

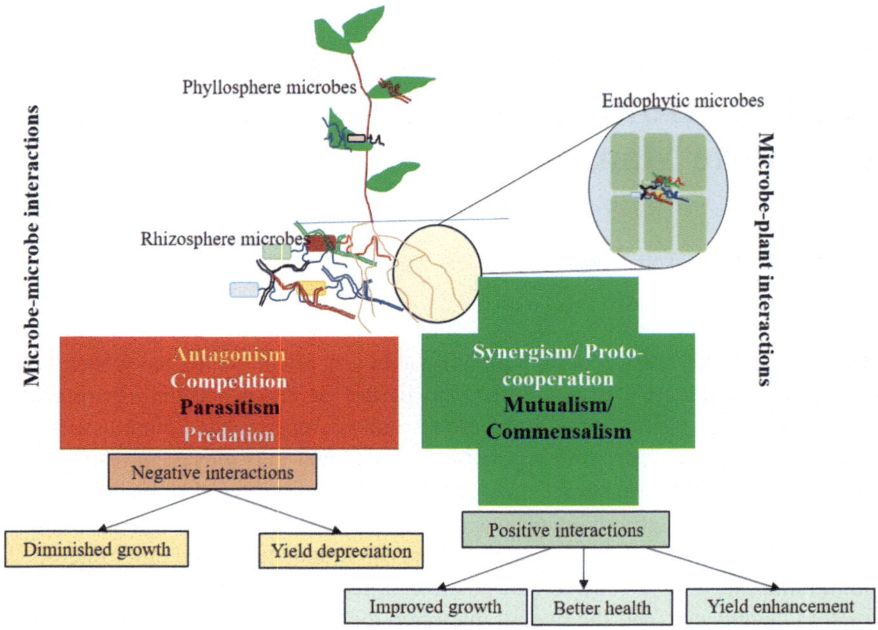

Fig. 3.1 Schematic representation of the plant as meta-organism and the various inter- and intra-microbial and microbe-plant interactions emphasizing the distinct plant-associated microbiome niches or hot-spots and the repercussions of these interactions on the holobiont plant

complex and intricate interdependencies for harnessing improved survival benefits and thus have led to emergence of specialized associations such as mutualism, parasitism, and so forth (Imam et al. 2016). Plants can be considered as the meta-organism harboring diverse microbiomes due to their ability to associate with different types and classes of microorganisms (Ho et al. 2017). These microbiomes are affected by and may vary according to several abiotic and biotic factors as well as the agronomic crop cultivation practices (Hartman et al. 2018; Prasad et al. 2020). Most likely, there exists spatial as well as temporal variabilities among the microbial communities inhabiting different organ-tissue systems of the meta-organism leading to occurrence of specific niches, the "hot-spots," or the regions of higher microbial diversity (Gopal and Gupta 2018) (Fig. 3.1).

One of the most complex niches exhibiting diverse associations among the fundamental biological and nonbiological components in the microbe-plant-soil continuum is the "rhizosphere" (Shrivastava et al. 2014; Igiehon and Babalola 2018). It is designated as the soil environment zone comprised of plant root cell exudates, sloughed root cap and border cells, and the complex polysaccharides or mucilage released by the roots which act as chemoattractant and also serve as food for the rhizospheric microorganisms (Babalola 2010). Therefore, the root-derived biochemical and microbiological entities can collectively regulate each other (Igiehon and Babalola 2018). The rhizo-microbiome, complete rhizospheric microbes and extensive intercellular chemical talks among themselves and plant cells and the relationships they develop, is considered the external functional genome of the crop plant (Turner et al. 2013).

3.2 Plant-Microbe Interaction System

The belowground soil microbial diversity can be considered as an index of the agroecosystem productivity (Kalia and Gosal 2011), and thus the plant-microbe interactome is critical to determine the crop productivity (Singh et al. 2019; Varma et al. 2019a, b, 2020). The response patterns of microbial communities and their assembly vary according to the agronomic crop cultivation practices (Hartmann et al. 2015). Moreover, the agronomic interventions may influence few key individual microbiome members that may initiate a cascade of events affecting the overall microbial community structure and composition (Finkel et al. 2017). These microbial communities form the core microbiome of the host plant and are predominated by bacterial and fungal genera, besides archaea, protozoan, and viruses (Ahkami et al. 2017). Among these primarily, the bacteria and fungi are largely and collectively responsible for enhancing growth of the host plant through numerous direct and indirect mechanisms and are referred to as plant growth-promoting rhizomicrobes (PGPMs) (Orozco-Mosqueda et al. 2018).

The plant probiotic and plant growth and yield enhancing microbes have the ability to colonize the rhizosphere, rhizoplane and sometimes even the endorhizosphere of the plant's roots, phyllosphere, endosphere, spermosphere, and other possible micro-niches in the plant (Van Der Heijden et al. 2008; Babalola 2010; Prasad et al. 2015). The microbes associated with these microclimates around, on, and inside the plant roots and aerial organs exhibit very diverse and complex relationships well-regulated and orchestrated through specific low-molecular weight compounds, the signal molecules, which ensure proper monitoring of the surroundings, cell–cell communication, and recognition of the compatible partners (Schlaeppi and Bulgarelli 2015; Yamazaki and Hayashi 2015; Igiehon and Babalola 2018).

3.2.1 Interaction System Variants: Factors Affecting the Plant-Associated Microbiome

Several factors affect the interaction(s) among the microbiont and its macrobiont partner. The microbial habitation and interactions are necessary for the plant's well-being and can be traced back as spermosphere microbiome (Flandroy et al. 2018). There are two major factors, the plant specific or related features which can affect the associated microbiome (due to organ, species, genotype/variety/cultivar) and the crop age besides the environment (Wintermans et al. 2016; Flandroy et al. 2018).

Thus, the plant microbiome may vary according to plant organs and tissues as well as according to certain temporal variations which occur due to the growth phase of the plant, i.e., vegetative or reproductive phase or crop age (Mitter et al. 2017; Hamonts et al. 2018). For an instance, the rhizobiome or the root microbiome composition and structure depends on the root physiology and the diversity and quantity of the root exudates secreted by the plant roots (Sasse et al. 2018). This "rhizosphere effect," the occurrence of distinct rhizobiome than the bulk soil, is critical. However, it is of less relevance in crops such as rice as the rhizobiome will be quite similar to bulk soil microbiome.

Further, the crop genotypes, wild versus domesticated, exhibit profound variations in their PAM particularly the composition of the root microbiome which is altered on domestication imparting an adverse effect on the diversity of the possible microbial interactions particularly agronomically important arbuscular mycorrhizal and N-fixer associations with landraces of wheat/maize and soybean, respectively (Pérez-Jaramillo et al. 2018). Domestication of crop plants has led to a shift from elaborate root traits mostly inhabited by Bacteroidetes group of bacteria to predominance of Proteobacteria and Actinobacteria.

The soil microbial diversity too gets altered broadly by factors including the edaphic, i.e., soil physicochemical characteristics, and climatic conditions and the anthropogenic management or interventions (Imam et al. 2016). The soil physical and chemical characteristics which may affect rhizobiome include the soil structure and type, soil aggregate potential, moisture content, pH, organic matter content, EC, soil nutrient status, and temperature (Fierer 2017; Santoyo et al. 2017). Contrarily, a study on Andean potato (*Solanum tuberosum* L.) grown in variable soil and climatic conditions exhibited rather uniform taxonomic composition of the prokaryotic bacterial genera at a particular stage of the potato plant development (Pfeiffer et al. 2017). Likewise, an interesting report depicted a plant genera dependent response with soybean rhizobiome which gets significantly affected by the variable soil pH, P, and K significantly while only pH and N content affected the alfalfa rhizobiome (Xiao et al. 2017).

Globally, the major arable land is under intensive agriculture to obtain two or more than two crops annually. However, the high cropping intensity and the region specific cropping system patterns have aroused concerns due to possible environmental footprints of the modern agricultural practices (Hartman et al. 2018). One of the key issues is the alterations in the soil microbial community structure and diversity due to agrichemical-based crop cultivation techniques (Sergaki et al. 2018). Mareque et al. (2018) have reported pronounced effect of rate of application of N fertilizer on the community structure and relative number/abundance of the sweet sorghum endophytic bacteria. Further, they have reported the exclusive effect of the plant organ on the structure and abundance with relative increase or decrease in the occurrence of certain key groups or genera of these endophytes. Another field study by Xue et al. (2018) on potato crop revealed that the soil texture and chemistry largely affected the microbial communities in terms of diversity and abundance of microbial groups or phyla. However, the agri-management practices, conventional versus organic, had subtle effects, while the composition of the microbial communities does not vary among the three test varieties, viz., red, yellow, and russet, of potato. Likewise, van der Heijden and Hartmann (2016) have also reported variation in microbial community structure of the land undergoing organic versus conventional cultivation practices. The cropping practices or regimes can also influence the microbiome structure and characteristics. Granzow et al. (2017) have observed distinct variations in the microbiota richness and diversity among the two cropping regimes—row and mixed intercropping with significantly higher bacterial and fungal diversity in the bulk soil samples of two monocultured wheat and fababean crops. Therefore, such studies can be instrumental in custom designing of the agricultural biologicals for sustainable productivity. The environmental quality- and equity-related problems can be properly resolved through ecological intensification aiming at maintenance, restoration, or at least minimization of the ill effects on the soil microbial biodiversity (Hartman et al. 2018).

3.2.2 Beneficial Plant-Microbe Interactions

The beneficial plant-microbial interactions fall in two fundamental categories on the basis of the extent of the interaction among the two partners and the mechanism of benefit imparted by the microbiont to the host plant (Ahkami et al. 2017). The former group may include the soil microbes that may actively inhabit the outer and/or internal tissues of various plant organs to develop commensal and mutualistic relationships such as nitrogen fixation by diazotrophic microbes in specialized nodules or the inhabitation of epidermal/cortical cells by endophytes (Van Der Heijden et al. 2008; Prasad et al. 2020). Therefore, these microbes actively facilitate nutrient acquisition such as nitrogen, phosphorus, and essential micronutrients or can modulate phytohormone concentration levels, increased auxins, gibberellins, and cytokinins, while preferential decrease in ethylene through activation of the 1-ACC deaminase enzyme activity, in the plant (Orozco-Mosqueda et al. 2018). This may lead to improved resource partitioning and nutrient supply primarily through nutrient mineralization processes (Van Der Heijden et al. 2008).

Otherwise, the plant partner can harvest indirect benefits due to both active and passive ways. The active mechanism involves the secretion of two fundamental groups of compounds, i.e., diffusible and secretory low-molecular weight compounds such as antibiotics, bacteriocins, small lipoproteins, lytic enzymes (proteases, chitinases), volatile organic compounds, and metal-chelating siderophores, which offer cidal/killing (antibiosis) effect or pose competition to the phytopathogens, thereby curbing their attack (Tikhonovich and Provorov 2011). The passive mechanism of plant probiotic action involves competition and occupancy of the receptor ligand sites on the surfaces of the plant tissues and organs and formation of homogenous or heterogeneous biofilms on the exterior as well as internal plant tissues (Yamazaki and Hayashi 2015). Furthermore, Panke-Buisse et al. (2017) have observed regulation and induction of early flowering trait in model plant *Arabidopsis thaliana* to be governed through the cultivable microbiome teased out of the whole microbiome of the test plant. This illustrates the positive role of the cultivable microbial groups as the key microbial genera for retention of the early flowering trait.

3.2.2.1 Plant Probiotic and Growth-Enhancing Microbes

As discussed in the above Sect. 3.2.2, the plant probiotic and plant growth-promoting (PGP) microbes which form the associated microbiome of a crop plant for regulation, optimization, and thus improvement in the plant vegetative characteristics can ensure provision of a durable immunity to the plant and thus can enhance the yield (Hunter 2016; Prasad et al. 2018). Since the interactions are two-way, more microbiome-responsive varieties or cultivars of the crop plants can be developed to harness both direct and indirect benefits of the associated microbiome such as enhanced nutrient use efficiency (NUE) and competitive edge for the test plant (Ahkami et al. 2017). Therefore, the application of agricultural biologicals is gaining impetus due to their positive effect on crop health and productivity besides no harm to the ecosystem (Schlaeppi and Bulgarelli 2015; Singh et al. 2018).

3.2.2.1.1 Improvement of Plant Growth and Health

A comprehensive information is provided in the Table 3.1 indicating the various positive benefits on the growth and health of the crop plants due to occurrence of PGPR in the microbiome of the host plant and due to their activities.

3.2.2.1.2 Counteracting Abiotic and Biotic Stresses

Agricultural productivity can be severely affected by both abiotic and biotic stresses indirectly affecting the host PAM. Conversely, the presence of certain microbial communities in the PAM can invariably reduce or manage if not totally eradicate the negative effects of the temperature, salt, alkalinity, and moisture extremes besides attack by pests and phytopathogens (Vejan et al. 2016). This way the PAM may provide a competitive benefit to develop tolerance to these stresses afflicted on the growing crop plants (Joshi et al. 2018). The contribution of PAM towards the overall growth and health of the plant can be identified as direct involving induction of a response, be it local or systemic, in the host plant towards a particular stress (Kalia et al. 2020). However, indirectly the PAM PGP substances may modulate for sustenance of overall improved growth through improved nutrient uptake and mobilization, improved cell division and elongation mediated through microbe secreted or plant induced higher phytohormone production and triggering of the reactive oxygen species-scavenging machinery in the host (Joshi et al. 2018).

Considering the abiotic stress such as drought and salinity conditions, the primary protective mode of action of the PAM is to impart water and salinity tolerance to the

Table 3.1 Effect of various microbiome alterations on the vegetative growth and total health of the test crop plants

Host crop plant	Microbiont(s) affected	Plant organ/ tissue inhabited	Key parameters improved	Possible mechanism of action	Remarks
Arabidopsis thaliana	• Decreased abundance of Actinobacteria, Acidobacteria, Bacteroidetes, Proteobacteria, and Verrucomicrobia • Increased abundance in Spirochaetae, Firmicutes, and the Crenarchaeota archaebacteria (for early flowering trait)	Rhizosphere and rhizoplane	Vegetative growth and early flowering	Inoculation of mixed strains or taxa from rhizosphere soil has modified the plant trait	Panke-Buisse et al. (2015)
Arabidopsis thaliana	Occurrence of predominance of genera *Xanthomonas*, *Microbacterium*, and *Stenotrophomonas*	Rhizosphere and rhizoplane	Defense activation against downy mildew pathogen *Hyaloperonospora arabidopsidis*	Foliar challenge with pathogen	Berendsen et al. (2018)
Oryza sativa accessions M206 (subsp. *japonica*), IR20 (subsp. *indica*) and *Oryza glaberrima* accessions TOg7102 and CG14	• Changes in terms of enrichment of multiple Actinobacteria and Chloroflexi • Depletion of several Acidobacteria and Deltaproteobacteria	Rhizosphere and endosphere microbiome	Improved drought tolerance by the rice plants	Compartment specific enrichment of drought-resistant microbiota probably due to modification of the concentration and composition of root exudates	Santos-Medellín et al. (2017)
Phyllostegia kaalaensis	Uncultured fungal community predominated with mycoparasitic *Pseudozyma aphidis*	Phyllosphere microbiota	Effective reduction in disease caused by *Neoerysiphe galeopsidis* and increased survival to an outplanted population	Total foliar microbiome transplants from healthy wild relatives altered the colonization and microbiome of the treated plants	Zahn and Amend (2017)
Saccharum sp. var. SP 80–3280	*Asticcacaulis*, *Burkholderia*, *Chitinophaga*, *Ensifer*, *Lysobacter*, *Pedobacter*,	Rhizosphere and root endosphere	3.4-fold increase in biomass of inoculated vs. uninoculated plants	• Efficient colonization of the plant organs • Displacement of the native	Armanhi et al. (2018)

(continued)

Table 3.1 (continued)

Host crop plant	Microbiont(s) affected	Plant organ/ tissue inhabited	Key parameters improved	Possible mechanism of action	Remarks
	Rhizobium, *Stenotrophomonas*, and two unknown genera of the families Comamonadaceae and Streptomycetaceae			microbiota on inoculation of the synthetic cultures • Predomination of the rhizospheric microbial abundance (53.9%)	
Solanum lycopersicum 'money maker' variety (Park Seed, 05851-PK-P1)	Multiple *Pseudomonas* spp.	Phyllosphere	Leaf-associated microbiota conferred protection against pathogen *Pseudomonas syringae* pv. tomato in dose-dependent manner	Decrease in pathogen colonization	Berg and Koskella (2018)
Solanum lycopersicum cv. Marmande	Habitat extension of *Pseudomonas* species	Rhizospheric microbiota and tomato root rhizoplane	Root microbiome composition shifts due to occurrence of pathogenic oomycete *P. parasitica*	Commensal interactions of *Pseudomonas* spp. with fungal oomycete leading to their preferential colonization of the surface biofilm of the fungal pathogen	Larousse et al. (2017)
Theobroma cacao (cacao tree)	*Colletotrichum tropicale* (fungal endophyte)	Rhizosphere microbiome	Leaf litter foliar application improved resistance to *Phytophthora palmivora*	Exposure of seedlings with healthy adult plant leaf litter enriched their microbiome with endophytic fungus, *Colletotrichum tropicale*, by upregulation of host defense pathways	Christian et al. (2017)
Vitis vinifera cv. Pinot gris	Bacterial genera *Haemophilus*, *Swaminathania*, *Paracoccus*, *Roseomonas*, *Kineosporia*, and *Porphyromonas* and fungal genera *Epicoccum*, *Teratosphaeria*, *Exophiala*, *Claviceps*, and *Chalastospora*	Phyllosphere	Effect of penconazole (chemical fungicide) and *Lysobacter capsici*AZ78 (biological control agent) on grapevine leaf microbiota	Improvement in the biocontrol communities through enrichment of the beneficial phyllosphere microbial population by agronomic practices	Perazzolli et al. (2014)

host plant and involves improvement of the leaf water status particularly through management of the stomatal conductance behavior of the probiotic microbiome positive host plants and hence the increased photosynthetic efficiency translating in enhanced growth (Vejan et al. 2016). Salinity stress is another common abiotic stress. Plants growing in high salinity conditions exhibit a variable root microbiome predominated by halophytic microorganisms including the salt-tolerant bacteria *Halobacterium*, *Halococcus*, *Halomonas*, and others besides fungal genera which include members of the phylum Glomeromycota (Ruppel et al. 2013). These microbes improve the nutritional status of the plant, enhance the antioxidant enzymatic processes, and also increase the plant phytohormone concentrations. Even certain bacterial isolates having ice nucleation proteins (INPs) can improve the tolerance to low and freezing temperatures (Chialva and Bonfante 2018).

The continuous mono-cropping practice has been one of the primary causes for the development of diseased soils due to buildup of the population of the pathogen or causative agent. Santhanam et al. (2015) have reported significant reduction in incidence of the *Fusarium-Alternaria* disease complex and mortality on inoculation with consortia of native bacterial isolates which were the component of the core microbiome in *Nicotiana attenuata* grown under sick plot field conditions. Chialva et al. (2018) have demonstrated the role of soil microbiota to curb the invasion and attack of *Fusarium oxysporum* f sp. *lycopersici* (FoL) in disease-suppressive/disease-conducive soil for both resistant and susceptible genotype of tomato plants. They elaborated that irrespective of the taxonomic composition, both the test soil microbiomes elicited two distinct responses, one the primary level pathogen-associated molecular patterns (PAMP)-triggered immunity pathway, besides increase in lignin synthesis leading to inhibition of further ramification by FoL. Likewise, Snelders et al. (2018) have discussed about decline in the incidence and severity of the wheat monoculture-derived take-all disease caused by *Gaeumannomyces graminis* var. *tritici* due to elevated occurrence of *Pseudomonas* spp., a take-all fungus antagonist in the soil microbiome of the disease-suppressive soils. However, they have argued for the use of effector molecules, low-molecular weight compounds, peptides, and sRNA, to modulate the rhizobiome of the plant. The disease severity can also be affected by the occurrence of pathogen complementing metabolic properties of the co-resident nonpathogenic microbes in the disease affecting tissue probably through quorum-sensing phenomena (da Silva et al. 2014). Therefore, inoculation of disease-suppressive soils to correct the pathogenic attack in plants growing in ailing or disease-conducive soils can possibly be one of the techniques to remediate disease incidence and severity in susceptible genotypes of crop plants (Gopal et al. 2013; Massart et al. 2015). Another report of Berendsen et al. (2018) advocated that a pathogen memory is developed in disease-afflicted soil and this legacy (soil-mediated) can confer increased protection against the test pathogen in the next plant population on cultivation in the same soil. Mechanistically, this may be traced back to shifts or variations in the composition of the rhizobiome caused by the pathogen invasion due to alterations in the exudates secreted by the roots (Gu et al. 2016).

3.3 Agri-waste Management

Prudent management of the agricultural waste can be another effective technique to improve the sustainability of the agroecosystems. This waste should be rather treated as the crop-derived biomass which can either function as a soil conditioner (straw mulches), nutrient supplement for the next crop (biochar, compost, manure), or can also be transformed to bioenergy (biogas, biochar, bioethanol, and more) (Nadeem et al. 2015).

3.3.1 What Is Agri-waste?

Agri-waste encompasses a huge diversity of crop residues such as cereal straws, stovers, aerial leaves/stem, residues, hulls, chaffs, weeds, and grasses which are rich in lignocellulosic substances, saccharide rich molasses, fruit and vegetable wastes as peels and outer coverings, other solid biomass (Timsina 2018). These waste biomass can be harnessed for several possible purposes such as production of energy (bioethanol), phytochemical extraction, and as organic carbon material to enhance the fertility of the soils on amendment (Zhang et al. 2012).

3.3.1.1 Why There Is a Need for Prudent Management of Agri-waste?

The conventional disposal techniques to get rid of the agricultural waste are not full proof and considered to be faulty as these approaches lead to significant environmental problems. The rampant burning of the paddy stubbles and straw to quickly clear the fields for sowing of wheat and other *rabi* crops has become an important source of emission of particulate matter and greenhouse gas pollutants into the ambient air. Gupta et al. (2004) estimated that 1 ton of straw when burnt releases 3 kg of particulate matter and 199 kg ash besides the gaseous components comprised of 60 kg CO, 1460 kg CO_2, and 2 kg SO_2. Substantial increase in particulate matter leads to severe respiratory disorders in humans and livestock. Therefore, this waste must be carefully managed rather than disposed off in landfills or burnt under open conditions.

3.3.1.2 How to Manage Agri-waste?

The practical ways of agri-waste management include fruitful transformation of the waste straw and other aboveground biomass to plant growth-promoting or soil fertility-enriching product. Paddy straw, the most abundant agri-waste generated in significant quantities in South Asia, can either be burnt in situ, incorporated in soil, used as mulch for following crop, baled for usage in energy generation, fed to cattle, or removed from field. The kind of management chosen determines the overall nutrient balance in the field. The most convenient way to manage agri-waste is through in situ management techniques. In general, the rice residue is high in potassium, besides having marginal to low levels of nitrogen, phosphorus, and sulfur. Therefore, it has a potential for improving the soil fertility on incorporation in the rice fields (Borah et al. 2016). The horticultural crop-derived waste generated due to processing of the fruits and vegetables can be better tackled by extraction of the phytochemicals from the waste and amelioration of other food products with these extracted compounds (Sagar et al. 2018). The cellulosic substrates extracted from the horticultural waste biomass can be utilized as a substrate to grow microorganisms and generate single-cell proteins, industrially useful enzymes, and other products (Das and Singh 2004). Therefore, a diversity of secondary industries can be supported on agri-waste as the basic input or substrate.

Conversion of lignocellulosic biomass to bioethanol or bioenergy is another possible way of deriving wealth from waste. As the prices of the fossil fuel always fluctuate and enhance over time, it is necessary to scout for alternative energy sources to replace the conventional fossil fuels (Nadeem et al. 2015). Alshammari et al. (2011) have reported the use of banana waste for the generation of bioethanol. The rice straw and husk can also be converted into bioethanol through a series of breakdown and saccharification processes (Isikgor and Becer 2015; Kaur et al. 2020). Treatment of the paddy straw amended with other agri-wastes such as pod husk of soybean and pigeon pea, peanut shells, corn stalks, and sugarcane bagasse with *Aspergillus fumigatus* CTS2 culture enhanced the saccharification of the paddy straw with maximum enhancement recorded on soybean husk supplementation (Singla et al. 2018).

3.4 Microbes and Agri-waste Management

Microbial decomposition of the waste material is one of the fundamental phenomena governing the recycling of the essential and trace elements (Xuan 2007). Microbes exhibit enormous physiological versatility. They actively degrade the lignocellulosic content of the agri-waste through production and secretion of lignocellulase, cellulase, hemicellulase, and lignase enzymes which are of great importance (Chandra et al. 2015). Another known mechanism for the microbial degradation of agriculture-derived lignocellulosic material is through the use of soil macrofauna, earthworms. However, the basic mechanism of biotransformation of soil and agri-waste to vermicompost in the earthworm gut is mediated via the enrichment of the cellulose-degrading microorganisms in its gut environment (Sharma et al. 2005). Therefore, the cellulase-producing microbes are the key for rapid degradation of the agri-waste particularly the soil-incorporated cereal straw.

3.4.1 Can Microbes Be a Possible Solution for Sustainable Straw Waste Management?: Factors Affecting Decomposition

The straw waste can be converted through several possible manners, but the microbial interventions can effectively transform straw to either useful products such as single-cell protein (Spalvins et al. 2018), bioethanol (Isikgor and Becer 2015), organic acids, and vitamins (ex situ straw management techniques) or can also help to degrade the straw in situ in the field. However, for the latter aspect, the primary limitation of the cellulose-degrading microbes will be the nonuniformity for the degradation potential as it is governed by a myriad of factors including the temperature, moisture content, organic matter content, soil C:N ratio, pH, relative physiologically active cellulose-degrading microbial populations, and micro- to mesofauna bacterial grazer populations in soil. Nawaz et al. (2013) compared the effect of incorporation of rice straw and burnt rice straw ash on soil physical and chemical properties. They have reported that irrigation and fertilization practices predominantly affect biogeochemical reactions. Further, they have observed occurrence of greater reduced environment and higher soil temperatures which are required for enhanced seed germination on incorporation of rice residues in comparison with burnt rice residue incorporation treatment. The role of changes in the soil organic carbon content on incorporation of agri-waste and its retrospective impact on microbial diversity and abundance has been elaborated by a field experiment performed by Wang et al. (2015). They have reported strong positive correlation of total soil organic carbon accumulation with bacterial abundance and enzymatic activity or enhanced microbial activity on C-stabilization. Likewise, the respiration rate, soil phosphatase, and urease enzyme activities were observed to be improved by incorporation of rice straw and helped in counteracting the effect of long-term and continuous application of pesticides (Rahmansyah et al. 2009).

A recent report by Li et al. (2018b) indicated the role of incubation temperature, moisture content, and soil tillage management systems to affect the decomposition of the incorporated straw besides affecting the CO_2-C evolution rates. They have summarized that high incubation temperature and moisture content will lead to significantly enhanced release of CO_2 from soil managed through conventional tillage technique compared to subsoil and no till management techniques. Likewise, the rate of straw decomposition is governed by two prominent factors, viz., litter quality and climatic conditions (Wang et al. 2012). However, irrespective of the climatic conditions, the decomposition chemistry remained similar with a decrease in the *O*/*N*-alkyl-*C* and di-*O*-alkyls, while a simultaneous increase in alkyl- and aromatic-Cs and their oxide groups was observed. Further, Curtin et al. (2008) have observed that well distribution of the straw in soil leads to faster decomposition. Thus, sustainable in situ agri-waste management demands appropriate and gradual decomposition of the stubbles and straw.

Fertilization may also affect the extent of decomposition of the straw incorporated in soil under field conditions. Zhan et al. (2018) have observed that organic mode of fertilization decelerates the population of straw degrading Firmicutes and Acidobacteria-like genera due to improvement in the fertility of the test soil. However, the control and NPK-amended soils exhibited enhanced prevalence of straw degrading microbes.

3.4.1.1 Microbial Genera Responsible for In Situ Degradation of Agri-waste

Agri-waste degradation potential has been reported for diverse group of microbes including bacteria, archaebacteria, and fungi. In a 1-year field experiment performed by Li et al. (2018a) in three contrasting soil types and six different organic wastes, incorporation of wheat straw and maize stovers significantly improved the microbial biomass carbon besides their functional diversity indices in all the three test soils in comparison to the unincorporated control treatment. Therefore, organic waste incorporation can be an effective technique to improve the soil microbial biomass, thereby indirectly affecting the overall soil properties. A terminal restriction fragment length polymorphism (T-RFLP) study indicated the prominence of Methanosarcinaceae and Methanobacteriaceae family members of archaebacteria to be responsible for emission of methane on incorporation of rice straw in rice field soil under anoxic environment, i.e., stagnant water conditions (Weber et al. 2001).

The common fungal genera known to effectively biotransform sugarcane waste include *Aspergillus flavipes*, *Chrysogenum*, *Cochliolous speifer*, *Penicillium*, *Rhizopus oryzae*, *Rhizopus stolonifera*, and *Trichoderma viride* (Sánchez 2009). The other most common group of fungi designated as white-rot fungi including the basidiomycetous *Phanerochaete chrysosporium* have several commercial applications as lignocellulose-producing genera. The white-rot fungi can degrade lignocellulosic waste by production of extracellular enzymes and hydrolases responsible for degradation of polysaccharides exclusively while an oxidative ligninolytic system having phenyl ring opening and lignin degrading properties (Sánchez 2009). Likewise, the rice straw degrading bacterial genera can also be useful if inoculated as consortial inoculant (Stella and Emmyrafedziawati 2015). Conventionally, microbes can be utilized for the conversion of the rice straw piles into compost, but this process is both time and energy intensive. Therefore, it will be prudent to perform in situ incorporation followed by low-energy composting of the straw and stubbles through application of efficient bacterial and fungal decomposer to achieve accelerated decomposition (Choudhary et al. 2016).

3.4.1.2 Microbial Inoculations for Agri-waste Management: Merits of Single, Dual, and Consortia Inoculants: Ab Initio Lab/Field Studies on Agri-waste Degradation

The application of a dual inoculant is generally beneficial to obtain faster degradation of the incorporated straw. Kausar et al. (2010) have developed a lingocellulolytic fungal inoculant comprised of *Aspergillus niger* and *Trichoderma viride* and tested it for in vitro biodegradation potential on rice straw. They have observed significantly higher decomposition of cellulose, hemicelluloses, lignin, and total carbon compared to the control such that the C/N ratio reduced substantially within 3 weeks of incubation of rice straw with the culture. Shruti et al. (2015) have isolated 19 bacterial and fungal isolates by rice straw enrichment technique under lab conditions and observed enhanced degradation of the rice straw on inoculation of the microbial cultures. Enhanced degradation of the rice straw through inoculation of a microbial consortium comprised of 30 different bacterial species leading to decrease in the ADF, NDF, and CF of the rice straw indicating rapid decomposition under lab conditions (Stella and Emmyrafedziawati 2015). Similarly, Choudhary et al. (2015) have observed accelerated decomposition of rice straw (up to 30% degradation) on inoculation of *Aspergillus flavus* and *Aspergillus terreus* under lab conditions.

Zeng et al. (2013) have reported enhanced degradation of the wheat straw on inoculation with an Actinobacteria, *Streptomyces viridosporus* T7A. They have further elucidated the ligno-cellulose degradation mechanism of these microbes which is primarily governed through modification of the carbonyl and methoxyl functional groups and deduction of the guaiacyl moieties in the lignin. Yu et al. (2007) have observed that the quinone Q-9(H_2) containing thermophilic Actinobacteria can possibly be the key gram-positive bacteria involved in degradation of lignin component in the agricultural waste during compost preparation from straw. In a metagenomics study performed for identification of microorganisms involved in delignification of the wood and other litter materials in the forest soil, Wilhelm et al. (2018) have observed a clear-cut high degradation of the lignin particularly by the members of the Comamonadaceae and Caulobacteraceae families of the Gram-negative bacteria. However, they have observed that cellulose was preferentially degraded by fungal communities. In a recent report by Li et al. (2017), a very efficient lignin depolymerization through breakage of recalcitrant C–C bonds in the neutral gut pH of the young worker termites facilitated polysaccharide cleavage by the ectosymbiotic fungi *Termitomyces* spp. Mimicking of such beneficial symbiotic natural associations can possibly provide the practical solutions to resolve the riddle of straw in situ management.

3.5 Future Prospects

The plant-associated microbiome studies can be epitomized as the ready information about the variabilities and functional role of the key microbial communities colonizing the plant which can be harnessed to identify and select the microbial groups or genera among the whole microbiome which have the capability to constructively alter the host plant phenology, physiology, and its ability to interact with the surrounding environment (Sanchez-Barrios et al. 2017). The use of informatics technologies can help in identification of the core microbiome members and linking these key groups to certain functional characteristics so as to organize the resident microbiome dynamics (Toju et al. 2018). Agler et al. (2016) have suggested for the use of a system biology approach for deciphering the complex microbe-microbe and microbe-plant interactions by simultaneous study of diverse group of microbial genera inhabiting a particular plant niche/organ. Further, core microbiomes can be capitalized for improving the agri-input use efficiency and tolerance to a variety of stress conditions of the present-day-intensive cultivation and monoculture-affected agroecosystems through the use of novel throughput, automated techniques such as microfluidics (Grossmann et al. 2011), and live root imaging through TRIS (tracking root interaction systems) (Massalha et al. 2017; Poole 2017). However, the extent of field translations of these results can be a bit tricky because majority research reports involve in vitro or controlled environment studies which can't be effectively extrapolated to obtain concrete conclusions for the identification of key groups on application under field conditions (Finkel et al. 2017).

The environmental factors complicate the scenario in comparison to the controlled environment studies. Further, studies have been performed for specific crop plants, and mostly the model plant microbiome studies have been published giving less clues for microbiome structure and the potential microbial genera of functional importance for other crops. This lab to land gap for the microbiome research probably exists due to variability in responses among different hosts and even for a single host in response to nutrient status, environment, and the crop age (Sessitsch et al. 2018). This ends up in a minima for the development of some universal and even designer plant probiotic culture(s) that can invariably perform and exhibit durable establishment over long periods (Sergaki et al. 2018).

These lacunae can be effectively addressed through application of consortia inoculum containing multiple strains or genera of the key players required for better establishment and colonization of the micro-niches or hot-spots. Gopal and Gupta (2018) have further suggested for the development of microbiome vaults equipped with preserved microbial genome, soil, and plant samples to configure minimum effective microbiome sets (MEMS) to harness short- to long-term benefits. Specifically, the realization of the rhizosphere engineering for the maintenance of the sustainable agriculture concept to ensure global food security can happen through putting know-how of rhizosphere interaction mechanisms in action (Ahkami et al. 2017). Furthermore, cognition of the plant host centric strategies integrating both microbial and plant traits can help in development of practical and effective solutions (Oyserman et al. 2018). The characterization and dissection of the individual and distinct roles of the plant genotype × environment × microbiome × management interactions have to be identified (Busby et al. 2017). Likewise, a holistic microbiome determination techniques or the microbiome deciphering toolbox has to be developed such that both cultivable and culture-independent protocols should be utilized to help us to avoid either under- or overestimation of the core microbiomes for ease of teasing out the cultivable fraction which can be utilized for development of the consortia or inoculant microbiota (Armanhi et al. 2018).

Acknowledgments Authors are thankful to the Head of the Department of Plant Breeding and Genetics as well as Soil Science, Punjab Agricultural University, Ludhiana, Punjab, India, for providing the necessary infrastructural facilities.

References

Agler MT, Ruhe J, Kroll S, Morhenn C, Kim ST, Weigel D, Kemen EM (2016) Microbial hub taxa link host and abiotic factors to plant microbiome variation. PLoS Biol 14:1–31. https://doi.org/10.1371/journal.pbio.1002352

Ahkami AH, Allen White R, Handakumbura PP, Jansson C (2017) Rhizosphere engineering: enhancing sustainable plant ecosystem productivity. Rhizosphere 3:233–243. https://doi.org/10.1016/j.rhisph.2017.04.012

Alshammari AM, Adnan FMA, Mustafa H, Hammad N (2011) Bioethanol fuel production from rotten banana as an environmental waste management and sustainable energy. J Microbiol 5:586–598. https://doi.org/10.5897/AJMR10.231

Armanhi JSL, de Souza RSC, Damasceno NDB, de Araújo LM, Imperial J, Arruda P (2018) A community-based culture collection for targeting novel plant growth-promoting bacteria from the sugarcane microbiome. Front Plant Sci 8:1–17. https://doi.org/10.3389/fpls.2017.02191

Babalola OO (2010) Beneficial bacteria of agricultural importance. Biotechnol Lett 32:1559–1570. https://doi.org/10.1007/s10529-010-0347-0

Berendsen RL, Vismans G, Yu K, Song Y, De Jonge R, Burgman WP, Burmølle M, Herschend J, Bakker PAHM, Pieterse CMJ (2018) Disease-induced assemblage of a plant-beneficial bacterial consortium. ISME J 12:1496–1507. https://doi.org/10.1038/s41396-018-0093-1

Berg M, Koskella B (2018) Nutrient- and dose-dependent microbiome-mediated protection against a plant pathogen. Curr Biol 28:2487–2492.e3. https://doi.org/10.1016/j.cub.2018.05.085

Borah N, Barua R, Nath D, Hazarika K, Phukon A, Goswami K, Barua DC (2016) Low energy rice stubble management through in situ decomposition. Procedia Environ Sci 35:771–780. https://doi.org/10.1016/j.proenv.2016.07.092

Busby PE, Soman C, Wagner MR, Friesen ML, Kremer J, Bennett A, Morsy M, Eisen JA, Leach JE, Dangl JL (2017) Research priorities for harnessing plant microbiomes in sustainable agriculture. PLoS Biol 15:1–14. https://doi.org/10.1371/journal.pbio.2001793

Chandra R, Yadav S, Kumar V (2015) Microbial degradation of lignocellulosic waste and its metabolic products. In: Chandra R (ed) Environmental waste management. CRC, Boca Raton, pp 249–298. https://doi.org/10.1201/b19243-11

Chialva M, Bonfante P (2018) Plant microbiota: from model plants to Mediterranean crops. Phytopathol Mediterr 57:123–145. https://doi.org/10.14601/Phytopathol

Chialva M, Zhou Y, Spadaro D, Bonfante P (2018) Not only priming: soil microbiota may protect tomato from root pathogens. Plant Signal Behav 13:1–3. https://doi.org/10.1080/15592324.2018.1464855

Choudhary M, Sharma PC, Garg N (2015) Crop residue degradation by autochthonous fungi isolated from cropping system management scenarios. BioResources 10:5809–5819. https://doi.org/10.15376/biores.10.3.5809-5819

Choudhary M, Sharma PC, Jat HS, Nehra V, McDonald AJ, Garg N (2016) Crop residue degradation by fungi isolated from conservation agriculture fields under rice–wheat system of north-West India. Int J Recycl Org Waste Agric 5:349–360. https://doi.org/10.1007/s40093-016-0145-3

Christian N, Herre EA, Mejia LC, Clay K (2017) Exposure to the leaf litter microbiome of healthy adults protects seedlings from pathogen damage. Proc R Soc B Biol Sci 284:1–8. https://doi.org/10.1098/rspb.2017.0641

Curtin D, Francis GS, McCallum FM (2008) Decomposition rate of cereal straw as affected by soil placement. Aust J Soil Res 46:152–160

da Silva DP, Castañeda-Ojeda MP, Moretti C, Buonaurio R, Ramos C, Venturi V (2014) Bacterial multispecies studies and microbiome analysis of a plant disease. Microbiol (UK) 160:556–566. https://doi.org/10.1099/mic.0.074468-0

Das H, Singh SK (2004) Useful byproducts from cellulosic wastes of agriculture and food industry – a critical appraisal. Crit Rev Food Sci Nutr 44:77–89. https://doi.org/10.1080/10408690490424630

Fierer N (2017) Embracing the unknown: disentangling the complexities of the soil microbiome. Nat Rev Microbiol 15:579–590. https://doi.org/10.1038/nrmicro.2017.87

Finkel OM, Castrillo G, Herrera Paredes S, Salas González I, Dangl JL (2017) Understanding and exploiting plant beneficial microbes. Curr Opin Plant Biol 38:155–163. https://doi.org/10.1016/j.pbi.2017.04.018

Flandroy L, Poutahidis T, Berg G, Clarke G, Dao MC, Decaestecker E, Furman E, Haahtela T, Massart S, Plovier H, Sanz Y, Rook G (2018) The impact of human activities and lifestyles on the interlinked microbiota and health of humans and of ecosystems. Sci Total Environ 627:1018–1038. https://doi.org/10.1016/j.scitotenv.2018.01.288

Gopal M, Gupta A (2018) Building plant microbiome vault: a future biotechnological resource. Symbiosis 77:1. https://doi.org/10.1007/s13199-018-0574-z

Gopal M, Gupta A, Thomas GV (2013) Bespoke microbiome therapy to manage plant diseases. Front Microbiol 4:10–13. https://doi.org/10.3389/fmicb.2013.00355

Granzow S, Kaiser K, Wemheuer B, Pfeiffer B, Daniel R, Vidal S, Wemheuer F (2017) The effects of cropping regimes on fungal and bacterial communities of wheat and faba bean in a greenhouse pot experiment differ between plant species and compartment. Front Microbiol 8:1–22. https://doi.org/10.3389/fmicb.2017.00902

Grossmann G, Guo W-J, Ehrhardt DW, Frommer WB, Sit RV, Quake SR, Meier M (2011) The RootChip: an integrated microfluidic chip for plant science. Plant Cell 23:4234. https://doi.org/10.1105/tpc.111.092577

Gu Y, Wei Z, Wang X, Friman V, Huang J, Wang X, Mei X, Xu Y, Shen Q, Jousset A (2016) Pathogen invasion indirectly changes the composition of soil microbiome via shifts in root exudation profile. Biol Fertil Soils 52:997–1005

Gupta PK, Sahai S, Singh N, Dixit CK, Singh DP, Sharma C, Tiwari MK, Gupta RK, Garg SC (2004) Residue burning in rice-wheat cropping system: causes and implications. Curr Sci 87:1713–1717

Hamonts K, Trivedi P, Garg A, Janitz C, Grinyer J, Holford P, Botha FC, Anderson IC, Singh BK (2018) Field study reveals core plant microbiota and relative importance of their drivers. Environ Microbiol 20:124–140. https://doi.org/10.1111/1462-2920.14031

Hartman K, van der Heijden MGA, Wittwer RA, Banerjee S, Walser JC, Schlaeppi K (2018) Cropping practices manipulate abundance patterns of root and soil microbiome members paving the way to smart farming. Microbiome 6:1–14. https://doi.org/10.1186/s40168-017-0389-9

Hartmann M, Frey B, Mayer J, Mäder P, Widmer F (2015) Distinct soil microbial diversity under long-term organic and conventional farming. ISME J 9:1177–1194. https://doi.org/10.1038/ismej.2014.210

Ho Y, Mathew D, Huang C (2017) Plant-microbe ecology: interactions of lants and symbiotic microbial communities. In: Zubaida Y (ed) Plant ecology – traditional approaches to recent trends. INTECH Open Science, pp 93–119

Hunter P (2016) Plant microbiomes and sustainable agriculture. EMBO Rep 17:1696–1699. https://doi.org/10.15252/embr.201643476

Igiehon NO, Babalola OO (2018) Rhizosphere microbiome modulators: contributions of nitrogen fixing bacteria towards sustainable agriculture. Int J Environ Res Public Health 15:E574. https://doi.org/10.3390/ijerph15040574

Imam J, Singh PK, Shukla P (2016) Plant microbe interactions in post genomic era: perspectives and applications. Front Microbiol 7:1–15. https://doi.org/10.3389/fmicb.2016.01488

Isikgor FH, Becer CR (2015) Lignocellulosic biomass: a sustainable platform for the production of bio-based chemicals and polymers. Polym Chem 6:4497–4559. https://doi.org/10.1039/c5py00263j

Joshi R, Singla-Pareek SL, Pareek A (2018) Engineering abiotic stress response in plants for biomass production. J Biol Chem 293:5035–5043. https://doi.org/10.1074/jbc.TM117.000232

Kalia A, Gosal SK (2011) Effect of pesticide application on soil microorganisms. Arch Agron Soil Sci 57:569–596. https://doi.org/10.1080/03650341003787582

Kalia A, Sharma SP, Kaur S, Kaur H (2020) Bacterial inoculants: how can these microbes sustain soil health and crop productivity? In: Giri B, Varma A (eds) Soil health. Soil biology, vol 59. Springer, Cham, pp 337–372

Kaur P, Taggar MS, Kalia A (2020) Characterization of magnetic nanoparticle–immobilized cellulases for enzymatic saccharification of rice straw. Biomass Conv Biorefinery. https://doi.org/10.1007/s13399-020-00628-x

Kausar H, Sariah M, Mohd Saud H, Zahangir Alam M, Razi Ismail M (2010) Development of compatible lignocellulolytic fungal consortium for rapid composting of rice straw. Int Biodeterior Biodegrad 64:594–600. https://doi.org/10.1016/j.ibiod.2010.06.012

Larousse M, Rancurel C, Syska C, Palero F, Etienne C, Industri B, Nesme X, Bardin M, Galiana E (2017) Tomato root microbiota and *Phytophthora parasitica*-associated disease. Microbiome 5:1–11. https://doi.org/10.1186/S40168-017-0273-7

Li H, Yelle DJ, Li C, Yang M, Ke J, Zhang R, Liu Y, Zhu N, Liang S, Mo X, Ralph J, Currie CR, Mo J (2017) Lignocellulose pretreatment in a fungus-cultivating termite. Proc Natl Acad Sci 114:4709–4714. https://doi.org/10.1073/pnas.1618360114

Li L, Xu M, Ali ME, Zhang W, Duan Y, Li D (2018a) Factors affecting soil microbial biomass and functional diversity with the application of organic amendments in three contrasting cropland soils during a field experiment. PLoS One 13:1–18. https://doi.org/10.1371/journal.pone.0203812

Li XS, Han HF, Ning TY, Lal R (2018b) CO2-C evolution rate in an incubation study with straw input to soil managed by different tillage systems. RSC Adv 8:12588–12596. https://doi.org/10.1039/c8ra00708j

Mareque C, Freitas T, Vollú RE, Beracochea M, Seldin L, Battistoni F (2018) The endophytic bacterial microbiota associated with sweet sorghum (*Sorghum bicolor*) is modulated by the application of chemical N fertilizer to the field. Int J Genomics 2018:1–10

Massalha H, Korenblum E, Malitsky S, Shapiro OH, Aharoni A (2017) Live imaging of root–bacteria interactions in a microfluidics setup. Proc Natl Acad Sci 114:4549. https://doi.org/10.1073/pnas.1618584114

Massart S, Margarita MM, Jijakli MH (2015) Biological control in the microbiome era: challenges and opportunities. Biol Control 89:98–108. https://doi.org/10.1016/j.biocontrol.2015.06.003

Mitter B, Pfaffenbichler N, Flavell R, Compant S, Antonielli L, Petric A, Berninger T, Naveed M, Sheibani-Tezerji R, von Maltzahn G, Sessitsch A (2017) A new approach to modify plant microbiomes and traits by introducing beneficial bacteria at flowering into progeny seeds. Front Microbiol 8:1–10. https://doi.org/10.3389/fmicb.2017.00011

Nadeem A, Dinesh G, Arun G (2015) Biodegradation of cellulose and agricultural waste material. In: Chandra R (ed) Advances in biodegradation and bioremediation of industrial waste. Taylor & Francis Group, New York, pp 211–234

Nawaz MF, Bourrie G, Trolard F, Mouret JC, Henry P (2013) Effects of agronomic practices on the physico-chemical properties of soil waters in rice culture. Turkish J Agric For 37:195–202. https://doi.org/10.3923/ijps.2014.70.75

Orozco-Mosqueda MDC, Rocha-Granados MDC, Glick BR, Santoyo G (2018) Microbiome engineering to improve biocontrol and plant growth-promoting mechanisms. Microbiol Res 208:25–31. https://doi.org/10.1016/j.micres.2018.01.005

Oyserman BO, Medema MH, Raaijmakers JM (2018) Road MAPs to engineer host microbiomes. Curr Opin Microbiol 43:46–54. https://doi.org/10.1016/j.mib.2017.11.023

Panke-Buisse K, Poole AC, Goodrich JK, Ley RE, Kao-Kniffin J (2015) Selection on soil microbiomes reveals reproducible impacts on plant function. ISME J 9:980–989. https://doi.org/10.1038/ismej.2014.196

Panke-Buisse K, Lee S, Kao-Kniffin J (2017) Cultivated sub-populations of soil microbiomes retain early flowering plant trait. Microb Ecol 73:394–403. https://doi.org/10.1007/s00248-016-0846-1

Perazzolli M, Antonielli L, Storari M, Puopolo G, Pancher M, Giovannini O, Pindo M, Pertot I (2014) Resilience of the natural phyllosphere microbiota of the grapevine to chemical and biological pesticides. Appl Environ Microbiol 80:3585–3596. https://doi.org/10.1128/AEM.00415-14

Pérez-Jaramillo JE, Carrión VJ, de Hollander M, Raaijmakers JM (2018) The wild side of plant microbiomes. Microbiome 6:4–9. https://doi.org/10.1186/s40168-018-0519-z

Pfeiffer S, Mitter B, Oswald A, Schloter-Hai B, Schloter M, Declerck S, Sessitsch A (2017) Rhizosphere microbiomes of potato cultivated in the high Andes show stable and dynamic core microbiomes with different responses to plant development. FEMS Microbiol Ecol 93:1–12. https://doi.org/10.1093/femsec/fiw242

Poole P (2017) Shining a light on the dark world of plant root–microbe interactions. Proc Natl Acad Sci 114:4281–4283. https://doi.org/10.1073/pnas.1703800114

Prasad R, Kumar M, Varma A (2015) Role of PGPR in soil fertility and plant health. In: Egamberdieva D, Shrivastava S, Varma A (eds) Plant growth-promoting rhizobacteria (PGPR) and medicinal plants. Springer International Publishing, Switzerland, pp 247–260

Prasad R, Gill SS, Tuteja N (2018) Crop improvement through microbial biotechnology. Elsevier, Amsterdam, Netherlands. ISBN 9780444639882. https://www.elsevier.com/books/crop-improvement-through-microbialbiotechnology/prasad/978-0-444-63987-5

Prasad R, Chhabra S, Gill SS, Singh PK, Tuteja N (2020) The microbial symbionts: potential for the crop improvement in changing environments. In: Tuteja N, Tuteja R, Passricha N, Saifi SK (eds) Advancement in crop improvement techniques. Elsevier, Amsterdam, Netherlands, pp 233–240

Rahmansyah M, Antonius S, Sulistinah N (2009) Phosphatase and urease instability caused by pesticides present in soil improved by grounded Rice straw. J Agric Biol Sci 4:56–62

Ruppel S, Franken P, Witzel K (2013) Properties of the halophyte microbiome and their implications for plant salt tolerance. Funct Plant Biol 40:940–951. https://doi.org/10.1071/FP12355

Sagar NA, Pareek S, Sharma S, Yahia EM, Lobo MG (2018) Fruit and vegetable waste: bioactive compounds, their extraction, and possible utilization. Compr Rev Food Sci Food Saf 17:512–531. https://doi.org/10.1111/1541-4337.12330

Sánchez C (2009) Lignocellulosic residues: biodegradation and bioconversion by fungi. Biotechnol Adv 27:185–194. https://doi.org/10.1016/j.biotechadv.2008.11.001

Sanchez-Barrios A, Sahib M, DeBolt S (2017) I've got magic in me: the microbiome of conventional vs organic production systems. In: Singh DP, Singh HB, Prabha R (eds) Plant-microbe interactions in agro-ecological perspectives. Springer Nature, Singapore, pp 85–95

Santhanam R, Luu VT, Weinhold A, Goldberg J, Oh Y, Baldwin IT (2015) Native root-associated bacteria rescue a plant from a sudden-wilt disease that emerged during continuous cropping. Proc Natl Acad Sci 112:E5013–E5020. https://doi.org/10.1073/pnas.1505765112

Santos-Medellín C, Edwards J, Liechty Z, Nguyen B, Sundaresan V (2017) Drought stress results in a compartment-specific restructuring of the rice root-associated microbiomes. MBio 8:1–15. https://doi.org/10.1128/mBio.00764-17

Santoyo G, Hernández-Pacheco C, Hernández-Salmerón J, Hernández-León R (2017) The role of abiotic factors modulating the plant-microbe-soil interactions: toward sustainable agriculture. A review. Spanish J Agric Res 15:1–15. https://doi.org/10.5424/sjar/2017151-9990

Sasse J, Martinoia E, Northen T (2018) Feed your friends: do Plant exudates shape the root microbiome? Trends Plant Sci 23:25–41. https://doi.org/10.1016/j.tplants.2017.09.003

Schlaeppi K, Bulgarelli D (2015) The plant microbiome at work. Mol Plant-Microbe Interact 28:212–217. https://doi.org/10.1094/MPMI-10-14-0334-FI

Sergaki C, Lagunas B, Lidbury I, Gifford ML, Schäfer P (2018) Challenges and approaches in microbiome research: from fundamental to applied. Front Plant Sci 9:1–12. https://doi.org/10.3389/fpls.2018.01205

Sessitsch A, Brader G, Pfaffenbichler N, Gusenbauer D, Mitter B (2018) The contribution of plant microbiota to economy growth. Microb Biotechnol 11:801. https://doi.org/10.1111/1751-7915.13290

Sharma S, Pradhan K, Satya S, Vasudevan P (2005) Potentiality of earthworms for waste management and in other uses – a review. Am J Sci 1:4–16

Shrivastava S, Prasad R, Varma A (2014) Anatomy of root from eyes of a microbiologist. In: Morte A, Varma A (eds) Root engineering, vol 40. Springer, Berlin, pp 3–22

Shruti, Sharma A, Kumar Malik D (2015) Lignocellulose biomass degradation by microbial consortium isolated from harvested rice field. Int J Curr Microbiol App Sci 4:274–280

Singh D, Singh J, Kalia A (2018) Yield, nutrient availability and uptake in wheat (*Triticum aestivum*) as influenced by *Azotobacter* and nitrogen levels. Indian J Agron 63(3):326–330

Singh D, Raina TK, Kumar A, Singh J, Prasad R (2019) Plant microbiome: a reservoir of novel genes and metabolites. Plant Gene 18:100177. https://doi.org/10.1016/j.plgene.2019.100177

Singla D, Taggar M, Kocher G, Kalia A (2018) Cellulase production by *Aspergillus fumigatus* using different plant-based agricultural biomass for paddy straw saccharification. Cellul Chem Technol 52:803–813

Snelders NC, Kettles GJ, Rudd JJ, Thomma BPHJ (2018) Plant pathogen effector proteins as manipulators of host microbiomes? Mol Plant Pathol 19:257–259. https://doi.org/10.1111/mpp.12628

Spalvins K, Ivanovs K, Blumberga D (2018) Single cell protein production from waste biomass: review of various agricultural by-products. Agron Res 16:1493–1508. https://doi.org/10.15159/AR.18.129

Stella M, Emmyrafedziawati AK (2015) Identification of rice straw degrading microbial consortium. J Trop Agric Fd Sc 43:119–127

Tikhonovich IA, Provorov NA (2011) Microbiology is the basis of sustainable agriculture: an opinion. Ann Appl Biol 159:155–168. https://doi.org/10.1111/j.1744-7348.2011.00489.x

Timsina J (2018) Can organic sources of nutrients increase crop yields to meet global food demand? Agronomy 8:214. https://doi.org/10.3390/agronomy8100214

Toju H, Peay KG, Yamamichi M, Narisawa K, Hiruma K, Naito K, Fukuda S, Ushio M, Nakaoka S, Onoda Y, Yoshida K, Schlaeppi K, Bai Y, Sugiura R, Ichihashi Y, Minamisawa K, Kiers ET (2018) Core microbiomes for sustainable agroecosystems. Nat Plants 4:247–257. https://doi.org/10.1038/s41477-018-0139-4

Turner TR, James EK, Poole PS (2013) The plant microbiome. Genome Biol 14:1–10. https://doi.org/10.1016/B978-0-12-417163-3.00011-1

van der Heijden MGA, Hartmann M (2016) Networking in the plant microbiome. PLoS Biol 14:1–9. https://doi.org/10.1371/journal.pbio.1002378

Van Der Heijden MGA, Bardgett RD, Van Straalen NM (2008) The unseen majority: soil microbes as drivers of plant diversity and productivity in terrestrial ecosystems. Ecol Lett 11:296–310. https://doi.org/10.1111/j.1461-0248.2007.01139.x

Varma A, Swati T, Prasad R (2019a) Plant microbe interface. Springer International Publishing, Switzerland. ISBN 978-3-030-19831-2. https://www.springer.com/gp/book/9783030198305

Varma A, Swati T, Prasad R (2019b) Plant biotic interactions: state of art. Springer International Publishing, Switzerland. ISBN 978-3-030-26657-8. https://www.springer.com/gp/book/9783030266561

Varma A, Swati T, Prasad R (2020) Plant microbe symbiosis. Springer International Publishing, Switzerland. ISBN 978-3-030-36247-8. https://www.springer.com/gp/book/9783030362478

Vejan P, Abdullah R, Khadiran T, Ismail S, Nasrulhaq Boyce A (2016) Role of plant growth promoting rhizobacteria in agricultural sustainability-a review. Molecules 21:1–17. https://doi.org/10.3390/molecules21050573

Wang X, Sun B, Mao J, Sui Y, Cao X (2012) Structural convergence of maize and wheat straw during two-year decomposition under different climate conditions. Environ Sci Technol 46:7159–7165. https://doi.org/10.1021/es300522x

Wang P, Liu Y, Li L, Cheng K, Zheng J, Zhang X, Zheng J, Joseph S, Pan G (2015) Long-term rice cultivation stabilizes soil organic carbon and promotes soil microbial activity in a salt marsh derived soil chronosequence. Sci Rep 5:1–13. https://doi.org/10.1038/srep15704

Weber S, Lueders T, Friedrich MW, Conrad R (2001) Methanogenic populations involved in the degradation of rice straw in anoxic paddy soil. FEMS Microbiol Ecol 38:11–20. https://doi.org/10.1016/S0168-6496(01)00168-4

Wilhelm RC, Singh R, Eltis LD, Mohn WW (2018) Bacterial contributions to delignification and lignocellulose degradation in forest soils with metagenomic and quantitative stable isotope probing. ISME J 13:413. https://doi.org/10.1038/s41396-018-0279-6

Wintermans PCA, Bakker PAHM, Pieterse CMJ (2016) Natural genetic variation in Arabidopsis for responsiveness to plant growth-promoting rhizobacteria. Plant Mol Biol 90:623–634. https://doi.org/10.1007/s11103-016-0442-2

Xiao X, Fan M, Wang E, Chen W, Wei G (2017) Interactions of plant growth-promoting rhizobacteria and soil factors in two leguminous plants. Appl Microbiol Biotechnol 101:8485–8497. https://doi.org/10.1007/s00253-017-8550-8

Xuan D (2007) Functional and molecular diversity of rice straw decomposing bacteria and fungi. SLU, Department of Soil Sciences, Uppsala

Xue D, Christenson R, Genger R, Gevens A, Lankau RA (2018) Soil microbial communities reflect both inherent soil properties and management practices in Wisconsin potato fields. Am J Potato Res 95:696–708. https://doi.org/10.1007/s12230-018-9677-6

Yamazaki A, Hayashi M (2015) Building the interaction interfaces: host responses upon infection with microorganisms. Curr Opin Plant Biol 23:132–139. https://doi.org/10.1016/j.pbi.2014.12.003

Yu H, Zeng G, Huang H, Xi X, Wang R, Huang D, Huang G, Li J (2007) Microbial community succession and lignocellulose degradation during agricultural waste composting. Biodegradation 18:793–802. https://doi.org/10.1007/s10532-007-9108-8

Zahn G, Amend AS (2017) Foliar microbiome transplants confer disease resistance in a critically-endangered plant. Peer J 5:e4020. https://doi.org/10.7717/peerj.4020

Zeng J, Singh D, Laskar D, Chen S (2013) Degradation of native wheat straw lignin by Streptomyces viridosporus T7A. Int J Environ Sci Technol 10:165–174. https://doi.org/10.1007/s13762-012-0085-z

Zhan Y, Liu W, Bao Y, Zhang J, Petropoulos E, Li Z, Lin X, Feng Y (2018) Fertilization shapes a well-organized community of bacterial decomposers for accelerated paddy straw degradation. Sci Rep 8(7981):1–10. https://doi.org/10.1038/s41598-018-26375-8

Zhang Z, Gonzalez A, Davies E, Liu Y (2012) Agricultural wastes. Water Environ Res 84:1386–1406. https://doi.org/10.2175/106143099X133767

Chapter 4
Plant-Microbe-Metal Interactions: A Biochemical and Molecular Analysis for Phytoremediation

Namrata Budhiraja, Priyanshi Srivastava, Sakshi Agrahari, Divyanshu Shukla, Bhawna Mudgil, Shikha Saxena, Rajesh Dahiya, and Siddharth Vats

Abstract Plants in nature face abiotic and biotic stress, attacked by phytopathogens and phyto-pests, which leads to loss in the productivity and health of the plants. Interaction among plant with rhizospheric microbes and soil metal contents involves various biochemical pathways which are exploited for phytoremediation. The plant-microbe interactions, plant-metal interaction, microbe-metal interactions and their involvement with each other, their role in phytoremediation and the molecular level at which all these activities are regulated, and how the plant immune system respond to it are of high significance. Plant immune system orchestrates interactions with beneficial microbes, pathogens, and insects, and this interaction is influenced by the presence of metals and provides a rational basis for developing sustainable strategies for phytoremediation. In this chapter all the aspect of plant-microbe-metals interactions in plants has been reviewed.

N. Budhiraja · P. Srivastava · D. Shukla · S. Saxena · S. Vats (✉)
Faculty of Biotechnology, Institute of Bio-Sciences and Technology, Shri Ramswaroop Memorial University, Barabanki, Uttar Pradesh, India

S. Agrahari
Division of Biological Sciences and Engineering, Netaji Subhash University of Technology, Delhi, India

B. Mudgil
TGT, Natural Science, Government Co-Ed, Sarvodaya Vidyalaya, Delhi, Rohini, India

R. Dahiya
University Institute of Engineering and Technology, Kurukshetra University, Kurukshetra, Haryana, India

A. Varma et al. (eds.), *Plant Microbiome Paradigm*,
https://doi.org/10.1007/978-3-030-50395-6_4

4.1 Introduction

The soil is the important component of the biosphere which supports life and provides vital components, essential for plantation, for the regulation process of some gases, climate, biochemical, and water cycling (Kamal et al. 2010). Soil is important for the biodiversity maintenance (Maurya et al. 2013, 2014; Vats et al. 2014; Vats and Mishra 2016; Goel et al. 2017; Magdoff and van Es 2000; Chen et al. 2018). Metals are always there in the soil, unprotected or in protected forms, like Arsenic, Cadmium, Copper, Mercury, Nickel, Lead, and Zinc which have the high level of accumulation and also led to toxicity via many farming activities and the residues of domestic household sewage sludge wastes and from various other sources (Kaur et al. 2010; Kumar et al. 2018; Bhargava et al. 2017; Behera and Prasad 2020a). These metals are generally toxic and harmful to plants and animals, which have the abilities to cause DNA damage and may lead to mutagenic animals and plants (Padmavathiamma and Li 2007; Hwang et al. 2018; Painuly et al. 2019; Tandon and Vats 2016). Phytoremediation is the plant-based and cost-effective technology which uses flora and fauna for environmental cleanup process, i.e., it has a role to play in removing organometallic wastes from soils and water (Tsao 2003; Gupta et al. 2018; Vats and Kumar 2015; Vats et al. 2019). In today's scenario phytoremediation acts as a tool with applications for decontamination of soil, water, and air by the help of detoxification, extraction, or hyper-accumulation of contaminants (Heinekamp and Willey 2007; Vats et al. 2011; Behera and Prasad 2020b).

4.2 Plant-Microbes Interaction in Phytoremediation

Rhizospheric plants show the interaction among soil with soil microorganism efficiently by extruding chemicals' components and signals' molecules (Saxena et al. 2019; Bhargava et al. 2019a, b). The microbes, which are in symbiotic association with the host plant, form an effective association by colonization and chemotaxis (Doornbos et al. 2012; Drogue et al. 2012; Bulgarelli et al. 2013; Kumar et al. 2020a, b; Prasad et al. 2020). The root-released phytochemicals (exudates) and microorganisms interaction in the rhizosphere has been recognized as an important segment of the involvement of the plants in phytoremediation (Badri et al. 2009). Root exudates/root-released phytochemicals have a significant action in phytoremediation, by helping in induction, the ability of host plants to instantaneously adapt and survive with various stress (physical) by regulating the growth of microbes living in association with roots of the host plant or by regulating the process like transformation, adsorption, chelation, and inactivation of metals. Organic molecules released by plants especially organic acids have better ability to form complexes and binding with metal ions, affecting solubility, mobility, and their bioavailability in the soil (Chiang et al. 2011). Root discharges enhance the absorption and mobility of nutrients and metals, the process known as acidification due to release of proton followed by intracellular binding compounds, electron transfer with the help of enzymes in the root zones of the plants, and lastly by indirectly stimulating the microbes of rhizosphere and their activity to enhance the efficiency of phytoremediation (Ström et al. 2002; Pérez-Montaño et al. 2013). Metals from the soil are accumulated by the plants, which required the metals to be in the solution form with the soil. pH plays a vital role in controlling the mobilization process (Gadd 2004; Wawra et al. 2018).

Variation in the pH affects the metal speciation in solution (Shaheen et al. 2016). Soil borne microorganism may lead to the increase in bioavailability of metal ions including Fe^{2+}, Mn^{2+}, and Cd^{2+} for absorption of the root (Poggenburg et al. 2018; Tricker et al. 2018; Burges et al. 2018). Metals can be immobilized by the help of various microorganisms in different ways which include accumulation in their biomass or on cell walls, or through precipitation or adsorption (Vats and Negi 2013; Vats et al. 2013a, b; Gadd 2018; Wawra et al. 2018). Microbes may enhance the phenomena of phytoremediation in various different ways: by accelerating biomass of plants, regulating metal availability, bioaccumulation, and translocation. There are several pros of employing plants and plant growth promoting microorganisms (PGPMs) instead of chemicals for bioremediation, as the metabolites released by the rhizobia (in situ) are biodegradable, nontoxic in nature (Rajkumar et al. 2012). Although, not all the metabolites released have role in metal absorption, some are neutral, and some have no negative effects in metal mobilization (Zhao et al. 2001).

An organic acid such as oxalate which has low molecular weight is secreted by non-mycorrhizal and ectomycorrhizal when microbes come in contact of metals and form a stable complex with them (Johansson et al. 2008). Microbes of the endophytic origin (endophytic bacteria, EB) locations and rhizobacteria degrade the toxic compounds in the rhizosphere. EB are those groups of bacteria which reside/colonize the plant internally, its tissues, and with no sign of infection and not harm or negative effect to the plants (Prasad and Dagar 2014; Vats and Miglani 2011; Ojha et al. 2013).

The most common cultivable endophytic species are *Pseudomonadaceae*, *Burkholderiaceae*, and *Enterobacteriaceae* (Mastretta et al. 2006). Microorganism presents in the rhizosphere plays a vital role in the ecological fitness of their plant hosts. Many essential processes of the microbes occur in the rhizosphere, like providing protection to the plants, pathogenesis, promotion in the growth of the plants, recycling of the minerals, production of antibiotics, and geochemical plants colonization (Kent and Triplett 2002). Interaction between microbes and plants can be beneficial neutral, or harmful to the plant, depending on the specific microorganisms and host plants involved and prevailing environmental conditions (Bais et al. 2006).

4.3 Rhizosphere Microbiome: The Second Genome of the Plant

Microbes play a vital role in the functioning of plants by enhancing their physiology and development. To enhance and influence the plant growth and health, it is mandatory to know which microorganism is present in the rhizosphere microorganisms and their impact on health and diseases. The diversity of microbes which is associated with plant roots is enormous, and the microbial community formed of this association between plants and microbes is also known as plant's second genome and supports plant's health (Berendsen et al. 2012). Figure 4.1 represents the interaction among plants, microbes, and metals. Soil microbial community represents the greatest reservoir of biological diversity. These microbes of diverse types (*Achromobacter*, *Acetobacter*, *Anabaena*, *Azoarcus*, *Azotobacter*, *Bacillus*, *Clostridium*, *Enterobacter*, *Frankia*, *Hydrogenophaga*, *Kluyvera*, *Microcoleus*, *Phyllobacterium*, *Pseudomonas*, *Serratia*, *Streptomyces*, *Vibrio*), and legume symbiont are known as Rhizobium (Vassilev and de Oliveira Mendes 2018).

The root also forms the system part and allows colonization of microbes inside and outside of it. Microbes which grow inside the plant tissues are termed as

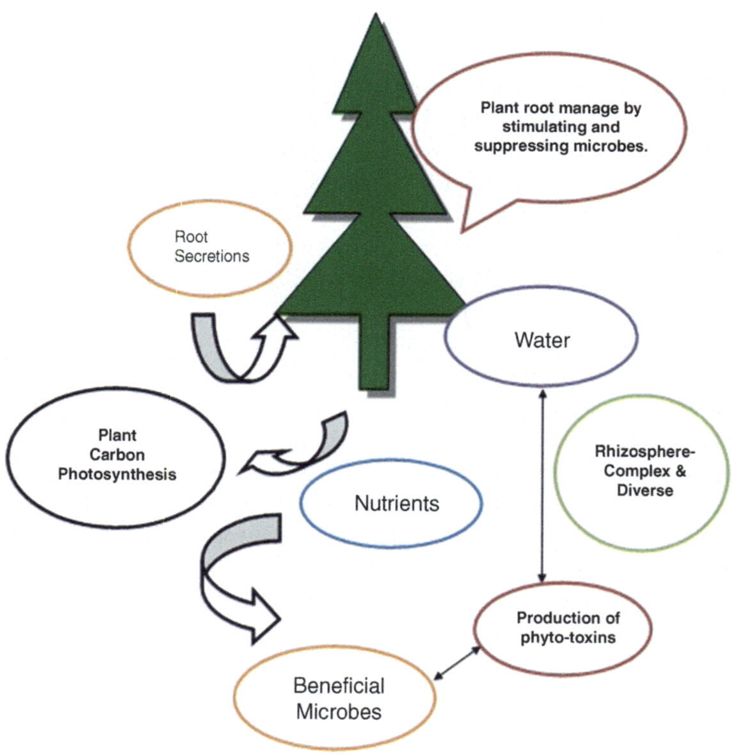

Fig. 4.1 Interaction among plant, microbes, and soil components

endophytes (Jeffries et al. 2003; do Amaral et al. 2017). It is the interaction of microbes of roots and plants that brings changes in the chemical environment of the rhizosphere to that of the soil (Marschner 1995; Hinsinger 2001). Exudates of roots are taken by microbes as nutrition. 30%–60% of the total photosynthetic carbons are taken by the roots for its use, and microbes in the roots provide 10–20% of the overall need (Marschner 1995; Salt et al. 1998). Exudates are acids (organic acids) of low as well as high molecular weight. The total concentrations of organic acids in roots generally range from 10 to 20 Mm, which usually comprises succinate, isocitrate, acetate, lactate, oxalate, fumarate, malate, citrate, and aconitate. The leftover of organic solutes in roots is consists of sugars (90 mM) and amino acids (10–20 mM) (Jones 1998).

Microbial cells have the ability to produce and recognize signal molecules; due to this, the whole populations are allowed to produce biofilms over the large areas of the root surface. This phenomenon is generally referred to as quorum sensing. It is processed and involves the cell–cell communication mechanism which is able to synchronize and helps in gene expression in response to population cell density.

4.4 Stimulation of Plant Growth Through the Microbial Community of the Rhizosphere

The rhizosphere is the term given to the area of plant roots and its surrounding soil. This is the region where exchange of signals and biochemical interactions between plants and soil microbes is studied and described (Pinton et al. 2007; Shrivastava et al. 2014), and it is the systems where roots are able to release metabolites in the large quantities from the root hairs. These metabolites act as a chemical signal for the mobility of the bacteria to the root surface. After being activated and attracted to these biochemical signals, bacteria colonize very efficiently with the rhizospheric soil of the crop plants. These bacteria are called plant growth promoting rhizobacteria (PGPR) (Prasad et al. 2005, 2015, 2020). PGPR helps in the process of phytoremediation in a profitable way; this partially depends upon the ability of the plants to withstand the toxicity of metal and to yield biomass inadequate amount (Novo et al. 2018; Vats and Bhargava 2017; Rosenkranz et al. 2018; Ma and Wang 2010; Mesa et al. 2015). Rhizospheric microbes with plants perform rhizoremediation (Lacalle et al. 2018). PGPR are of two groups: firstly, nutrient cycling and photostimulation which include atmospheric nitrogen fixation which synthesizes siderophores that are able to extract Fe and provide it to producer cells of auxin, cytokinins, and gibberellins which solubilize phosphorous and enzymes, namely, ACC deaminase which reduces ethylene levels; secondly which have biocontrol actions, by antibiotic production (Sharma et al. 2014; Jain et al. 2011; Bashan and Holguin 1998; Glick et al. 2007; Bhargava et al. 2020). The bacteria which are associated with the roots having ACC deaminase activity help the plants to withstand toxic and polluted soil (Ghosh et al. 2018). Under high stress caused by toxic soil, ethylene concentration increases which leads to inhibition of root growth and proliferation. Bacterial ACC deaminase can significantly decrease ACC levels by metabolizing its ethylene precursor ACC into a ketobutyric acid and ammonia. Widely rhizosphere can be divided into three different components: the rhizosphere soil, the rhizoplane, and the root. The rhizosphere is thus the zone of soil which is influenced by roots itself by the release of substrates which affect the microbial activity.

The rhizoplane is the surface of roots which includes the strong adherent root particles. The root is a part of the system as certain rhizo-endophytic microorganisms (Bowen and Rovira 1999). The rhizospheric environment provides a dynamic condition where microbes can sustain as well as interact with each other and host plant. Organic acid produces by soil fungi (Richardson and Simpson 2011) and rhizobacteria (Goldstein et al. 1999; Nautiyal 2003) helps in promoting the metals availability and lowering of pH which accompany anions. It also helps in metal solubilization, bioavailability, and mobility by supplying metal-complexing organic acid ligands (Kamnev and Van der Lelie 2000). The constituents of heavy toxic metals are obtained from dissolving minerals which increase the bioavailability and phytoavailability in soils by these microbial-driven processes which change the soil fertility. The exudates of root, in the resolver zone, act as a source of abundant energy for the microbial transformation of organic compounds (Table 4.1). Microorganisms present in the soil produces biosurfactants, for facilitating the removal of organic pollutants (Chen et al. 2018). Detoxification of soil contaminated with heavy metals takes place directly by the utilization of root exudates which can also be carried out in soils which are contaminated with heavy metals. The PGPR plays a vital role in enhancing the growth of the plants even in the soil with heavy metal contamination and finds application in phytoremediation.

4.5 Plant-Microbe-Metal Interactions

Plants and microbes coexist, and their cohesive interactions play an important role in adapting to the environments which are producing metals need to be explored and exploited to improve and enhance their ability to phytoremediate metals. Phytochemicals released by plant's root find application as nutrient and also provide energy sources for soil microorganism. Many beneficial rhizobia or fungi acting are PGPM, which is able to make metal phyto-toxicity less strong and stimulate the growth of plant indirectly with the help of the induction of defense mechanisms against the phytopathogens and are also done directly through the mineral solubilization of nutrients such as N, K, Fe, P, etc. and release of enzymes (Specific) such as 1-aminocyclopropane-1-carboxylate deaminase and production of phytohormones. PGMP may also modify the bioavailability of metals in soil by various mechanisms like precipitation, acidification, redox reactions, chelation, and complexation. Plant-microbe interaction enhances the phytoremediation and supports the biogeochemical cycling of metals (Basu et al. 2020). The proper combination of both plant and microbes involved in applied processes for the enhancement of phytoremediation

Table 4.1 Components of root exudates and their roles in the rhizosphere

Components	Species	Rhizosphere function
Amino acids	α alanine, β alanine, arginine, asparagines, aspartic acid, cysteine, cystine, glutamic acid, glutamine, histidine, homoserine, isoleucine, leucine, lysine, methionine, phenylalanine, proline, serine, threonine, tryptophan	Nutrient source, the energy source Chelators of insoluble mineral nutrients, chemoattractant signals to microbes
Enzymes	Amylase, DNase, phosphates, polygalacturonase, RNase, sucrose, urease, xylanase	Catalysts for phosphorous release from organic molecules, biocatalysts for organic matter transformation in soil
Organic acids	Acetic, acoustic, adipic, butyric, citric, cyclic, formic, gluconic, glutaric, glyoxylic, indole-3-acetic, isocitric, lactic, maleic, malic, oxalic, piscidic, propionic, pyruvic, succinic, tartaric, valeric	Nutrient source Chemoattractant signals to microbes Chelators of poorly soluble mineral nutrient Acidifiers of soil, detoxifiers of AI, *nod* gene inducers, antibacterial agents
Phenols	Caffeic acid, ferulic acid, flavonoids, isoflavonoids, neoflavonoids, pyrocatechol, quercetin, strigolactones, styrene	Nutrient source, energy source, chemoattractant signals to microbes, chelators of insoluble minerals, microbial growth promoters, inhibitors in rhizobia, inductors of resistance against phytopathogens, detoxifiers of AI, phytoalexins against soil pathogens
Purines	Hypoxanthine, isoguanine uric acid, theobromine, xanthine	Nutrient source, the energy source
Root border cells		Produce signals that control mitosis, produce signals controlling gene expression, stimulate microbial growth, release chemoattractants, synthesize defense molecules for the rhizosphere, act as decoys that keep root cap infection-free, release mucilage and proteins
Saccharides	Arabinose, fructose, fucose, galactose, glucose, lactose, mannose, raffinose, rhamnose, ribose, sucrose, xylose	Nutrient source, energy source, anchoring to bacteria to plant surfaces
Vitamins	Ascorbic acid, biotin, folic acid, niacin, pantothenate, pyridoxine, riboflavin, thiamine, thioctic acid, tocopherol, vitamin B12	Promoters of plant and microbial growth, nutrient source, resistance to soil pathogens, facilitation of organic pollutant degradation, induction of plant-microbe symbiosis
Others	Billneurine, bradyoxetin, glomalin, inositol, nicotinic acid, rhamnolipids, somatropin, surfactants	Stimulation of plant and microbial growth, regulators of symbiotic expression of nodulation genes

efficiency is to be considered, so as to maximize the performance from conventional microbe-based technologies. Plant-microbe interactions benefit both toward the metal tolerance and their detoxification. Together with their functioning includes mobilization of metal ions as well as their immobilization, translocation, and transformation, which lead to a better strategy for bioremediation processes (Bruins et al. 2000; Ma et al. 2011). The activities of PGPM which includes metal bioaccumulation, bioleaching, and bioexclusion are involved in adaptation microbes toward heavy metals and making them resistant and tolerant to heavy metal-rich environments. Processes like chelation with acidification and protonation make metals to get mobilized, whereas precipitation, alkalization, and complexation cause metal mobilization or immobilization (Tandon and Vats 2016; Alford et al. 2010).

4.5.1 Bioaccumulation

The arrangement made to increase and improve the efficiency of remediation techniques is to increase the process of bioaccumulation and biotransformation potential of both plants and microbes for the detoxification of toxic metals. If metals such as Cr, Cd, Pb, As, etc. are present in amount higher which is toxic, soil will also start showing toxic effects on overall growth of the plants and metabolism microbes and host. And on bioaccumulation these heavy metals pose serious risk to the health and life of the humans and its livestock and other animals (Vats et al. 2012, 2017; Vats 2017; Negi and Vats 2013; Sharma et al. 2018; Singh and Vats 2019; Vats and Bhargava 2017; Ajmani et al. 2019). To remove such excess of metal ions from the contaminated site is done by the use of some chemicals as well as by biological means. Bioaccumulation is a process of intracellular accumulation of metals that are comprised of two stages:

1. Metabolism-independent passive biosorption like chemical and physical adsorption, ion exchange (metal), chelation, micro precipitation, coordination, surface complexation
2. Metabolism-dependent active bioaccumulation like transportation of metal ions or biosorption of ions into the microbial cells by carrier-mediated ion pumps, metal ion-based complex formation and permeation, and endocytosis (Chojnacka 2010)

The process of bioaccumulation is more complex than that of biosorption because it needs the metabolic pathways and utilizes cells metabolic activities that involve intracellular sequestration, metal accumulation, extracellular precipitation, and complex formation (Gadd 2004). Numerous microbes undergo metal bioaccumulation, which can be considered for the uptake characteristics as well as reduction in metal toxicity of plants (Ma et al. 2011; Deng and Wang 2012).

4.5.2 Bioleaching

Bioleaching is a simple and effective technique for the extraction of metals. Some of the mesophilic bacteria, iron-oxidizing bacteria, thermophilic bacteria, as well as thermophilic bacteria and fungi are able to bioleach heavy metals from the contaminated sludge, sediments, and soils (Wong et al. 2004; Kletzin 2006). Dissolution, complexation, reduction, oxidation, and adsorption metabolic processes are responsible for removing metals including Zn, Fe, Cd, Cu, Pb, and Cr (Pathak et al. 2009) with the help of *Acidithiobacillus thiooxidans* which creates favorable acidic conditions. The capability of bioleaching is depended upon the bacterial species in comparison with the neutrophilic bacteria. Acidophilic bacteria are more common as well as more capable of metal bioleaching (Navarro et al. 2013).

4.5.3 Bioexclusion

Nonessential metals like Cd and As utilize transport pathways for passage of ions into the cell and can be ATPase-linked or non-ATPase-linked which are selective for the transported anion or cation (Nies and Silver 1995), while ATPase efflux technique is utilized by bacterial cells for essential metal (e.g., Cu^{2+}) transport through active transport (Bruins et al. 2000). Prokaryotes exhibit resistance from heavy metals due to the distribution pattern, the physiological functions, and the action-exporting proteins like P-type ATPase, chromate proteins including resistance elements (i.e., CnrT and NreB), and cation diffusion facilitator. Possessions of the highly specialized mechanisms make a metal bacterial resistant.

4.5.4 Metal Mobilization

Metal mobilization is often known as strong binding of metals to soil particles or precipitation for the countable or uncountable of a significant amount of metals in soil. The probability and percolation (mobility and solubilization) of metals have certified as a fundamental term in affecting the effectiveness of phytoextraction (Ma et al. 2009). Metal-mobilizing microbes are commonly used to modify rhizo deposition (soil habitat), although inducing soil mobility using biogeochemical cycle processes (Basu et al. 2020), i.e., a phenomenon which involves circulation of chemical element, circulates through the biotic and abiotic factor of the ecosystem and speciation of metals. Metals, which includes acidification, protonation and chelation (Argueso et al. 2007; Ma et al. 2011; Rajkumar et al. 2012; Sessitsch et al. 2013).

4.5.4.1 Acidification

pH of the soil is an indication of soil's alkalinity/acidity which is calculated in terms of pH units. pH of soil is relevant factor which affects the quantity, quality, and mobility/probability of metals in soil. pH and mobility of metal are inversely proportional (Richards et al. 2000). Soil pH is extensively induced by activities of both plants and many other microorganisms. Plant roots excrete hydrogen ions which undergo adsorption on the soil granules as well as transpose heavy metal cations and eventually lead to acidification of the rhizosphere. In comparison to the bulk soil, the pH of rhizosphere is reduced by 1 or 2 units due to the root latex (Sheoran et al. 2011), therefore amplifying mobility of soil and biological availability of the soil sample (Alford et al. 2010; Kim et al. 2010). Chem et al. (2014) indicated that the amount of exudation by roots can be regulated by inoculation of Sphingomonas SaMR12, endophytic bacterium from *S. alfredii*, thus fundamentally upgrading plant's absorption capability and bioavailability of Cd. Recently it was reported that *P. myrsinacearum* RC6b notably elevated the uptake of metals by *S. plumbizincicola*. This was credited due to the organic acid production capacity and solubilization of insoluble tricalcium phosphate (Ma et al. 2011).

4.5.4.2 Protonation

Microorganisms present in soil are capable of acidification of their habitat by transporting proton by substituting the cationic heavy metals present at the sorption site (Rajkumar et al. 2012). In order to understand, acknowledge, and advance such processes, there have been immense experiments to illustrate the synergy among bacterial surfaces, metal ions, and protons, as well as to distinguish them by undergoing techniques like spectroscopy. Giotta et al. (2011) examined the communication of proton sets present on the surface of Rhodobactersphaeroides with Ni^{2+} by employing lessened total reflection Fourier transform infrared (ATR-FTIR) spectroscopy. The result disclosed that carboxylate moieties that exist on the bacterial surface play a crucial role in extracellular biosorption of Ni^{2+}.

4.5.4.3 Chelation

Sequestering agents or organic chelators' compounds liberated from plants and rhizobacteria perform rummage of metal ions around it, consequently protecting or preserving from resorption (Gadd 2004). Natural chelators of organic nature are often recognized as organic acid anions, metallophores, siderophores, metal-binding agents, and biosurfactants (Sessitsch et al. 2013). Upon induction of binding of metals by peptides (MTs and PCs), the chelating metal ion can abolish the impact of phytotoxins from free ions, permitting for transport in plants, metal uptake, and xylem loading (Cai and Ma 2002). Tripeptide glutathione produces phytochelatins, peptides with the ability to attach heavy metals. Also been synthesized PCs synthase upon catalysis by enzymes (Solanki and Dhankhar 2011).

Heavy metal exposure induces the production/manufacturing of PCs immediately; heavy metal exposures are positively or empathetically interrelated with deposition of metals in plant tissues (Pal and Rai 2010). In distinction to small cysteine-rich, PCs, metal binding proteins and MTs, plays a significant role in various organisms for e.g. Eukaryotic, Prokaryotic, animals, plants, and many other microbes, including homeostasis and metal detoxification by rummage reactive oxygen species (Leitenmaier and Küpper 2013). Bolchi et al. (2011) labeled the polypeptides of fungal mycorrhiza *Tuber melanosporum*, i.e., MTs (TmelMT) and PC synthase (TmelPCS), were essential of consulting an enhance tolerance to stress. In spite of this, it is known that MTs transpire in AMF and genes encode several enzymes for the synthesis of PCs which can be triggered in mycorrhizal roots; thus, there is an increase in mycorrhizal photosynthesis unsheltered to metal stress. Although, there has been no remarkable implementation of the metal-binding phenomena to regulate metal tolerance in AM symbiosis (Rivera-Becerril et al. 2005). Iron (Fe) is one of the important micronutrients, and due to its low solubility, its concentration in soil is below the level necessary to support microbial life. Thus, plants overcome challenges to grasp iron (Fe) by three mechanisms:

Strategy 1: refers to the solubilization of ferrous (Fe), by the plants of dicots and monocots type, through acidification at rhizosphere.
Strategy 2: involves release of phytosiderophores (PSs) or absorption of Fe^{3+}-PS.
Strategy 3: the plants are involved in the absorption of Fe^{3+}-microbial siderophores.

According to various studies, it has been either mentioned or demonstrated that PSs have metal solubilizing and transporting ability by chelation, and however they are transferred into rhizosphere using a potassium mutagenic acid symporter (Sakaguchi et al. 1999). Microorganisms that form siderophores show more absorption for metals over PSs.

Thus, microbes may develop and produce their own strategies to solubilize metals for a systematic uptake by plants. Not long ago, but Yuan et al. (2014) also illustrated that solubilization of Cd enhanced due to liberation of siderophores, upon inoculation of endophyte, i.e., Rahnella sp. JN27, into metal chelated soil, thus assisting Cd-hyperaccumulators, i.e., *Amaranthus mangostanus* and *A. hypochondriacus*. Microbes and plants that release organic acids usually participate in varying pathways which occur in the rhizosphere, which even include detoxification of heavy metals, assession, mobilization, and mineral weathering in soil (Rajkumar et al. 2012). Plant roots that excrete organic acids, such as citrate, acetate, and oxalate, are broadly recalled to be responsible for dissolving the solid phase metal through complexation reaction in soil and though make obtainable for plants uptake. Mucha et al. (2005) establish that the malonate and oxalate exhibited complexation properties with trace metals.

Although, the organic acids produced by microbes consequently led to rise in metal release rather than the direct change by secreted roots (Amir and Pineau 2003). Percolates of LMWOAs from microbial populations comprise of acid derivatives of phenols and aliphatic compounds, with immense perspective to improve metal solubilization processes (Rajkumar et al. 2012). A recent study by Chem et al. (2014) exemplified that *Pseudomonas* sp. Lk9, endophytes with organic acid production capability, plays a significant role in enhancing the availability of metal ions and minerals (like Fe and P) into the soil by LMWOAs secretion, thereby significantly affecting the *Solanum nigrum's* production of shoot biomass as well as metal accumulation in aerial plant parts. However, AMF produces the specific protein glomalin which appears to be systematic as the heavy metals are sequestered outside mycelial boundaries (Gonzalez-Chavez et al. 2004).

The BSs (biosurfactants) structures comprise one or combination of more than one compound, namely, complexes of polysaccharide with lipids, mycolic acids, glycolipids, or cell surfaces of microbes (Pacheco et al. 2011). BSs can create their complexes with metals due to their amphiphilic structures, and soil matrix is responsible to desorb metals to the soil solution. Hence increase metal solubility of soils is contaminated with high metal ions (Sheng et al. 2008). Currently, studies on surfactant from *Bacillus subtilis*, di-rhamnolipids, and sophorolipids produced from *Torulopsis bombicola* have been implemented for the removal of metal ions from tainted soil (Mulligan et al. 2001; Juwarkar et al. 2007; Venkatesh and Vedaraman 2012).

4.5.5 Metal Immobilization

Reduction of microbes can be done by plant-metal uptake or translocation to aerial plant parts by a decrease of metal bioavailability in soil through some processes like precipitation, alkalization, and complexation processes (Fig. 4.2).

4.5.5.1 Precipitation

The ability of microorganism is to publicize the enzymatically catalyzed precipitation of radionuclides and toxic metals through a reduction process carried out by

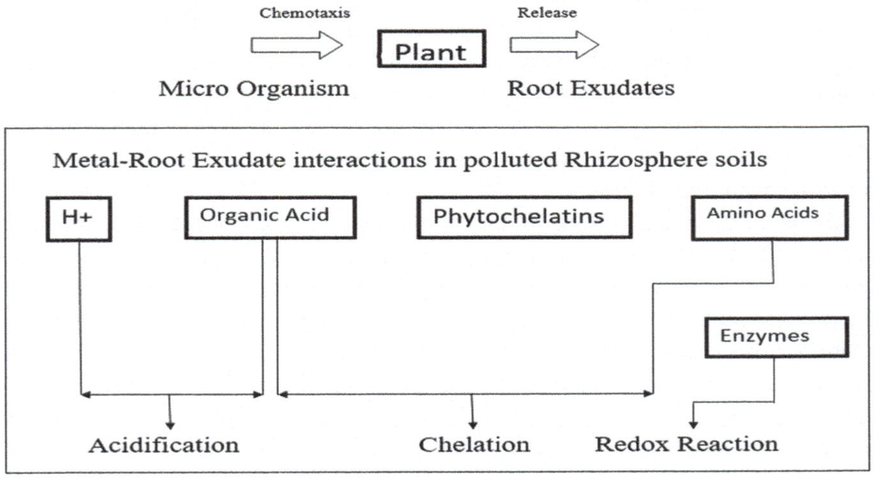

Fig. 4.2 Basic of phytoremediation

microbes. This proves substantial for phytoremediation of soil contaminated with metals (Payne and DiChristina 2006). Oves et al. (2013) described that the inoculation of Cr reduces OSG41 sp. of *P. aeruginosa* over chickpea gown in Cr^{6+}. Cr uptake is consequently decreased by a percentage of 38, 40, and 36 in shoots, grains, and roots, respectively. The result denotes that bacteria procure the ability in contrast to the inhibition pattern of high concentrates of Cr^{6+} as to protect host plant. Hence, an insoluble mineral formed by metals and radionuclides can be immobilized straight away else by actions of enzymes (Pagnanelli et al. 2010), or it can be indirectly immobilized by ferrous oxidation of a microbial inorganic acid (Park et al. 2011). Park et al. (2011) identified that the significance of PSB reduced PB availability by releasing P from insoluble P compounds in soil contamination. In this perception, metal ions present in contaminated soil are phytostabilized using bacteria due to their ability to solubilize minerals.

4.5.5.2 Alkalinization

Bacteria like cyanobacteria are able to pass through substratum alkalization to absorb metals, although affect the metal stability in soil. AMF act as a metal sink for the reduction of mobile and metal cations that are accessible in the soil, hence fabricating a more appropriate habitat for growing plants in contaminated soil. Hou et al. (2013) perceived that upon inoculating *S. alfredii* and *Lolium perenne* with AMF *G. mosseae* and *G. caledonium* led to decrease in soil DTPA-extractable Cd up to 21–38% through the process of alkalization.

4.6 Metal Transformation, Transportation, and Distribution

Se, Hg, As, Mn, Cr, and Fe constitute the common heavy metals which undergo microbial redox reaction by modulating the soil mobility and speciation as well as reduction in phytotoxicity of metals (O'Loughlin et al. 2003; Kashefi and Lovley 2000). Hg and Cu exhibit greater solubility at lesser oxidation state, while metals like Cr, As, and Se have higher oxidation state with higher solubility and toxicity. Reduction of Cr metal is another aspect for metal ion precipitation present in soil or in aqueous solutions. Microorganisms with redox reactions of metals have the ability to minimize the phytotoxicity of metals by conversion of metal ions present in the rhizosphere to a non-bioavailable state, thus depicting the phytotransformation property of microbes.

The mobility of different metals differs from each other like Cadmium and Zinc which are more mobile as compare to Copper and Lead. The metallic translocation from plant roots depends on plant species and metals. During the process of transportation through plants, metals are largely bounded on the cell walls of roots that result in increase in concentration of metals in plant roots. Ligands, namely, amino acids, organic acids, and thiols chelate with the metal ions and favor metal transport into the shoots from roots (Zacchini et al. 2009). Most of the heavy metals can only be able to transfer by the formation of complexes of metal-organic acids (Maser et al. 2001). Organic acids secreted by the microorganisms alter the existing form as well as metal dispersion by combining metal ions in plants, thereby leading to the transport of ions to shoots from root cells, and can also be able to improve the efficiency of phytoextraction (Sheng et al. 2008).

4.7 Signaling and Communication in Plant-Microbe Interaction

Vast communication between plants and microbes takes place through various different signaling molecules that are involved in maintenance of growth. It includes chemical signaling through plants that is perceived through the microbes and microbial signals due to which it recognizes the change in plant physiology (Fig. 4.3).

The major signaling phytochemicals and constituents for microbe and plants interaction are done via root exudates flavonoids including mycorrhizal synthesis, symbiotic relation among legume, and rhizobium (Steinkellner and Mammerler 2007). The significant roles of flavonoids are colonizing roots, growth of hyphae, germination of spores and differentiation in arbuscular mycorrhizal fungi and plant interactions (Mandal et al. 2010; Badri et al. 2009). Another aspect involving

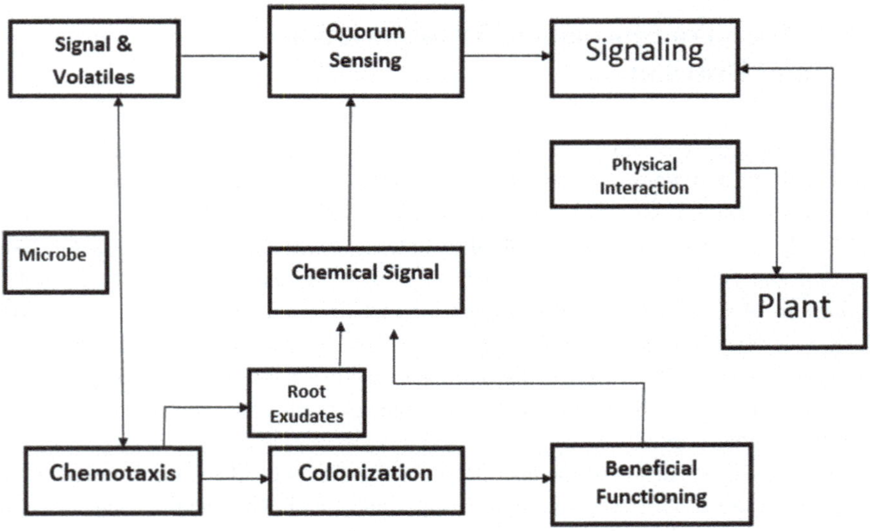

Fig. 4.3 Signaling components in plant-microbe interaction

enhancement of growth in host-specific rhizobium by flavonoids due to their function as chemo-attractants induction of nodule formation (nod) genes that play role in the release of lipochitin- oligosaccharides signaling components, the Nod factors (Perret et al. 2000; Mandal et al. 2010; Singh et al. 2019). Plant roots release phytochemicals like flavonoids that are recognized by the nod proteins of the rhizobium, transcriptional regulators which are able to bind directly to signaling molecules and help in the synthesis and also to export nod gene. This is the reason why specific flavonoids not only induce expression of nod genes but also chemotaxis of rhizobium and growth of bacteria (Bais et al. 2006).

The chemical components of the root exudates are altered by the free-living microbes including rhizobium and fungus by which release of various signaling molecules takes places like volatile organic compounds, Myc factor, exopolysaccharides, nod factors, and microorganism-related molecular patterns (Goh et al. 2013). The functions of these signaling molecules may contribute to improving growth of plants that is the basic requirement for the success of phytoremediation. Quorum sensing is the mechanism by which bacteria such as *Pseudomonas aeruginosa* regulate gene expression according to the population density. Basically, it is a cell–cell communication process which monitors the population density, cumulative alterative in bacterial gene expressions which are controlled by the help of diffusible signals which is synthesized by single cells of bacteria (Daniels et al. 2004). Root exudates such as flavonoids and genistein play a vital role in bacterial quorum sensing communication because of their ability to attract the rhizobium by chemotaxis in the direction of leguminous roots as well as to adhere and favour colonization which regulates the expression of nod genes in the tissue of rhizobial plant (Loh et al. 2002).

4.8 Conclusion

This chapter has focused on the studies regarding the most vital properties of plants and microbes as well as mechanisms coexisting together and competing for survival and their cohesive interactions that play an important role in adapting to metalliferous environments. Plant root exudates are useful nutrient and energy sources for soil microorganisms, with which they establish intricate communication systems. Bacteria and fungi work together with plants and reduce the damage that can be caused by the metals, and these microbes are termed as growth promoting microorganisms (PGPMs) and also help the plants by activating their defense mechanisms directed toward phytopathogens. PGPMs help plants in solubilizing minerals present in the soil and release of enzymes. This review helps in understanding the biochemical and molecular mechanisms involved in phytoremediation of heavy metals and their detoxification, biotransformation, transportation, and distribution. Plant and microbe's interaction-based-modulated phytoremediation is a viable technology for the cleanup of contaminated environments.

References

Ajmani A, Kumar R, Bhargava P, Vats S (2019) Mathematically optimized production, purification and characterization of penicillin G Acylase from AA17A and AA17B: an industrial biocatalyst for production of aminopenicillanic acid a core moiety of different penicillins. Indian J Biotechnol 18:260–268

Alford EA, Pilon-Smits EAH, Paschke M (2010) Metallophytes – a view from the rhizosphere. Plant Soil 337:33–50

Amir H, Pineau R (2003) Release of Ni and Co by microbial activity in new Caledonian ultramafic soils. Can J Microbiol 49:288–293

Argueso CT, Hansen M, Kieber JJ (2007) Regulation of ethylene biosynthesis. J Plant Growth Regul 26(2):92–105

Badri DV, Weir TL, van der Lelie D, Vivanco JM (2009) Rhizosphere chemical dialogues: plant-microbe interactions. Curr Opin Biotechnol 20:642–650

Bais HP, Weir TL, Perry LG, Gilroy S, Vivanco JM (2006) The role of root exudates in rhizosphere interactions with plants and other organisms. Annu Rev Plant Biol 57:233–266

Bashan Y, Holguin G (1998) Proposal for the division of plant growth-promoting rhizobacteria into two classifications: biocontrol-PGPB (plant growth-promoting bacteria) and PGPB. Soil Biol Biochem 30:1225–1228

Basu S, Kumar G, Chhabra S, Prasad R (2020) Role of soil microbes in biogeochemical cycle for enhancing soil fertility. In: Verma JP, Macdonald C, Gupta VK, Podile AR (eds) New and future developments in microbial biotechnology and bioengineering: phytomicrobiome for sustainable agriculture. Elsevier, Amsterdam, pp 149–157

Behera BK, Prasad R (2020a) Strategies for soil management. In: Behera BK, Prasad R (eds) Environmental technology and sustainability. Elsevier, Amsterdam, Netherlands, pp 143–167

Behera BK, Prasad R (2020b) Environmental technology and sustainability. Elsevier, Amsterdam, Netherlands. ISBN 9780128191033. https://www.elsevier.com/books/environmental-technology-and-sustainability/behera/978-0-12-819103-3

Berendsen RL, Pieterse CM, Bakker PA (2012) The rhizosphere microbiome and plant health. Trends Plant Sci 17(8):478–486

Bhargava P, Gupta N, Vats S, Goel R (2017) Health issues and heavy metals. Austin J Environ Toxicol 3(1):3018

Bhargava P, Khan M, Verma A, Singh A, Singh S, Vats S, Goel R (2019a) Metagenomics as a tool to explore new insights from plant-microbe interface. In: Plant microbe interface. Springer, Cham, pp 271–289

Bhargava P, Vats S, Gupta N (2019b) Metagenomics as a tool to explore mycorrhizal fungal communities. In: Varma A, Choudhary DK (eds) Mycorrhizosphere and pedogenesis. Springer, Singapore, pp 207–219

Bhargava P, Gupta N, Kumar R, Vats S (2020) Plants and microbes: bioresources for sustainable development and biocontrol. In: Plant microbe symbiosis. Springer, Cham, pp 153–176

Bolchi A, Ruotolo R, Marchini G, Vurro E, di Toppi LS, Kohler A, Tisserant E, Martin F, Ottonello S (2011) Genome-wide inventory of metal homeostasis-related gene products including a functional phytochelatin synthase in the hypogeous mycorrhizal fungus *Tuber melanosporum*. Fungal Genet Biol 48(6):573–584

Bowen GD, Rovira AD (1999) The rhizosphere and its management to improve plant growth. Adv Agron 66:1–102

Bruins MR, Kapil S, Oehme FW (2000) Microbial resistance to metals in the environment. Ecotoxicol Environ Saf 45:198–207

Bulgarelli D, Schlaeppi K, Spaepen S, Ver Lorenvan Themaat E, Schulze-Lefert P (2013) Structure and functions of the bacterial microbiota of plants. Annu Rev Plant Biol 64:807–838

Burges A, Alkorta I, Epelde L, Garbisu C (2018) From phytoremediation of soil contaminants to phytomanagement of ecosystem services in metal contaminated sites. Int J Phytoremediation 20 (4):384–397

Cai Y, Ma LQ (eds) (2002) Metal tolerance, accumulation, and detoxification in plants with emphasis on arsenic in terrestrial plants. In: Biogeochemistry of environmentally important trace elements. American Chemical Society, Washington, DC, pp 95–114

Chem B, Zhang Y, Rafiq MT, Khan KY, Pan F, Yang X et al (2014) Improvement of cadmium uptake and accumulation in *Sedum alfredii* by endophytic bacteria *Sphingomonas* SaMR12: effects on plant growth and root exudates. Chemosphere 117:367–373

Chen W, Teng Y, Li Z, Liu W, Ren W, Luo Y, Christie P (2018) Mechanisms by which organic fertilizer and effective microbes mitigate peanut continuous cropping yield constraints in a red soil of South China. Appl Soil Ecol 128:23–34

Chiang PN, Chiu CY, Wang MK, Chen BT (2011) Low-molecular-weight organic acids exuded by millet (*Setaria italica* (L.) Beauv.) roots and their effect on the remediation of cadmium-contaminated soil. Soil Sci 176:33–38

Chojnacka K (2010) Biosorption and bioaccumulation–the prospects for practical applications. Environ Int 36(3):299–307

Daniels R, Vanderleyden J, Michiels J (2004) Quorum sensing and swarming migration in bacteria. FEMS Microbiol Rev 28:261–289

Deng X, Wang P (2012) Isolation of marine bacteria highly resistant to mercury and their bioaccumulation process. Bioresour Technol 121:342–347

do Amaral FP, Agtuca BJ, Stacey G (2017) Setaria root–microbe interactions. In: Doust A, Diao X (eds) Genetics and genomics of Setaria. Springer, Cham, pp 239–250

Doornbos RF, vanLoon LC, Bakker PAHM (2012) Impact of root exudates and plant defense signaling on bacterial communities in the rhizosphere. A review. Agron Sustain Dev 32:227–243

Drogue B, Dore H, Borland S, Wisniewski-Doré F, Prigent-Combaret C (2012) Which specificity in cooperation between phytostimulating rhizobacteria and plants? Res Microbiol 163:500–510

Gadd GM (2004) Microbial influence on metal mobility and application for bioremediation. Geoderma 122:109–119

Gadd GM (2018) Geomycology: fungi as agents of biogeochemical change. In: Biology and environment: proceedings of the Royal Irish Academy. Royal Irish Academy, August 2018

Ghosh PK, De TK, Maiti TK (2018) Role of ACC deaminase as a stress ameliorating enzyme of plant growth-promoting rhizobacteria useful in stress agriculture: a review. In: Meena VS (ed) Role of rhizospheric microbes in soil. Springer, Singapore, pp 57–106

Giotta L, Mastrogiacomo D, Italiano F, Milano F, Agostiano A, Nagy K et al (2011) Reversible binding of metal ions onto bacterial layers revealed by protonation-induced ATR-FTIR difference spectroscopy. Langmuir 27:3762–3773

Glick BR, Cheng Z, Czarny J, Duan J (2007) Promotion of plant growth by ACC deaminase containing soil bacteria. Eur J Plant Pathol 119:329–339

Goel R, Bhargava P, Gupta N, Vats S (2017) Health issues and heavy metals. Austin J Environ Toxicol 3(1):1018

Goh HH, Sloan J, Malinowski R, Fleming A (2013) Variable expansion expression in *Arabidopsis* leads to different growth responses. J Plant Physiol 171:329–339

Goldstein AH, Braverman K, Osorio N (1999) Evidence for mutualism between a plant growing in a phosphate-limited desert environment and a mineral phosphate solubilizing (MPS) rhizobacterium. FEMS Microbiol Ecol 30(4):295–300

Gonzalez-Chavez MC, Carrillo-Gonzalez R, Wright SF, Nichols KA (2004) The role of glomalin, a protein produced by arbuscular mycorrhizal fungi, in sequestering potentially toxic elements. Environ Pollut 130(3):317–323

Gupta N, Vats S, Bhargava P (2018) Sustainable agriculture: role of metagenomics and metabolomics in exploring the soil microbiota. In: Choudhary DK, Kumar M, Prasad R (eds) In silico approach for sustainable agriculture. Springer, Singapore, pp 183–199

Heinekamp Y, Willey N (2007) Using real-time polymerase chain reaction to quantify gene expression in plants exposed to radioactivity. In: Willey N (ed) Phytoremediation: Methods and reviews. Humana Press, Totowa, pp 59–70

Hinsinger P (2001) Bioavailability of soil inorganic P in the rhizosphere as affected by root-induced chemical changes: a review. Plant Soil 237:173–195

Hou W, Ma Z, Sun L, Han M, Lu J, Li Z et al (2013) Extracellular polymeric substances from copper-tolerance *Sinorhizobium meliloti* immobilize Cu2C. J Hazard Mater 261:614–620

Hwang JH, Park H, Choi DW, Nam KT, Lim KM (2018) Investigation of dermal toxicity of ionic liquids in monolayer-cultured skin cells and 3D reconstructed human skin models. Toxicol In Vitro 46:194–202

Jain P, Miglani K, Vats S (2011) Aptamers-potential applications in diagnostics and therapeutics. Everyman's Sci XLV(6):361

Jeffries P, Gianinazzi S, Perotto S, Turnau K, Barea JM (2003) The contribution of arbuscular mycorrhizal fungi in sustainable maintenance of plant health and soil fertility. Biol Fertil Soils 37(1):1–16

Johansson EM, Fransson PMA, Finlay RD, van Hees PAW (2008) Quantitative analysis of exudates from soil-living basidiomycetes in pure culture as a response to lead, cadmium and arsenic stress. Soil Biol Biochem 40:2225–2236

Jones D (1998) Organic acids in the rhizosphere—a critical review. Plant Soil 205:24–44

Juwarkar AA, Nair A, Dubey KV, Singh SK, Devotta S (2007) Biosurfactant technology for remediation of cadmium and lead contaminated soils. Chemosphere 10:1996–2002

Kamal S, Prasad R, Varma A (2010) Soil microbial diversity in relation to heavy metals. In: Sherameti I, Varma A (eds) Soil heavy metals. Springer-Verlag, Berlin-Heidelberg, pp 31–64

Kamnev AA, Van der Lelie D (2000) Chemical and biological parameters as tools to evaluate and improve heavy metal phytoremediation. Biosci Rep 20:239–258

Kashefi K, Lovley DR (2000) Reduction of Fe(III), Mn(IV), and toxic metals at 100 degrees C by *Pyrobaculum islandicum*. Appl Environ Microbiol 66:1050–1056

Kaur A, Vats S, Rekhi S, Bhardwaj A, Goel J, Tanwar RS, Gaur KK (2010) Physico-chemical analysis of the industrial effluents and their impact on the soil microflora. Procedia Environ Sci 2:595–599

Kent AD, Triplett EW (2002) Microbial communities and their interactions in soil and rhizosphere ecosystems. Annu Rev Microbiol 56(1):211–236

Kim S, Lim H, Lee I (2010) Enhanced heavy metal phytoextraction by *Echinochloa crus-galli* using root exudates. J Biosci Bioeng 109:47–50

Kletzin A (2006) Metabolism of inorganic sulfur compounds in archea. In: Garret RA, Klenk HP (eds) Archea: evolution, physiology and molecular biology. Blackwell, Oxford, pp 262–274

Kumar A, Singh N, Pandey R, Gupta VK, Sharma B (2018) Biochemical and molecular targets of heavy metals and their actions. In: Rai M, Ingle A, Medici S (eds) Biomedical applications of metals. Springer, Cham, pp 297–319

Kumar N, Balamurugan A, Shafreen MM, Rahim A, Vats S, Vishwakarma K (2020a) Nanomaterials: emerging trends and future prospects for economical agricultural system. In: Biogenic nano-particles and their use in agro-ecosystems. Springer, Singapore, pp 281–305

Kumar N, Srivastava P, Vishwakarma K, Kumar R, Kuppala H, Maheshwari SK, Vats S (2020b) The rhizobium–plant symbiosis: state of the art. In: Plant microbe symbiosis. Springer, Cham, pp 1–20

Lacalle RG, Gómez-Sagasti MT, Artetxe U, Garbisu C, Becerril JM (2018) *Brassica napus* has a key role in the recovery of the health of soils contaminated with metals and diesel by rhizoremediation. Sci Total Environ 618:347–356

Leitenmaier B, Küpper H (2013) Compartmentation and complexation of metals in hyperaccumulator plants. Front Plant Sci 4:374

Loh J, Russell WC, William SY, Stacey G (2002) Bradyoxetin, a unique chemical signal involved in symbiotic gene regulation. Proc Natl Acad Sci USA 99:14446–14451

Ma X, Wang C (2010) Fullerene nanoparticles affect the fate and uptake of trichloroethylene inphytoremediation systems. Environ Eng Sci 27:989–992

Ma Y, Rajkumar M, Freitas H (2009) Improvement of plant growth and nickel uptake by nickel resistant-plant growth promoting bacteria. J HazardMater 166:1154–1161

Ma Y, Prasad MNV, Rajkumar M, Freitas H (2011) Plant growth promoting rhizobacteria and endophytes accelerate phytoremediation of metalliferous soils. Biotechnol Adv 29:248–258
Magdoff F, van Es H (2000) Building soils for better crops. In: Sustainable agriculture network handbook series, book 4, 2nd edn. University of Vermont, Burlington
Mandal SM, Chakraborty D, Dey S (2010) Phenolic acids act as signaling molecules in plant-microbe symbioses. Plant Signal Behav 5:359–368
Marschner H (1995) Mineral nutrition of higher plants, 2nd edn. Academic, San Diego
Maser P, Thomine S, Schroeder JI, Ward JM, Hirschi K, Sze H et al (2001) Phylogenetic relationships within cation transporter families of *Arabidopsis*. Plant Physiol 126:1646–1667
Mastretta C, Barac T, Vangronsveld J et al (2006) Endophytic bacteria and their potential application to improve the phytoremediation of contaminated environments. Biotech Gen Eng Rev 23:175–207
Maurya DP, Vats S, Rai S, Negi S (2013) Optimization of enzymatic saccharification of microwave pretreated sugarcane tops through response surface methodology for biofuel. Indian J Exp Biol 51(11):992–996
Maurya DP, Singh D, Vats S (2014) Cellulase production and utilization. In: Jian A (ed) Category-chemical technology. LAP LAMBERT Academic, Saarbrücken, p 80
Mesa J, Mateos-Naranjo E, Caviedes MA, Redondo-Gómez S, Pajuelo E, Rodríguez-Llorente ID (2015) Scouting contaminated estuaries: heavy metal resistant and plant growth promoting rhizobacteria in the native metal rhizoaccumulator *Spartina maritima*. Mar Pollut Bull 90 (1–2):150–159
Mucha AP, Marisa C, Almeida R, Bordalo AA, Teresa M, Vasconcelos SD (2005) Exudation of organic acids by a marsh plant and implications on trace metal availability in the rhizosphere of estuarine sediments. Estuar Coast Shelf Sci 65:191–198
Mulligan CN, Yong RN, Gibbs BF (2001) Surfactant-enhanced remediation of contaminated soil: a review. Eng Geol 60:371–380
Nautiyal CS (2003) Plant growth-promoting rhizobacteria from stressed ecosystems. In 6th International PGPR workshop, October 2003, pp 5–10
Navarro CA, von Bernath D, Jerez CA (2013) Heavy metal resistance strategies of acidophilic bacteria and their acquisition: importance for biomining and bioremediation. Biol Res 46:363–371
Negi S, Vats S (2013) Pine forest litter based bio-refinery for biofuels and value-added phytochemicals. In: Singh RS, Pandey A, Larroche C (eds) Advances in industrial biotechnology. IK International, Delhi, pp 98–116
Nies DH, Silver S (1995) Ion efflux systems involved in bacterial metal resistances. J Ind Microbiol 14:186–199
Novo LA, Castro PM, Alvarenga P, da Silva EF (2018) Plant growth–promoting rhizobacteria-assisted phytoremediation of mine soils. In: Prasad MNV, Favas PJC, Maiti SK (eds) Bio-geotechnologies for mine site rehabilitation. Elsevier, Amsterdam, pp 281–295
O'Loughlin EJ, Kelly SD, Kemner KM, Csencsits R, Cook RE (2003) Reduction of AgI, AuIII, CuII, and HgII by FeII/FeIII hydroxysulfate green rust. Chemosphere 53:437–446
Ojha AK, Forster S, Kumar S, Vats S, Negi S, Fischer I (2013) Synthesis of well–dispersed silver nanorods of different aspect ratios and their antimicrobial properties against gram positive and negative bacterial strains. J Nanobiotechnol 11(1):42
Oves M, Saghir Khan M, Zaidi A (2013) Chromium reducing and plant growth promoting novel strain *Pseudomonas aeruginosa* OSG41 enhance chickpea growth in chromium amended soils. Eur J Soil Biol 56:72–83
Pacheco GJ, Ciapina EMP, Gomes EB, Pereira N Jr (2011) Biosurfactant production by *Rhodococcus erythropolis* and its application to oil removal. Braz J Microbiol 41:685–693
Padmavathiamma PK, Li LY (2007) Phytoremediation technology: hyper-accumulation metals in plants. Water Air Soil Pollut 184(1–4):105–126
Pagnanelli F, Viggi C, Toro L (2010) Development of new composite biosorbents from olive pomace wastes. Appl Surf Sci 256:5492–5497

Painuly AS, Gupta R, Vats S (2019) Bio-accumulation of arsenic (III) using *Nelumbo nucifera* Gaertn. J Health Pollut 9(23):190902
Pal R, Rai JPN (2010) Phytochelatins: peptides involved in heavy metal detoxification. Appl Biochem Biotechnol 160:945–963
Park JH, Bolan N, Megharaj M, Naidu R (2011) Isolation of phosphate solubilizing bacteria and their potential for lead immobilization in soil. J Hazard Mater 185:829–836
Pathak A, Dastidar MG, Sreekrishnan TR (2009) Bioleaching of heavy metals from sewage sludge: a review. J Environ Manag 90:2343–2353
Payne AN, DiChristina TJ (2006) A rapid mutant screening technique for detection of technetium [Tc(VII)] reduction-deficient mutants of *Shewanella oneidensis* MR-1. FEMS Microbiol Lett 259:282–287
Pérez-Montaño F, Jiménez-Guerrero I, Sánchez-Matamoros RC, López-Baena FJ, Ollero FJ, Rodríguez-Carvajal MA et al (2013) Rice and bean AHL-mimic quorum-sensing signals specifically interfere with the capacity to form biofilms by plant-associated bacteria. Res Microbiol 164:749–760
Perret X, Staehelin C, Broughton WJ (2000) Molecular basis of symbiotic promiscuity. Microbiol Mol Biol Rev 64:180–201
Pinton R, Varanini Z, Nannipieri P (2007) The rhizosphere: biochemistry and organic substances at the soil-plant interface. CRC, Boca Raton
Poggenburg C, Mikutta R, Liebmann P, Koch M, Guggenberger G (2018) Siderophore-promoted dissolution of ferrihydrite associated with adsorbed and coprecipitated natural organic matter. Org Geochem 125:177–188
Prasad R, Garg AP, Varma A (2005) Interaction of medicinal plants with plant growth promoting rhizobacteria and symbiotic fungi. In: Podila GK, Varma A (eds) Basic research and applications of Mycorrhizae, vol 1. IK International, Delhi, pp 363–407
Prasad MP, Dagar S (2014) Identification and characterization of endophytic bacteria from fruits like avocado and black grapes. Int J Curr Microbiol Appl Sci 3:937–947
Prasad R, Kumar M, Varma A (2015) Role of PGPR in soil fertility and plant health. In: Egamberdieva D, Shrivastava S, Varma A (eds) Plant growth-promoting rhizobacteria (PGPR) and medicinal plants. Springer International Publishing, Switzerland, pp 247–260
Prasad R, Chhabra S, Gill SS, Singh PK, Tuteja N (2020) The microbial symbionts: potential for the crop improvement in changing environments. In: Tuteja N, Tuteja R, Passricha N, Saifi SK (eds) Advancement in crop improvement techniques. Elsevier, Amsterdam, Netherlands, pp 233–240
Rajkumar M, Sandhya S, Prasad MNV, Freitas H (2012) Perspectives of plant-associated microbes in heavy metal phytoremediation. Biotechnol Adv 30:1562–1574
Richards BK, Steenhuis TS, Peverly JH, McBride MB (2000) Effect of sludge-processing mode, soil texture and soil pH on metal mobility in undisturbed soil columns under accelerated loading. Environ Pollut 109(2):327–346
Richardson AE, Simpson RJ (2011) Soil microorganisms mediating phosphorus availability update on microbial phosphorus. Plant Physiol 156(3):989–996
Rivera-Becerril F, van Tuinen D, Martin-Laurent F, Metwally A, Dietz KJ, Gianinazzi S et al (2005) Molecular changes in *Pisum sativum* L. roots during arbuscular mycorrhiza buffering of cadmium stress. Mycorrhiza 16:51–60
Rosenkranz T, Kidd P, Puschenreiter M (2018) Effect of bacterial inoculants on phytomining of metals from waste incineration bottom ash. Waste Manag 73:351–359
Sakaguchi T, Nishizawa NK, Nakanishi H, Yoshimura E, Mori S (1999) The role of potassium in the secretion of mugineic acids family phytosiderophores from iron-deficient barley roots. Plant Soil 215:221–227
Salt DE, Smith RD, Raskin I (1998) Phytoremediation. Annu Rev Plant Physiol 49:643–668
Saxena P, Srivastava J, Pandey S, Srivastava S, Maurya N, Kaushik NC, Mishra S, Asthana G, Bhargava P, Kumar R, Vats S (2019) Plants for biocontrol and biological control of plant pathogens. In: Plant biotic interactions. Springer, Cham, pp 147–179

Sessitsch A, Kuffner M, Kidd P, Vangronsveld J, Wenzel WW, Fallmann K et al (2013) The role of plant-associated bacteria in the mobilization and phytoextraction of trace elements in contaminated soils. Soil Biol Biochem 60:182–194

Shaheen SM, Tsadilas CD, Ok YS, Rinklebe J (2016) 13 potential mobility, bioavailability, and plant uptake of toxic elements in temporary flooded soils. In: Rinklebe J, Knox A, Paller M (eds) Trace elements in waterlogged soils and sediments. CRC; Taylor & Francis Group, New York, pp 287–312

Sharma KM, Kumar R, Vats S, Gupta A (2014) Production, partial purification and characterization of alkaline protease from *Bacillus aryabhattai* K3. Int J Adv Pharm Biol Chem 3(2):290–298

Sharma D, Javed S, Arshilekha, Saxena P, Babbar P, Shukla D, Srivastava P, Vats S (2018) Food additives and their effects: a mini review. Int J Curr Res 10(06):69999–70002

Sheng XF, He LY, Wang QY, Ye HS, Jiang C (2008) Effects of inoculation of biosurfactant producing *Bacillus* sp. J119 on plant growth and cadmium uptake in a cadmium amended soil. J Hazard Mater 155:17–22

Sheoran V, Sheoran AS, Poonia P (2011) Role of hyperaccumulators in phytoextraction of metals from contaminated mining sites, a review. Crit Rev Environ Sci Technol 41:168–214

Shrivastava S, Prasad R, Varma A (2014) Anatomy of root from eyes of a microbiologist. In: Morte A, Varma A (eds) Root engineering, vol 40. Springer-Verlag, Berlin-Heidelberg, pp 3–22

Singh M, Vats S (2019) Mathematically designed bioprocess for release of value added products with pharmaceutical applications from wastes generated from spices industries. Int J Pharm Sci Res 10(1):130–138

Singh D, Raina TK, Kumar A, Singh J, Prasad R (2019) Plant microbiome: a reservoir of novel genes and metabolites. Plant Gene 18:100177. https://doi.org/10.1016/j.plgene.2019.100177

Solanki R, Dhankhar R (2011) Biochemical changes and adaptive strategies of plants under heavy metal stress. Biologia 66:195–204

Steinkellner S, Mammerler R (2007) Effect of flavonoids on the development of *Fusarium oxysporum* f. sp. *Lycopersici*. J Plant Interact 2:17–23

Ström L, Owen AG, Godbold DL, Jones D (2002) Organic acid mediated P mobilization in the rhizosphere and uptake by maize roots. Soil Biol Biochem 34:703–710

Tandon S, Vats S (2016) Microbial biosynthesis of cadmium sulfide (Cds) nanoparticles and their characterization. Eur J Pharm Med Res 3(9):545–550

Tricker PJ, ElHabti A, Schmidt J, Fleury D (2018) The physiological and genetic basis of combined drought and heat tolerance in wheat. J Exp Bot 69(13):3195–3210

Tsao DT (2003) Overview of phytotechnologies. In: Tsao DT (ed) Phytoremediation. Springer, Berlin, pp 1–50

Vassilev N, de Oliveira Mendes G (2018) Solid-state fermentation and plant-beneficial microorganisms. In: Pandey A, Larroche C, Soccol CR (eds) Current developments in biotechnology and bioengineering. Elsevier, Cambridge, pp 435–450

Vats S (2017) Methods for extractions of value-added nutraceuticals from lignocellulosic wastes and their health application. In: Grumezescu AM, Holban A-M (eds) Ingredients extraction by physicochemical methods in food. Academic, London, pp 1–64

Vats S, Bhargava P (2017) Alternate energy: fuel for "Modi's India" and "smart cities". Int J Curr Res 9(04):49090–49097

Vats S, Kumar R (2015) Amylolytic-extremoenzymes: saviour of environments. Eur J Biomed Pharm Sci 2(5):694–702

Vats S, Miglani K (2011) Synergistic antimicrobial effect of cow urine and *Azadirachta indica* on infectious microbes. Int J Pharm Sci Res 2(7):1781

Vats S, Mishra A (2016) Soil agro-ecological management by vermicompost a potential organic nutrient source for the state of Uttar Pradesh. Eur J Pharm Med Res 3(9):604–609

Vats S, Negi S (2013) Use of artificial neural network (ANN) for the development of bioprocess using *Pinus roxburghii* fallen foliages for the release of polyphenols and reducing sugars. Bioresour Technol 140:392–398

Vats S, Kumar R, Miglani AK (2011) Isolation, characterization and identification of high salinity tolerant, heavy metal contaminant and antibiotics resistant amylolytic-thermophilic pseudomonas Sp. Int J Pharm Sci Rev Res 10(2):125–129
Vats S, Kumar R, Negi S (2012) Natural food that meet antibiotics resistance challenge: in vitro synergistic antimicrobial activity of *Azadirachta indica*, *Terminalia chebula*, *Piper nigrum* and photoactivated cow urine. Asian J Pharm Biol Res 2(2):122–126
Vats S, Maurya DP, Agarwal A, Shamoon M, Negi S (2013a) Development of a microbial consortium for the production of blend of enzymes for the hydrolysis of agricultural wastes into sugars. J Sci Ind Res 72:585–790
Vats S, Maurya DP, Jain A, Mall V, Negi S (2013b) Mathematical model-based optimization of physico-enzymatic hydrolysis of *Pinus roxburghii* needles for the production of reducing sugars. Indian J Exp Biol 51:944–953
Vats S, Kumar R, Maurya DP (2014) Alkaline amylase from multi resistant microbes and its applications. In: Alexei E (ed) Category-microbiology. Lambert Academic, Saarbrücken, p 100
Vats S, Singh M, Siraj S, Singh H, Tandon S (2017) Role of nanotechnology in theranostics and personalized medicines. J Health Res Rev 4(1):1
Vats S, Gupta N, Bhargava P (2019) Vulnerability of soil micro biota towards natural and anthropogenic induced changes and loss of pedospheric functionality. In: Varma A, Choudhary DK (eds) Mycorrhizosphere and pedogenesis. Springer, Singapore, pp 191–205
Venkatesh NM, Vedaraman N (2012) Remediation of soil contaminated with copper using rhamnolipids produced from *Pseudomonas aeruginosa* MTCC 2297 using waste frying rice bran oil. Ann Microbiol 62:85–91
Wawra A, Friesl-Hanl W, Jäger A, Puschenreiter M, Soja G, Reichenauer T, Watzinger A (2018) Investigations of microbial degradation of polycyclic aromatic hydrocarbons based on 13 C-labeled phenanthrene in a soil co-contaminated with trace elements using a plant assisted approach. Environ Sci Pollut Res 25(7):6364–6377
Wong JWC, Xiang L, Gu XY, Zhou LX (2004) Bioleaching of heavy metals from anaerobically digested sewage sludge using FeS2 as an energy source. Chemosphere 55:101–107
Yuan M, He HD, Xiao L, Zhong T, Liu H, Li SB et al (2014) Enhancement of Cd phytoextraction by two *Amaranthus* species with endophytic *Rahnella* sp. JN27. Chemosphere 103:99–104
Zacchini M, Pietrini F, Mugnozza GS, Iori V, Pietrosanti L, Massacci A (2009) Metal tolerance, accumulation and translocation in poplar and willow clones treated with cadmium in hydroponics. Water Air Soil Pollut 197:23–34
Zhao FJ, Hamon RE, McLaughlin MJ (2001) Root exudates of the hyperaccumulator *Thlaspi caerulescens* do not enhance metal mobilization. New Phytol 151(3):613–620

Chapter 5
Ecosystem Diversity as a Function of Plant and Soil-Microbe Interactions

Sanjukta Dey, Soumaryya Bhattacharyya, and Rabindranath Bhattacharyya

Abstract Microbial diversity in soil is overwhelming, and so is their interaction with roots of higher plants. Rhizosphere is the area of the soil around the roots of higher plants where amazingly intense cross talk occurs with soil microbes, and they form a continuous network in a particular ecosystem. It is now well established that underground microbial diversity determines community composition of plants and of the ecosystem in an area. This is possible due to the diverse form of plant and soil-microbe interaction occurring in the rhizosphere. In this review it has been our endeavour to elucidate ecosystem diversity as a function of plant and soil-microbe interactions. We have focussed primarily on soil microbial diversity determining plant community composition. However, literature concerning ecosystem diversity is scarce. To our understanding this could be a potential area of ecological research in years to come.

5.1 Introduction

Since beginning of study concerning diversity in ecosystem, ecologists have theorized and/or hypothesized the process as a function of anthropogenic activities (Nelson et al. 2006), ecosystem size (Blakely and Didham 2010), invasive species and infectious diseases (Crowl et al. 2008). It also includes various other aboveground abiotic factors like climate, age, environmental harshness, disturbance, environmental heterogeneity, as well as biotic interactions (Tilman and Pacala 1993). However, attention to the belowground drivers (soil microbes) of ecosystem

S. Dey
School of Life Sciences and Biotechnology, Adamas University, Kolkata, West Bengal, India

S. Bhattacharyya
Department of Botany (Post Graduate), Ramakrishna Mission Vivekananda Centenary College (Rahara), Kolkata, West Bengal, India

R. Bhattacharyya (✉)
Department of Life Sciences, Presidency University, Kolkata, West Bengal, India
e-mail: rabindranathbpc@yahoo.co.in

A. Varma et al. (eds.), *Plant Microbiome Paradigm*,
https://doi.org/10.1007/978-3-030-50395-6_5

diversity has been to a lesser extent (Thompson et al. 2001). It is reported that soil microbial diversity is overwhelming (Rudi et al. 2007). Though soil microbes are main drivers in shaping the belowground biodiversity, there are technical constraints for their study. The tendency to view microbial associates as simply extensions of the host plant and their outright exclusion from experimental designs, through use of artificial soil mixes or soil amendments, resulted in depleted microbial communities (Reynolds et al. 2003). Studies estimate that approximately 10^7 microbial species are present per gram of soil (Gans et al. 2005), and it implies that it would be a casual approach if we neglect the impact of soil micro biota in shaping diversity of ecosystem.

In recent years, however, ecologists have developed a growing body of research on plant-soil microbe interactions. These studies have helped us understand plant community structure and diversity from microbially based perspective. This chapter discusses the basics of plant-soil microbe interactions and how the process determines plant community structure and diversity.

5.2 Method of Studying Rhizosphere Microflora

Hiltner (1904) observed that microorganisms were more abundant in the soil surrounding plant roots than in soil remote from the root. This zone of soil in which the microflora was influenced by the plant root was called the "rhizosphere" by German physiologist and agronomist Hiltner. Since soil is a continuous system, a clear-cut demarcation of rhizosphere has not been possible, and ecologists have differing opinion while describing the same. Attempts to better define the zones of influence have led to such terms as "outer rhizosphere," "inner rhizosphere," "root surface," and "rhizoplane." The heterogeneous nature of soil makes it impossible to define precisely where each of these zones begins and ends. Of special interest is rhizoplane which consists of the outer surface of root and associated microbes and soil debris. However, the term is ambiguous as microbial population over the root surface is not present in a single plane.

Since the discovery of rhizosphere, many workers have shown that quantity and quality of microbes are diverse and richer in soil around plant roots from that in soil beyond the influence of roots. Ecologists have sampled soil to compare microbial population in soil influenced by plant roots and of soil distant from the zone. Experiments have been conducted from time to time with better techniques adapted every passing time. The traditional method of sampling the rhizosphere is to free the roots of much of the adhering soil by vigorous shaking, suspend the roots plus "firmly adhering soil" in a given volume of diluent, and prepare a dilution series from which aliquots are taken for counting. A comparison is made of the counts per gram of "firmly adhering soil" (R) with the counts per gram of soil taken some distance from the root (S). The ratio of these two counts was termed the R/S ratio (Katznelson 1946) and is the most widely accepted method of expressing the extent of the rhizosphere effect. However, the classical method has a drawback;

considering the greater concentration of organisms at the root surface, it is obvious the energy with which the sample is shaken will markedly affect the R/S ratio; when a small amount of soil is included with the roots in the rhizosphere sample, a erroneously higher R/S ratio is obtained compared to the results obtained when large amount of soil adheres to the roots. Experiments show that R/S ratio of identical plants varies considerably by simply varying the amount of soil adhering to the root surface, at the time of suspension in the diluent, and give faulty results (Clark 1949). While there is little doubt that the ratio is handy in comparing rhizosphere effect, we need to observe caution while applying force shaking the plant roots to remove the loosely attached soil particles as results may vary greatly with force and making manipulations are easy.

To reduce the significant differences that appear in results while following the traditional method of sampling rhizosphere population, Ishizawa et al. (1957) proposed a better method which requires washing of roots. Roots from the field are gently shaken in sterile water when the loose rhizosphere soil is washed away. It is then transferred to another flask where it is shaken vigorously to wash down firmly adhering soil particles. Samples taken from the first and second washing give results of rhizosphere and rhizoplane microbial population. Ishizawa et al. (1957) and Louw and Webley (1959), however, tried to make a clear distinction between rhizosphere and rhizoplane microflora when the latter is a part of the former. Chances prevail that during this method, there will be some washing off of organisms from the root surface into the rhizosphere sample, but even so the results obtained for the rhizosphere population will be reasonably accurate and should form the basis of comparison with control soil. Also, the rhizoplane microflora estimated by this method will include many rhizosphere organisms as the roots are not thoroughly washed before the final shaking for the rhizoplane count.

The serial washing techniques used by Harley and Waid (1955), Parkinson et al. (1963), and Brown et al. (1962), to assess the rhizoplane populations of fungi (Harley and Waid 1955; Parkinson et al. 1963) and *Azotobacter* (Brown et al. 1962), provide valuable information on the organisms which are tightly held to the root surface. Rouatt and Katznelson (1960) distinguished between the rhizosphere and rhizoplane microflora by shaking roots and soil to provide the rhizosphere sample and then, after several washings, macerated and suspended the roots to provide the rhizoplane sample. In the interpretation of these results, consideration should be given to the likelihood that, at least for bacteria, the successive vigorous washings of the roots will remove not only the outer layers of the rhizoplane population but wash off as well colonies and cells intimately associated with a root but having less adhesive properties than those persisting through all the treatments. The ability of an organism to penetrate the cortical cells of the host (Rouatt and Katznelson 1960) or produce gum would enhance its chances of being rated a rhizoplane inhabitant.

The true essence of the entire sample count is to find out diversity in microflora in the rhizosphere and compare it with non-rhizosphere soil (R/S ratio). However, in serial washing technique comparison is made between rhizosphere and rhizoplane microflora which are essentially integral part of the same system. And there is drawback of the technique as chances exist for erroneous and manipulative result. Therefore, the question arises, what is the best method of studying and comparing rhizosphere microflora? According to Rovira et al. (1974), expression both by numbers of organisms per gram of soil in the sample and by numbers per gram of root is probably the most satisfactory; on soil weight basis, comparisons may be made between sphere and non-rhizosphere soil (R/S ratio), while on a root weight basis, comparisons between different portions of the root systems and between roots of different plants are more valid. Such comparisons on a root weight basis can be made only between root systems of similar morphology. Clark (1949) found less variation when results were expressed on root weight basis than on a soil weight basis.

The sampling of rhizosphere microflora by techniques mentioned in the aforementioned paragraphs gave quantitative results. To obtain qualitative knowledge on the type of microcosms in the rhizosphere, we need to look at the root under microscope or thin-walled glass observation boxes (traditional method), and currently we can avail molecular tools in rhizosphere microbiology. Roots and adhering soil are often viewed under scanning electron microscopy (SEM) which requires extensive sample preparation (dehydration) and is destructive for specimen. To make thin sections of the roots, it is first imbedded in resin followed by sectioning in microtome. A better yet resource saving microscopic technique is environmental scanning electron microscopy (ESEM) which does not require extensive sample preparation and is less destructive for the specimen. ESEM has been extensively used to study mineral composition in rhizosphere. More recently the technique has been applied to study the interaction between wheat and plant growth-promoting rhizobacteria (PGPR) which revealed an excellent ability of bacteria to adhere to the surface of intact leaves and roots and to colonize both leaf mesophyll and root vascular tissues in aseptic conditions (Dal Cortivo et al. 2017).

Microorganisms in rhizosphere can be identified down to complete genera or functional group using fluorescence microscopy and molecular stains. Two types of staining techniques are usually used—general and specific cell staining. Acridine orange which stains DNA in soil and rhizosphere sample is a general stain that has been used extensively (Bottomley and Maggard 1990). The drawback of general staining is that it stains every kind of DNA without any species specificity. Therefore, it can be used for preliminary detection of rhizosphere microflora. Acridine orange can also serve as a counter stain by binding to the soil humic material, and this technical offshoot has been exploited to observe fungal (*Pythium aphanidermatum*) zoospores on cucumber roots (Sørensen et al. 2009). Specific cell stains are often used with confocal laser scanning microscopy (CLSM) and florescence microscopy and have been used extensively to study the structural and spatial composition of microbial communities in environmental samples (Dabral et al. 2017). Strain specific fluorescent-antibody staining has been used to study bacterial root colonization by *Azospirillum* in the rhizosphere of wheat (Scholter et al. 1993). Florescent in situ hybridization (FISH) probes which target rRNA oligonucleotides are used to detect rRNA hybridization signal in bulk soil of rhizosphere assuming that cellular RNA content correlate with growth activity (Assmus et al. 1997).

Insertion of reporter genes (*lux* and *gfp*) into bacteria can be used to track single cell in action. Construction of *Rhizobium meliloti* GFP mutants showed detailed CLSM images of their growth and behaviour during the early stages of infection and nodulation in living alfalfa roots (Gage et al. 1996).

Microscopy, staining and use of reporter genes are state-of-the art molecular techniques to study rhizosphere microbiology, but they study single cells. However, to gain deeper insight into associations in the rhizosphere, we need to study the community structure of microorganisms. It is best to make in situ analysis of community so that we can avoid biases from studying single cells. Therefore, from 1990s focus has been on studying microcosms by DNA- and PCR-based approaches. The rapid interest for these methods is requiring the DNA (or RNA) to be directly extracted from the environment to overcome biases in isolation and in vitro cultivation (Sørensen et al. 2009).

5.3 Different Types of Plant-Soil Microbe Interactions

Astronomical number of microorganisms dwells in soil. But their number is greater within, on and around plant roots (rhizoplane and rhizosphere) than remaining soil. This is because root exudates exert a selective pressure on underground microcosms. Rhizosphere microflora depends for their energy supply on organic substances provided by the roots, and the growth is thus intimately related to the metabolic activity of plants involved. Among these microorganisms, ability to metabolize the relatively simple compounds exuded by living roots is universal. A more limited segment of the population is capable of decomposing proteins, structural polymers and other complex products of plant metabolism. The energy yield from degradation of these materials is relatively low, and the reactions often proceed slowly (Gaskins et al. 1985).

To understand anything about how rhizosphere microorganisms shape ecological diversity, we have to first review the few types of plant-soil microbe interactions that exist in soil. Interactions between plant roots and soil microorganisms embrace such topics as plant root diseases, nodulation of legumes, mycorrhizal associations and the noninvasive associations between plant roots and microorganisms. Dealing with ecological diversity shaped by rhizosphere microflora, we will discuss mutual, symbiosis as well as commensalistic interactions between plant roots and soil microorganisms which occur in the noninvasive associations or in the preinvasive phases of the above associations. However, pathogenic interactions and ammensalism have not yet been reported to drive diversity in ecosystem. It is not unlikely that exclusion of a species due to attack by pathogen and ammensalism can determine the species richness and variation in ecosystem.

5.3.1 Plant Root-Fungi Association

Symbiotic mutualism between plant roots and members of kingdom fungi is termed as mycorrhizae. Two types of fungal propagule can be found in the rhizosphere—spores present on the surface of root and intra- and intercellular fungal hyphae. But for the association to be termed mycorrhiza, it is essential that the fungi extend hyphae within or between cells (Varma et al. 2017). Interactions of fungi in rhizosphere are diverse, and existence of every species impacts the other. However, while describing true association, we will disregard the spores which are superficially present on roots. Mycorrhizae is an extensive association with more than 85% vascular plants being a part of this symbiosis (Leake and Lead 2017; Prasad et al. 2017). The rhizosphere soil under the influence of mycorrhizal fungal hyphae is called mycorrhizosphere (Giri et al. 2005). There are six distinct groups of mycorrhizal association (Brundrett 2002; Smith and Read 1997). They have been classified as arbuscular, ecto-, ericoid, arbutoid, monotropoid and orchid.

The mechanism by which mycorrhizal fungi mediate rhizosphere process has two dimensions. Firstly, the mycorrhizal fungi and host plant share a cost-benefit relationship. The fungal partner helps its host to access nutrient pools otherwise inaccessible. The host provide carbon as energy source to the fungus. Besides interacting with host, the mycorrhizal fungi display an array of interaction with rhizosphere microflora. It has been experimentally shown that co-inoculation of *Medicago sativa* with *Glomus* spp. (arbuscular mycorrhizal fungi) and *Bacillus* spp. produced greater shoot biomass and root length over singly inoculated (with fungus) and un-inoculated control. Thymidine concentration (indicative of bacterial population) was greater around roots inoculated with *Glomus* spp. Therefore it is likely that mycorrhizal fungi facilitated bacterial growth in rhizosphere, and hence they must interact in some way to bring about good growth in host plant (Medina et al. 2003).

In a similar experiment by Azcón, he observed that tomato plants grown on sand-vermiculite medium dually inoculated with AM fungi (*Glomus mosseae* or *G. fasciculatum* or *Glomus* sp. E3 type) and *Azotobacter vinelandii* (A), and a strain of *Enterobacteriaceae* (P) showed variable degree of plant growth and nutrient uptake. He concluded that the effect of *Glomus* species on plant growth and nutrition is related to the associated bacterial groups. He called the interaction to be selective between the mycorrhizal fungi and bacteria. The rhizosphere bacteria assayed did not decrease plant growth and nutrient content in any *Glomus* treatment. A positive effect of bacterial-fungal inocula was evidenced in some cases. Mycorrhizas and rhizosphere microorganisms can influence the mutual development of each other. Enhanced plant growth did not arise from direct effect of percentage of infection as it was observed that present bacterial strain did not increase infection rate (Azcón 1989).

In an experiment conducted my Mar Vázquez et al. (2000), they attempted to study the interaction between mycorrhizal fungi and selected strains of bacteria. They observed that inoculating maize plant with arbuscular mycorrhizal fungi (three *Glomus* spp.), *Azospirillum* (phytostimulator), *Pseudomonas* and *Trichoderma* (fungal antagonists) enhanced plant growth. They found that upon none of the microbial inoculants interfered with mycorrhizal colonization. *Azospirillum* inoculum stimulated mycorrhizal colonization of maize roots by increasing the susceptibility of the host and increasing spore germination or the growth of mycelium, thereby increasing the chance of contacts between fungal hyphae and plant roots. The effects of biological control agents (*Pseudomonas* and *Trichoderma*) are of particular relevance because of the possibility that these antagonists of fungi could also negatively interfere with AMF. *P. fluorescens* WCS365 has not been shown to produce antifungal compounds (Bloemberg, personal communication). In spite of this, it has been considered an effective biological control agent by inducing systemic resistance in inoculated plants (Simons et al. 1996). AMF colonization has been demonstrated to induce similar defence mechanisms in mycorrhizal plants (Gianinazzi-Pearson 1996); however it seems logical that *P. fluorescens* WCS365 would have no effect on AM colonization. The mechanisms of action of *Trichoderma* are known to be based on antibiosis, fungistasis, and mycoparasitism. Rousseaeu et al. (1996) reported a mycoparasitism of the extramatical phase of *G. intraradices* by *T. harzianum*. Such results are difficult to generalize, because they may be linked to the aggressiveness of the *Trichoderma* strain used. Therefore, it could be that *T. harzianum* T12 was not aggressive enough to interact negatively with AMF.

Besides aiding the host plant access remote nutrient pools, mycorrhizal associations also help its host access organic nitrogen, increase heavy metal and aluminium tolerance, decrease disease susceptibility and, in some cases, increase water uptake (Marschner 2012). It also mediates plant community structure and diversity by selecting the quantity and quality of microbes in and around the rhizosphere. The mycorrhizal fungi and associated rhizobacteria exert a selection pressure on the plant community of the area and plants surviving the pressure are selected over the perishing.

5.3.2 *Plant-Rhizobacteria*

Soil bacteria that grow on/in vicinity of plant roots and promote good growth directly (assisting in resource acquisition or modulating plant hormones) or indirectly (decreasing the inhibitory effect of various pathogen on plant growth and development in forms of biocontrol agents) are known as plant growth-promoting rhizobacteria (PGPR) (Prasad et al. 2015).

So, principally, PGPR may have dual effect on plant growth and development; these are indirect and direct effect. Direct effect is the promotion of plant growth by the synthesizing phytohormones by the bacteria and also by uptake of nutrients from the environment (Glick 1995; Prasad et al. 2005). In indirect effect PGPR help in growth promotion by decreasing or preventing same of the harmful effects of phytopathogenic organisms by different mechanisms. Chemical substances which are used to control plant diseases are hazardous to human being and can be accumulated in the natural ecosystem. These hazardous chemicals can be successfully replaced by using PGPR which can also be used as biocontrol agent. At present day, the number of biocontrol strains is increasing a lot including the development of super quality genetically engineered biocontrol agents (Eden et al. 1996).

PGPR may directly help the host plants for the improved growth by a single way or by combination of (Penrose and Glick 2003) which may be fixation of atmospheric nitrogen fixation through production which give iron to host plants; by phosphate solubilization to uptake available form of phosphorus; enzyme synthesis for modulation of plant growth and development; and by synthesizing auxins and cytokines. However, to impart the above effects by PGPR, soil composition plays a vital role. Even the effect of PGPR may be negligible, if the plants are cultivated in the soil which is rich in nutrient.

Again, biosynthesis of plant growth regulators by *Rhizobium* sp. is involved in growth and development of root nodules in leguminous plants (Hirsch et al. 1997). However, regarding root morphogenesis, the concentration of auxin produced by the rhizobacteria is very much important.

Apart from direct effect of PGPR on plant growth and development, the PGPR also protect the plants from the unwanted effects of stresses of heavy metals (Burd et al. 1998), salt (Mayak et al. 2004) and phytopathogens (Wang et al. 2000).

5.4 Plant Community Structure with Its Diversity and Microbial Interaction

Ecological succession in particular secondary succession is an important aspect in the development of community structure and diversity which is influenced by microbial interactions. Zhao et al. (2019) reposted changes of bacterial community in sloped farmlands abandoned for different years (0–40 years). They reported that secondary succession greatly affected soil bacterial beta diversity.

There are several determinants for the study of plant community structure (mainly the determination of the diversity of plant species and their frequencies and relative abundance). Plant community diversity and microbial interactions have a notable effect on soil microbial community and can be altered with increasing plant community richness. They also reported that influence of plant species on microbial communities within the soil depends on the diversity of plant community, and this happens due to the plant-derived resources and antagonistic soil microbes. Sun et al. (2019) reported that higher plant evenness promotes a positive plant bacterial richness relationship. On the other hand, Cui et al. (2018) reported that there is no correlation between the diversity of species of weeds of aboveground with bacterial richness. On the contrary, Dassen et al. (2017) reported that there might be very marginal positive effect on the fungal richness but no such effect in respect of bacteria *Archaea*.

5.5 Conclusion

Although difference of opinion exists between ecologists regarding the effect soil microbes have in shaping plant community composition, it cannot however be neglected altogether, given the astronomical number of microbes dwelling in soil. With the development of advanced techniques to study microbial population of rhizosphere and rhizoplane, we are able to categorize them to the species level. New molecular techniques allow us to dissect the pathways by which soil microbes interact in the rhizosphere and help shape plant community composition. Advanced microbiological technology has enabled us to isolate individual rhizosphere microbial species and understand their physiology and metabolism. Recombinant DNA technology, as it is a boon for us, has made it feasible to develop new strains of soil dwelling microbes which help has been immensely helpful in agriculture.

Presently we have better understanding of how soil microbes and plant interactions are drivers of plant community diversity in ecosystem. Scanty literature on how this interaction might be a playing force in determining the entire ecosystem diversity makes us believe that this could be a potential area of future research for ecologists. We know that advanced techniques in research will be helpful in explaining how ecosystem diversity is a function of plant and soil-microbe interaction.

References

Assmus B, Schloter M, Kirchhof G, Hutzler P, Hartmann A (1997) Improved in situ tracking of rhizosphere bacteria using dual staining with fluorescence-labelled antibodies and rRNA-targeted oligonucleotides. Microb Ecol 33:32–40. https://doi.org/10.1007/s002489900005

Azcón R (1989) Selective interaction between free-living rhizosphere bacteria and vesicular arbuscular mycorrhizal fungi. Soil Biol Biochem 21(5):639–644. https://doi.org/10.1016/0038-0717(89)90057-6

Blakely TJ, Didham RK (2010) Disentangling the mechanistic drivers of ecosystem-size effects on species diversity. J Anim Ecol 79(6):1204–1214. https://doi.org/10.1111/j.1365-2656.2010.01729.x

Bottomley PJ, Maggard SP (1990) Determination of viability within serotypes of a soil population of *Rhizobium leguminosarum* bv. *trifolii*. Appl Environ Microbiol 56(2):533–540

Brown ME, Burlingham SK, Jackson RM (1962) Studies on Azotobacter species in soil. 1. Comparison of media and techniques for counting Azotobacter in soil. Plant Soil 17:309. https://doi.org/10.1007/BF01377670

Brundrett MC (2002) Coevolution of roots and mycorrhizas of land plants. New Phytol 154:275–304

Burd GI, Dixon DG, Glick BR (1998) A plant growth-promoting bacterium that decreases nickel toxicity in seedlings. Appl Environ Microbiol 64(10):3663–3668

Clark FE (1949) Soil microorganisms and plant roots. In: Norman AG (ed) Advances in agronomy, vol 1. Academic, New York, pp 241–288

Crowl TA, Crist TO, Parmenter RR, Belovsky G, Lugo AE (2008) The spread of invasive species and infectious disease as drivers of ecosystem change. Front Ecol Environ 6(5):238–246. https://doi.org/10.1890/070151

Cui H, Liu LL, Dai JR, Yu XN, Guo X, Yi SJ, Zhou DY, Guo WH, Du N (2018) Bacterial community shaped by heavy metals and contributing to health risks in cornfields. Ecotoxicol Environ Saf 166:259–269

Dabral S, Bhola D, Attri MK, Nath M, Prasad R, Singh S, Varma A (2017) Principles and application of confocal microscopy to understand symbiotic fungi. In: Varma A, Prasad R, Tuteja N (eds) Mycorrhiza. Springer International Publishing AG, Switzerland, pp 431–354

Dal Cortivo C, Barion G, Visioli G, Mattarozzi M, Mosca G, Vamerali T (2017) Increased root growth and nitrogen accumulation in common wheat following PGPR inoculation: assessment of plant-microbe interactions by ESEM. Agric Ecosyst Environ 247:396–408

Dassen S, Cortois R, Martens H, de Hollander M, Kowalchuk GA, van der Putten WH, De Deyn GB (2017) Differential responses of soil bacteria, fungi, archaea and protists to plant species richness and plant functional group identity. Mol Ecol 26(15):4085–4098

Eden MA, Hill RA, Stewart A (1996) Biological control of *Botrytis* stem infection of greenhouse tomatoes. Plant Pathol 45(2):276–284

Gage DJ, Bobo T, Long SR (1996) Use of green fluorescent protein to visualize the early events of symbiosis between *Rhizobium meliloti* and alfalfa (*Medicago sativa*). J Bacteriol 178:7159–7166

Gans J, Woilinsky M, Dunbar J (2005) Computational improvements reveal great bacterial diversity and high metal toxicity in soil. Science 309:1387–1390

Gaskins MH, Albrecht SL, Hubbell DH (1985) Rhizosphere bacteria and their use to increase plant productivity: a review. Agric Ecosyst Environ 12(2):99–116

Gianinazzi-Pearson V (1996) Plant cell response to arbuscular mycorrhizal fungi: getting to the roots of the symbiosis. Plant Cell 8:1871–1883

Giri B, Giang PH, Kumari R, Prasad R, Sachdev M, Garg AP, Oelmuller R, Varma A (2005) Mycorrhizosphere: strategies and functions. In: Buscot F, Varma A (eds) Microorganisms in soils: roles in genesis and functions, vol 3. Springer, Berlin, pp 213–252

Glick BR (1995) The enhancement of plant growth by free-living bacteria. Can J Microbiol 41 (2):109–117

Harley JL, Waid JS (1955) A method of studying active mycelia on living roots and other surfaces in the soil. Trans Br Mycol Soc 38(2):104–118. ISSN: 0007-1536. https://doi.org/10.1016/S0007-1536(55)80022-8

Hiltner L (1904) Uber neuere Erfahrungen und Probleme auf dem Gebiet der Bodenbakteriologie und unter besonderer Berucksichtigung der Grundungung and Brache. Arbeiten der Deutschen Landwirtschaftlichen Gesellschaft 98:59–78

Hirsch AM, Fang Y, Asad S, Kapulnik Y (1997) The role of phytohormones in plant-microbe symbioses. Plant Soil 194(1–2):171–184

Ishizawa S, Suzuki T, Sato O, Toyoda H (1957) Studies on microbial population in the rhizosphere of higher plants with special reference to method of study. Soil Plant Food 2(2):85–94

Katznelson H (1946) The rhizosphere effect of Mangels on certain groups of soil micro-organism. Soil Sci 62:343–354

Leake JR, Lead DJ (2017) Mycorrhizal symbioses and pedogenesis throughout Earth's history. In: Johnson NC, Gehring C, Jansa J (eds) Mycorrhizal mediation of soil. Elsevier, Amsterdam

Louw HA, Webley DM (1959) A study of soil bacteria dissolving certain mineral phosphate fertilizers and related compounds. J Appl Bacteriol 22:227–233. https://doi.org/10.1111/j.1365-2672.1959.tb00155.x

Mar Vázquez M, César S, Azcón R, Barea JM (2000) Interactions between arbuscular mycorrhizal fungi and other microbial inoculants (*Azospirillum, Pseudomonas, Trichoderma*) and their effects on microbial population and enzyme activities in the rhizosphere of maize plants. Appl Soil Ecol 15(3):261–272. https://doi.org/10.1016/s0929-1393(00)00075-5

Marschner P (2012) Rhizosphere biology. In: Marschner P (ed) Marschner's mineral nutrition of higher plants. Academic, New York, pp 369–388. https://doi.org/10.1016/b978-0-12-384905-2.00015-7

Mayak S, Tirosh T, Glick BR (2004) Plant growth-promoting bacteria confer resistance in tomato plants to salt stress. Plant Physiol Biochem 42(6):565–572

Medina A, Probanza A, Gutierrez Mañero F, Azcón R (2003) Interactions of arbuscular-mycorrhizal fungi and Bacillus strains and their effects on plant growth, microbial rhizosphere activity (thymidine and leucine incorporation) and fungal biomass (ergosterol and chitin). Appl Soil Ecol 22(1):15–28. https://doi.org/10.1016/s0929-1393(02)00112-9

Nelson GC, Bennett E, Berhe AA, Cassman K, DeFries R, Dietz T, Dobermann A, Dobson A, Janetos A, Levy M, Marco D, Nakicenovic N, O'Neill B, Norgaard R, Petschel-Held G, Ojima D, Pingali P, Watson R, Zurek M (2006) Anthropogenic drivers of ecosystem change: an overview. Ecol Soc 11(2). Retrieved from http://www.jstor.org/stable/26266018

Parkinson D, Taylor GS, Pearson R (1963) Studies on fungi in the root region. I. the development on fungi on young roots. Plant Soil 19:332–349

Penrose DM, Glick BR (2003) Methods for isolating and characterizing ACC deaminase-containing plant growth-promoting rhizobacteria. Physiol Plant 118(1):10–15

Prasad R, Garg AP, Varma A (2005) Interaction of medicinal plants with plant growth promoting rhizobacteria and symbiotic fungi. In: Podila GK, Varma A (eds) Basic research and applications of Mycorrhizae, vol 1. IK International, Delhi, pp 363–407

Prasad R, Kumar M, Varma A (2015) Role of PGPR in soil fertility and plant health. In: Egamberdieva D, Shrivastava S, Varma A (eds) Plant growth-promoting Rhizobacteria (PGPR) and medicinal plants. Springer International, Cham, pp 247–260

Prasad R, Bhola D, Akdi K, Cruz C, Sairam KVSS, Tuteja N, Varma A (2017) Introduction to mycorrhiza: historical development. In: Varma A, Prasad R, Tuteja N (eds) Mycorrhiza. Springer International, Cham, pp 1–7

Reynolds HL, Packer A, Bever JD, Clay K (2003) Grassroot ecology: plant–microbe–soil interactions as drivers of plant community structure and dynamics. Ecol Soc Am 84(9):2281–2291

Rouatt JW, Katznelson H (1960) Influence of light on bacterial flora of roots. Nature 186(4725):659

Rousseaeu A, Benhamou N, Chat I, Piché Y (1996) Mycoparasitism of the extramatical phase of *Glomus intraradices* by *Trichoderma harzianum*. Phytopathology 86:434–443

Rovira AD, Newman EI, Bowen HJ, Campbell R (1974) Quantitative assessment of the rhizoplane microflora by direct microscopy. Soil Biol Biochem 6(4):211–216

Rudi K, Zimonja M, Trosvik P, Naes T (2007) Use of multivariate statistics for 16S rRNA gene analysis of microbial communities. Int J Food Microbiol 120(1–2):95–99

Scholter M, Borlinghaus R, Bode W, Hartmann, A (1993) Direct identification and localization of *Azospirillum* in the rhizosphere of wheat using fluorescence labeled monoclonal antibodies and confocal scanning laser microscopy. J Microsc 171:173–177

Simons M, van der Bij AJ, Brand I, de Weger LA, Wijffelman CA, Lugtenberg BJ (1996) Gnotobiotic system for studying rhizosphere colonization by plant growth-promoting *Pseudomonas* bacteria. Mol Plant Microb Interact 9:600–607

Smith SE, Read DJ (1997) Mycorrhizal symbiosis. Academic, New York

Sørensen J, Nicolaisen MH, Ron E, Simonet P (2009) Molecular tools in rhizosphere microbiology—from single-cell to whole-community analysis. Plant Soil 321(1–2):483–512

Sun YQ, Wang J, Shen C, He JZ, Ge Y (2019) Plant evenness modulates the effect of plant richness on soil bacterial diversity. Sci Total Environ 662:8–14

Thompson JN, Reichman OJ, Morin PJ, Polis GA, Power ME, Sterner RW, Couch CA, Gough L, Holt R, Hooper DU, Keesing F, Lovell CR, Milne BT, Molles MC, Sharon, Roberts DW, Strauss Y (2001) Frontiers of ecology: as ecological research enters a new era of collaboration, integration, and technological sophistication, four frontiers seem paramount for understanding how biological and physical processes interact over multiple spatial and temporal scales to shape the earth's biodiversity. Bioscience 51(1):15–24

Tilman D, Pacala S (1993) The maintenance of species richness in plant communities. In: Ricklefs RE, Schluter D (eds) Species diversity in ecological communities: historical and geographical perspectives. University of Chicago Press, Chicago, pp 13–25

Varma A, Prasad R, Tuteja N (2017) Mycorrhiza: function, diversity and state-of-art. Springer International Publishing, Switzerland. ISBN 978-3-319-53064-2. http://www.springer.com/us/book/9783319530635

Wang C, Knill E, Glick BR, Défago G (2000) Effect of transferring 1-aminocyclopropane-1-carboxylic acid (ACC) deaminase genes into *Pseudomonas fluorescens* strain CHA0 and its gac a derivative CHA96 on their growth-promoting and disease-suppressive capacities. Can J Microbiol 46(10):898–907

Zhao FZ, Bai L, Wang JY, Deng J, Ren CJ, Han XH et al (2019) Change in soil bacterial community during secondary succession depend on plant and soil characteristics. Catena 173:246–252

Chapter 6
Plant Growth-Promoting Potentials of Endophytic Fungi for the Management of Agricultural Crops and Grasses

Siva Sundara Kumar Durairajan, Suchitra Rakesh, Barkavi Durairajan, Kaushik Rajaram, Nagarathinam Arunkumar, and Rajesh Jeewon

Abstract This chapter features the impetus to study endophytic fungi (EF) in order to invigorate the economic feasibility in agriculture production. EF occupy a dominant place in the habitat adaptation of plants leveraging better yields and protection from biotic as well as abiotic stresses. They secrete several secondary metabolites of organic nature, which, besides safeguarding plants from pathogens and pests, also valorize agricultural plants like rice and corn, and offer better postharvest management. It is emphasized that the methods currently employed in plant breeding, seed preparation, and agricultural practices are hindering the diversity of fungal endophytes due to the inadequate knowledge about them. Besides, they also make some endophytes to perform ineffectively in mitigating stresses. The objective of this chapter is to suggest how to overcome such impediments for the advancement of future agriculture practices.

6.1 Introduction

The term endophytic microorganisms include bacteria, fungi, and algae, which colonize within the internal structure of host plants without causing symptoms (Behie and Bidochka 2014; Schulz and Boyle 2005). Many recently published reports have outlined endophytic fungi (EF) and their biotechnological potentials. The analysis of EF in different climatic and geographic zones indicates that they are ubiquitous in internal plant tissues and diverse in nature (Ghimire et al. 2011; Li et al.

S. S. K. Durairajan (✉) · S. Rakesh · K. Rajaram · N. Arunkumar
Department of Microbiology, School of Life Sciences, Central University of Tamil Nadu, Thiruvarur, India
e-mail: d.sivasundarakumar@cutn.ac.in

B. Durairajan
Department of Microbiology, Lady Doak College, Madurai Kamaraj University, Madurai, India

R. Jeewon
Faculty of Science, Department of Health Sciences, University of Mauritius, Réduit, Mauritius

A. Varma et al. (eds.), *Plant Microbiome Paradigm*,
https://doi.org/10.1007/978-3-030-50395-6_6

2012; Petrini 1991). They perform activities benefitting the hosts by providing nutrients and adapting them to suit their environment, shielding them from biotic and abiotic stresses, and encouraging biodiversity of plant community (Berg 2009; Gond et al. 2010; Pandey et al. 2011).

Endophytic relationships may have evolved at the time when plants first evolved millions of years ago. Although the existence of endophytes within plants was first described in the nineteenth century, they received sufficient attention in the last 35 years. The evidence of plant-associated microbes dates back to the fossilized tissues of stems and leaves (Taylor and Taylor 2000). EF are classified under two broad groups such as clavicipitaceous (CE) and non-clavicipitaceous (NCE) based on different traits according to the taxonomical variations. CEs are ascomycetes belonging to family *Clavicipitaceae* and colonize systemically in shoots and rhizomes of slender grasses of Poaceae (Clay and Schardl 2002). CEs probably enter through seeds and colonize in the intercellular spaces of plant tissues (Schardl et al. 2004). Conversely, NCEs are phylogenetically diverse and belong to various orders of *Ascomycota* phylum and exhibit horizontal transmission. Owing to their facultative saprophytic nature, they have often been isolated from disinfected tissues of host plant bodies using culture media (Arnold and Lutzoni 2007; Schulz and Boyle 2005; Rodriguez et al. 2009; Chadha et al. 2014; Mishra et al. 2015). Molecular techniques have revealed a high biodiversity of NCE, which do not grow in standard agar media or exist as obligate biotrophs (Duong et al. 2006; Ko Ko et al. 2011; Porras-Alfaro and Bayman 2011; Huang et al. 2014; Vandenkoornhuyse et al. 2002).

The growth pattern of NCE depends on host varieties, nature of infected tissues, biodiversity, fitness advantage, and mode of transmission. Rodriguez classified NCE under three functional categories such as classes 2, 3, and 4 (Rodriguez et al. 2009) considering different fungal lifestyles and life strategies with r-K selection continuum.

The association of EF with host plants is not superficial as some species colonize within the plant either in the root system or aerial parts. This symbiotic association continues without exposing any disease symptom to the plant (Hyde and Doytong 2008; Schulz and Boyle 2005). A specific behavior of EF in inducing plant growth has often been reported (Schulz and Boyle 2005). This chapter focuses on class 2 fungal NCEs, cohabitating symbiotically with a vast range of plant species, even though their biodiversity has not been well analyzed (Schulz and Boyle 2005).

Our priority is to compile the current knowledge available on the diversity, pattern of colonization, and potential performance of EF that colonize vascular tissues and shoots (Fig. 6.1). This chapter will also consider the dark septate endophytes (DSEs) (Jumpponen and Trappe 1998), the endophytic nonmycorrhizal basidiomycetes species of order *Sebacinales* (Weiss et al. 2004) and endophytic yeast owing to their ubiquitous presence in healthy roots of the various plants thriving in different ecosystems. Root endophytic fungi (REF) in *Sebacinales* have drawn considerable focus in view of their benefits conferred to the plants (Weiß et al. 2016). However, many findings based on the evidence in agricultural crops (Franken 2012) are still unclear in about their ecological role in grasslands where they are mostly found. Many lines of evidence, however, have shown that ascomycetes REF

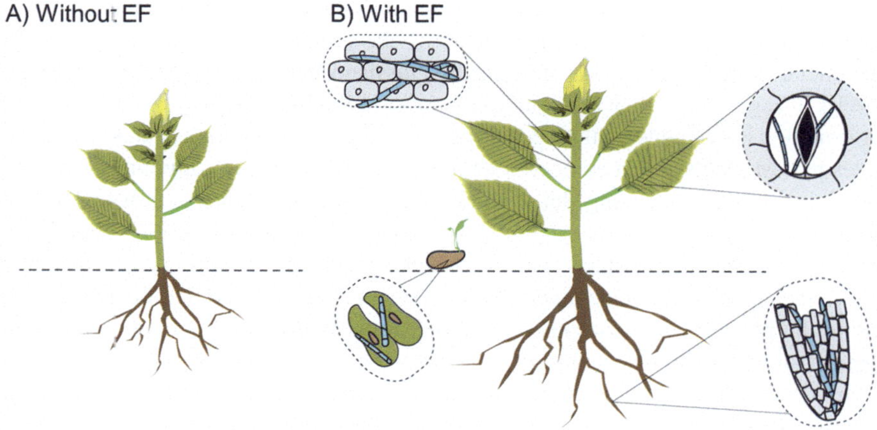

Fig. 6.1 (**a**) Absence of plant colonization by endophytic fungi (EF) shows normal development of plant growth (**b**) Plant inoculated with EF(+AMF). Growth promotion is often observed due to secretion of plant growth hormones and enhanced attainment of mineral nutrients through the EF's hyphal network (represented in blue)

may impact the structure of the plant community (Wehner et al. 2014; Hersh et al. 2011; Malcolm et al. 2013). The NCEs in roots vary phylogenetically and exhibit an inconsistent response toward host leading to a continuum in the interrelated activities of root-fungus association (Schulz and Boyle 2005). The consequential effects of EF, positive or otherwise, are controlled by the genotype of the host and the fungus (Knapp et al. 2012), in addition to the non-genetical aspects such as the developmental phase, prevailing ecological conditions at interactions stages, and nutritional standing of the host and fungus (Saikkonen et al. 1998). Unless the outcome of costs and mutual benefits resulting in a true mutualism is well determined, such activities cannot be well defined (Faeth and Fagan 2002). This chapter covers the EF that are found in plant roots but lack mycorrhizal associations like the formation of arbuscules.

6.2 Role of Endophytic Fungi in Promoting the Growth of Crops and Grassland Species

Endophytes are superior to mycorrhizal fungi in promoting plant growth. Various symbiotic interactions occurring in ecosystems have been classified into various categories based on the participatory advantage to both the hosts and the guests, such as mutualism, commensalism, parasitism, and amensalism, but this classification could not satisfactorily establish the relationship between host plants and root endophytes. EF are ancient as arbuscular mycorrhizal fungi (AMF) (Krings et al. 2007) and likely play an active role in selected ecosystems for plant survival, affecting structures of a single species or whole community (Saikkonen et al.

2004; Porras-Alfaro et al. 2008; Rodriguez and Redman 2008; Bultman et al. 2012; Knapp et al. 2012). In particular, REF of the *Ascomycota* group may prompt beneficial activities in plants thriving in grassland ecosystems (Rodriguez et al. 2009). The grassland REF are common in most of the fungal phyla, but only limited information is available on their performance regarding physiological and ecological roles in host plants, when compared with those of AMF. Many community surveys, involving DNA sequencing, have shown the ubiquitous manifestation of REF in *Ascomycota*, invariably almost five times greater in terms of species abundance collated to AMF in grassland species (Wehner et al. 2014). Further, they have the thriving capability in a wide range of hosts (Hersh et al. 2011; Malcolm et al. 2013), while the plant species associated in colonization express different reactions even against the same fungal genotype. For instance, when some plant species were administered with identical strains of *Microdochium* sp., under similar conditions, certain plant species revealed enhanced biomass production when other species showed no response despite similar colonization (Mandyam et al. 2012). These micromycetes flourish asymptomatically in living plant tissues (Petrini 1991; Saikkonen et al. 1998; Stone et al. 2000). They are identified by multiple taxonomic methods, predominantly belong to *Ascomycota* and *Basidiomycota* (Arnold et al. 2007; Arnold and Lutzoni 2007; Andrade-Linares et al. 2011; Weiss et al. 2004), and maintain a continuum of interactions with host organisms ranging from positive to neutral and neutral to negative.

In order to overcome the impediments in agriculture caused by the poor and conventional procedures (long-term application of inorganic fertilizers and pesticides), innovative microbial bioinoculant techniques are being designed and practiced. These new procedures gain momentum owing to their advantages in agriculture. Usually, microbes cohabitate with host plants symbiotically and create a congenial situation for mutual benefits of both organisms. Moreover, this association results in healthy growth of plants, ameliorating ineffective agricultural traits and improving the nutritional cycle and quality of the soil. Plant growth-promoting fungal endophytes (PGPFE) rooted in plant tissues and the coordination of endophytes within the plant tissue encourage rapid nutrients exchange and enzymes performance (Waller et al. 2005; Aly et al. 2011; Hiruma et al. 2016; Almario et al. 2017).

The wide prevalence of growth-promoting hormones produced by endophytic microbes in plant tissues favorably induces plant growth (Khan et al. 2015). Endophytes can mobilize insoluble phosphate and supply nitrogen to the hosts (Siddikkee Zereen et al. 2016; Almario et al. 2017; Malla et al. 2004).

PGPFE secrete different bioactive formulations performing multiple biological functions and are described as plant growth-promoting agents. Despite that most plants foster endophytes in their tissue interiors, the research on PGPE and their biological functions is limited. Further comprehension of the indigenous endophytes of plants will endorse their potential in improving plant growth and formulate an apt system for sustainable crop production.

REF are capable of influencing differential growth activities in host plants in different communities. Such influences considerably change the traits of plants according to different soil conditions adopted in the analysis, for instance, some strains of *Fusarium* responded differently when present in *Festuca brevipila* (Poaceae) than the herbaceous plant *Arrhenatherum elatius*. The endophytic fungi, *Alternaria elongate*, enhanced shoot mass considerably when used in a high sand condition compared to the plants reared in low sand condition (which possess greater nutrient accumulation than the plants in high sand treatment) even in the existence of rapidly growing *A. elatius* (Aguilar-Trigueros and Rillig 2016).

The weaker parasitism can retard the biomass synthesis in response to the colonization of root endophytes (Mandyam et al. 2010); otherwise, the induction of host resistance renders carbon allocation to the synthesis of antiherbivory compounds instead of facilitating vegetative growth (Aimé et al. 2013). The synthesis of plant growth hormones exerts many beneficial effects (Teale et al. 2006) or promotes nutrient transportation to the plants due to the mineralization of organic substances (under saprophytic capabilities of REF) (Newsham 2011).

6.3 Are the Plant-Endophyte Interactions Responsible to Produce Plant Hormones?

The role of endophytes and their relationship with plants are uncertain (Stone and Petrini 1997). Some EF appear to be ubiquitous (e.g., *Colletotrichum* spp., *Fusarium* spp., *Pestalotiopsis* spp., *Xylaria* spp.), while others may be host-specific or systemic within plants. Endophytes have evolved with plants over a long evolutionary time (Taylor and Taylor 2000), and they may have exchanged genetic information with the plants and vice versa (Stierle et al. 1993). Some plant-associated microorganisms, for example, can produce plant growth hormones in order to facilitate nutrient accumulation (Tudzynski 1997). Five classes of plant hormones (auxins, abscisins, ethylene, gibberellins, and kinetins) are produced by plant-associated fungi and bacteria (Tudzynski 1997).

Indole-3-acetic acid (IAA) is a major auxin that plays a key role in stimulating plant growth (Teale et al. 2006), inducing plant self-defense (Navarro et al. 2006), and serves as an effector molecule of the desired pathway (Spaepen et al. 2007). The EF producing indoleacetic acid (IAA) are considered as an efficient modulator of the colonization and growth of plant-endophyte association (Fu et al. 2015). The endophytes, *Penicillium chrysogenum*, *P. crustosum*, *Phoma glomerata*, *Pestalotiopsis neglecta*, and yeasts, play a vital role in the synthesis of phytohormones, favoring plants' in situ conservation (Fu et al. 2015; Fouda et al. 2015; Hassan 2017; Hoffman and Arnold 2010; Waqas et al. 2012). Nassar et al. (2005) showed that an endophytic yeast isolate, namely, *Williopsis saturnus* from sugarcane, can predominantly encourage the growth of rice, by producing indole-3-pyruvic acid and IAA. The growth enhancement features of *Trichoderma* significantly endorsed (Yeddia et al. 2001; Shoresh et al. 2010). The effectiveness of auxins in enhancing plant growth was proved in the case of *Arabidopsis* raised together with *T. virens* and *T. atroviride* (Contreras-Cornejo et al. 2009). The endophyte *Piriformospora indica*, a basidiomycete from the order *Sebacinales*, in a symbiotic association with plants roots (Prasad 2008; Franken 2012; Varma et al. 2012, 2014; Prasad et al. 2005, 2013; Gill et al. 2016; Chadha et al. 2015). Several studies reported that *P. indica* regulate the auxin production and contribute to root growth (Xu et al. 2018 and reference therein). The production of IAA by *P. indica* was first demonstrated by Sirrenberg et al. (2007) in vitro, and it has also been shown that *P. indica* can colonize the root of *A. thaliana* to promote root and shoot growth and lateral root development (Sirrenberg et al. 2007).

The fundamental inoculation tests revealed the capability of *P. indica* in colonizing the roots of plants (Verma et al. 1998; Prasad et al. 2008; Bagde et al. 2010), and till this date, *P. indica* is popular for its wide range of hosts enabling symbiotic interactions in roots of monocot and dicot plants inclusive of agricultural crops like barley, tobacco, and *Arabidopsis thaliana* (Varma et al. 2012, 2014; Johnson et al. 2014; Prasad et al. 2020).

The fungal root endophytes synthesize specific phytohormones that induce plant growth, as evidenced by the abundant adventitious growth in *Pelargonium* and *Poinsettia* cuttings, despite the absence of colonization by *P. indica* (Druege et al. 2007). However, this assumption of growth stimulation due to IAA synthesis was repudiated, as the elimination of the pathway in the endophyte did not result in the impairment of growth-promoting effects in plants (Hilbert et al. 2012).

Gibberellins (GAs) are one of the well-known classes of plant hormones, which regulate several plant developmental processes, including germination, stem elongation, leaf and flower development, and seed dormancy. Several species of fungi, like *Gibberella fujikuroi*, *Sphaceloma manihoticola*, *Phaeosphaeria* sp., and *Neurospora crassa* (Rademacher 1994) have been known to secrete GAs. Recent studies have established the capability of some strains of endophytic fungi, such as *Sesamum indicum* (Choi et al. 2005), *Phaeosphaeria* sp. *L487* (Kawaide 2006), *Penicillium citrinum* (Khan et al. 2008), *Chrysosporium pseudomerdarium* (Hamayun et al. 2009), *Aspergillus fumigatus* (Khan et al. 2011a), *Penicillium funiculosum* (Khan et al. 2011b), and *Paecilomyces formosus* LHL 10 (Khan et al. 2012) in producing GAs.

The endophytes, *P. formosus* LHL 10 and *P. janthinellum*, isolated from cucumber and tomato roots emanate gibberellins (GAs) in C rice cultivar (Waito-C) that are deficient in GA and in wild-type cultivar rice with normal GS synthesis (Dongjinbyeo) (Khan et al. 2012). The culture extract of *P. formosus* LHL 10 fungal isolate considerably induced the growth of Waito-C and Dongjinbyeo seedlings in comparison to a control culture extract. *P. formosus* was found to produce GAs (GA –1, –3, –4, –8, –9, –12, –20, and –24) and IAA (Khan et al. 2012). The inoculation of *P. formosus* was shown to increase shoot length of cucumber and related growth features in comparison to the uninoculated plants. The hyphae of *P. formosus* were found in the cortical and pericycle regions of the roots of the host, and the strain was confirmed by PCR techniques (Khan et al. 2012). Colonization of Chinese cabbage and barley seedlings by *P. indica* was shown to promote growth, and this is probably correlated with the increased level of GA in the colonized roots (Schäfer et al. 2009; Lee et al. 2011), and the genetic factor responsible for the inactivation of GA was decreased in *P. indica*-colonized barley roots (Schäfer and Kogel 2009).

Ethylene a common gaseous phytohormone plays a significant role in plant growth. A nonessential amino acid, 1-aminocyclopropane-1-carboxylic acid (ACC), which acts as a precursor of ethylene, can be metabolized by the enzyme ACC deaminase of endophytes to reduce ethylene levels in plants favoring growth potential. This enzyme synthesized by *T. asperellum* T 203 was shown to develop root-promoting capability in canola (Viterbo et al. 2010). In contrast, mutualistic fungus, *Phomopsis liquidambari*, increased root aerenchyma formation through auxin-mediated ethylene accumulation in rice (Hu et al. 2018). The endophyte *S. vermifera*, related closely to *P. indica*, was found to enhance plant growth in *Nicotiana attenuata* (Barazani et al. 2005) and *Panicum virgatum* (Ghimire et al. 2009). This enhancement effect seems to be related to the interference with ethylene synthesis occurring with the resistance retardation to herbivores (Barazani et al. 2005). Plant growth enhancement was also noticed during the generative performance. Tomato plants inoculated with DSEs and *P. indica* exhibited spurt in flowering and considerably higher fruit biomass during early stages of harvesting. However, the increase in fruit production receded at later phases of harvest (Andrade-Linares et al. 2011, 2013). In the root colonization studies, the fungus, *P. indica*, increases the expression of ACC synthase (Khatabi et al. 2012; Ansari et al. 2013) in the roots of *Arabidopsis* and barley probably via differentially modulating the expression of genes related to ethylene synthesis and signaling (Khatabi et al. 2012).

The influence of endophytes on these physiological performances cannot be predicted as in the case of plant-fungus symbiosis that mostly favors plant growth (Schulz and Boyle 2005; Andrade-Linares et al. 2013) and remains neutral (Brundrett 2006). These differences are due to intricate coordination between fungal symbionts and plants, which are regulated by a range of interaction and the nature and severity of stress along with other determinant factors. Hence it is possible to exemplify the results in a general context based on such studies. As such it is imperative to evaluate the central tendency and classify various modes of endophyte activities on plants under adverse conditions and appraise them under standard control.

6.4 Endophytic Fungi Promote the Supply of Plant Nutrition

The mechanism behind the uptake of mineral nutrients and the reciprocity of nutrients between plants and fungi has not so far been well established, as has been done in the case of mycorrhizal symbiotic association. Most of the experimental data collected so far have only led to the presumption of the role of endophytes in nutrient supply to plants. For example, instant inoculation of Chinese cabbage plants with DSE *Heteroconium chaetospira* prompted the host to use amino acids as a nitrogen source than inorganic N (NO_3^-) (Usuki and Narisawa 2007). The symbiotic association of *Phoma liquidambari* with rice plants, which rely on environmental N levels, facilitates N transformation by decomposing belowground straw (Sun et al. 2019). Thus fungal endophyte-plant systems are able to mobilize N by enhancing straw decomposition. *Trichoderma harzianum* T22, when associated with maize plants, increases growth and enhances crop yield even in low supply of nitrogen fertilizer (Harman 2000). Upson et al. (2009) revealed an interrelationship between plant growth enhancement in *Deschampsia antarctica* and utilization of organic resources by DSE. The infected roots of *D. antarctica* showed better plant growth with increased contents of nitrogen and phosphorous in root and shoot (Upson et al. 2009). Another feature of DSE, in increasing phosphorus content in *Carex* (Haselwandter and Read 1982) and *Pinus contorta*, has also been established. The meta-analysis of Newsham (2011) showed that DSE in addition to the supply of organic N, subjected to certain conditions, increased the biomass of plants and N and P content in shoots. Moreover, a direct evaluation process based on compartment experiments for solubilization of tricalcium phosphate and rock phosphate with *Atriplex canescens* (Pursh) plants infected by *Aspergillus ustus* showed that growth occurs in the phosphate source compartment and a sizeable increase occurs in plant shoot mass (Barrow and Osuna 2002). The influence of *P. indica* in plant nutritional aspect has also been studied recently. An enhancement in phosphate in shoots was observed in maize plants associated with fungus, and the colonization confirmed the inevitable role of fungal high affinity phosphate transporter (Yadav et al. 2010).

Colonizing *P. indica* to *Arabidopsis* roots enhances the activities of nitrate reductase and nitrate uptake in plants (Sherameti et al. 2005), which evidence the role of fungus in increasing nitrate supply to the plants. In such true mutualistic association, the fungal partner ought to derive benefits from the host, i.e., the transfer of carbohydrates from the host to the guest for the benefit of the latter. In the symbiotic association of fungus *H. chaetospira* with Chinese cabbage, the presence of carbohydrates, sucrose, and mannitol was noticed in plant roots by using $^{13}C\text{–}CO_2$ (Usuki and Narisawa 2007). In another study, the colonization of *P. indica* prompted an enzyme co-responsible for starch degradation (Sherameti et al. 2005). Appropriately, lower concentrations of hexose, starch, and certain amino acids in the colonized plant roots are indicative of likely transportation to the fungus (Schäfer et al. 2009). In contrast to the activities in mycorrhizal structure, this nutrient transfer could be regulated by increased CO_2 assimilation in plants concerned (Achatz et al. 2010).

Recent finding has shown a functional plant-fungal symbiosis in a nonmycorrhizal plant belongs to the Brassicaceae family. Almario et al. (2017) showed a new functional colonization between a *Helotiales* ascomycete fungus and the plant *Arabis alpina*, a nonmycorrhizal plant. The prospective significance of this group of fungi has been evidenced from cultivation-independent root microbiota profiling. Various comparative studies on growth condition have shown that this taxon is recurrent and abundant group of the root microbiota of *Arabis* exhibiting a specific high ampleness under low-P conditions. Later on,

Almario et al. (2017) conducted a functional analysis with a strain of *Helotiales*, isolated from *Arabis* root microbiota, which exhibited the colonization of the root interspaces without showing disease symptoms, carrying P to the host plant, enriching shoot biomass, and increasing shoot P content in native low-P soil, all such performances are the hallmarks of mycorrhizal symbiosis.

Several evidences show that nonmycorrhizal plants possess the capability of exhibiting advantageous symbiotic colonization with endophytic fungi, as in the case of the ascomycete *Colletotrichum tofieldiae* forming precise cellular compositions like certain epidermal and cortical cells appended with swollen hyphal cells, and conduct P to *Arabidopsis thaliana*, nonmycorrhizal host (Himura et al. 2016). *C. tofieldia* represents a native REF of *Arabidopsis* and promotes only growth under low-P condition. This trait is a contrary to the *Helotiales* fungus (transfer of P happens only in low and high-P conditions) or the *Basidiomycete P. indica* (Yadav et al. 2010) which enhances plant growth under low- and high-P conditions.

6.5 Possible Factors for the Loss of Endophytic Fungi and Their Plant Growth-Promoting Activities in Agricultural Plants

It can be postulated that the fungal endophytes colonized in a wild plant species cannot be sustained in the agricultural varieties of the same species. As under variable environmental and physiological conditions, there will be selective pressure to maintain the endophyte within the plant and to keep its beneficial activity for the plant in an optimal form. It seems reasonable to assume that plants can lose endophytes in the absence of any benefit by their presence and by the frequent application of pesticides and fungicides. Besides, in the absence of low selective pressure, the endophytes may be less effective due to the loss of secondary metabolites and/or might undergo mutations in the absence of any selection. The result is a reduction in the efficacy of the endophyte and/or loss of its beneficial activity.

It can be speculated that the ineffective or retarded performance of endophytes, besides losing some secondary metabolites, could be due to the following factors:

1. Plants growing with less or nil selective pressure.
2. Application of systemic fungicides to plants for avoiding fungal attack during plant breeding process and in vitro propagation.
3. Acquiring endophyte-mediated nutrients and pathogenic resistance may be lost due to the absence of selective pressure during plant breeding phases.
4. Seeds subjected to "pathogens cleaning off" process and heat treatment process may lose certain endophytes.
5. Commercially processed seeds might possibly be contaminated with fungicides that hamper the endophyte growth during the stages of storage and germination.
6. Application of systemic fungicides during plant growth.

6.6 Translation of Fungal Endophytes as Bioinoculant from Lab to Field Study

REF perform as valuable crop inoculants and also develop the abiotic and biotic stress tolerance in plants (Kumar et al. 2017). However, most of the related studies carried out in a conditioned system could not be extrapolated to the field (Nelissen et al. 2014). The commercialization of endophyte inoculants beneficial to agricultural crops is plagued by two pertinent issues. The first one is the lack of dependability and consistency, and the second is the uncertainty in the minds of ultimate users about the potentiality of such treatments. For instance, the case study perceived by Murphy et al. (2015, 2017) focused on resolving the inconsistency in the performance of endophytic inoculations in field crops by selecting a single crop variety as their sample plant and subjected the plant under different environmental stress factors culminating in a wide range of field trials. Their trials were developed for designing a de novo source endophyte inoculant from an inceptive concept originated from *Hordeum murinum*, a wild barley species. The group carried out several multiyear field experiments, combining rigorous methodologies inclusive of vigilant screening, appropriate selection procedures, elaborate testing of strains of fungal endophytes under controlled environments, and eventually validating an effective consortium of endophytes suitable for dry barley growing regions. The perspective followed by Murphy's group was targeted to develop a "pot-to-plot" approach. The problem of producing a really working inoculation could be solved by following this approach (Rosier et al. 2016). The group established the potentiality of a fungal endophyte inoculum that was capable of consistently enhancing the yield of barley grain fortified with several fertilizers continued for several seasons.

The analysis and identification of appropriate endophytes that support agricultural crop development is a continuous and open-ended endeavor, and the success rate depends upon the availability of resources and feasible methodologies. Further, the focus should be on integrating various approaches resulting in successful and consistent benefits to agriculture. For instance, marching toward a "no-till" farming activities carried through extended periods of rotations and adopting cover crops methods can convincingly reduce the cost of chemical inputs and wages (and may enhance the growth potential of crops when associated with an appropriate endophyte inoculant.

6.7 Conclusion

Endophytic fungi are more divergently expressed in a natural ecosystem than the broadly analyzed PGP fungi. The habitation of EF is ever-expanding despite adverse environmental factors thrust upon by inner plant tissues and other impediments imposed by the cultivation and independent cultivation methods.

In the ecosystem, plants are not self-sustaining bodies and coexist with different organisms encompassing microbes forming a complex community. These plant-associated organisms, especially EF, exert positive benefits to the plants existing even in agricultural lands. This review chapter fairly endorses the findings enunciating that the mutualistic association of EF confers benefits of PGP hormones and the supply of nutrition to the host plants. This review investigated the role of EF in growth, hormone synthesis, and nutrient supply occurring in plants and the interconnected parameters influenced by each other at the plant physiological stage. The mechanism behind the activities of the fungi influencing these parameters is still unknown. We believe that further analysis at molecular levels will explain the comprehensive response of plant-endophyte activities at various stages. In addition, an extensive analysis adopting integrated approaches encompassing biochemical, physiological, and molecular techniques and "omics" will throw light on the interrelated mechanisms of molecules and the metabolic routes of endophyte mediating growth promotion in plants.

In a nutshell, upgrading the cultivation circumstances and improving molecular tools will clarify how an endophyte-plant association works and could be used in the direction to improve crop yield. Further, in order to understand the influence of PGPEF, the activities occurring in the endophytes must be explored in detail.

References

Achatz B, von Ruden S, Andrade D, Neumann E, Pons-Kuhnemann J, Kogel KH, Franken P, Waller F (2010) Root colonization by *Piriformospora indica* enhances grain yield in barley under diverse nutrient regimes by accelerating plant development. Plant Soil 333:59–70

Aguilar-Trigueros CA, Rillig MC (2016) Effect of different root endophytic fungi on plant community structure in experimental microcosms. Ecol Evol 6:8149–8158

Aimé S, Alabouvette C, Steinberg C, Olivain C (2013) The endophytic strain *Fusarium oxysporum* Fo47: a good candidate for priming the defense responses in tomato roots. Mol Plant-Microbe Interact 26:918–926

Almario J, Jeena G, Wunder J, Langen G, Zuccaro A, Coupland G, Bucher M (2017) Root-associated fungal microbiota of nonmycorrhizal *Arabis alpina* and its contribution to plant phosphorus nutrition. Proc Natl Acad Sci USA 114:E9403–E9412

Aly AH, Debbab A, Proksch P (2011) Fungal endophytes: unique plant inhabitants with great promises. Appl Microbiol Biot 90:1829–1845

Andrade-Linares DR, Grosch R, Franken P, Karl HR, Kost G, Restrepo S, de Garcia MC, Maximova E (2011) Colonization of roots of cultivated *Solanum lycopersicum* by dark septate and other ascomycetous endophytes. Mycologia 103:710–721

Andrade-Linares DR, Anja Müller A, Fakhro A, Schwarz D, Franken P (2013) Impact of *Piriformospora indica* on tomato. In: Varma A (ed) *Piriformospora indica*, soil biology, vol 33. Springer, Berlin, pp 107–117

Ansari MW, Trivedi DK, Sahoo RK, Gill SS, Tuteja N (2013) A critical review on fungi mediated plant response with special emphasis to Piriformospora indica on improved production and protection of crops. Plant Physiol Biochem 70:403–410

Arnold AE, Lutzoni F (2007) Diversity and host range of foliar fungal endophytes: are tropical leaves biodiversity hotspots? Ecology 88:541–549

Arnold AE, Henk DA, Eells RA, Lutzoni F, Vilgalys R (2007) Diversity and phylogenetic affinities of foliar fungal endophytes in loblolly pine inferred by culturing and environmental PCR. Mycologia 99:185–206

Bagde US, Prasad R, Varma A (2010) Interaction of *Piriformospora indica* with medicinal plants and of economic importance. Afr J Biotechnol 9(54):9214–9226

Barazani O, Benderoth M, Groten K, Kuhlemeier C, Baldwin IT (2005) *Piriformospora indica* and Sebacina vermifera increase growth performance at the expense of herbivore resistance in *Nicotiana attenuata*. Oecologia 146:234–243

Barrow JR, Osuna P (2002) Phosphorus solubilization and uptake by dark septate fungi in fourwing saltbush, *Atriplex canescens* (Pursh) Nutt. J Arid Environ 51:449–459

Behie SW, Bidochka MJ (2014) Nutrient transfer in plant-fungal symbioses. Trends Plant Sci 19:734–740

Berg G (2009) Plant-microbe interactions promoting plant growth and health: perspectives for controlled use of microorganisms in agriculture. Appl Microbiol Biotechnol 84:11–18

Brundrett MC (2006) Understanding the roles of multifunctional mycorrhizal and endophytic fungi. In: Schulz BJE, Boyle CJC, Sieber TN (eds) Microbial root endophytes. Springer, Berlin, pp 281–293

Bultman TL, Aguilera A, Sullivan TJ (2012) Influence of fungal isolates infecting tall fescue on multitrophic interactions. Fungal Ecol 5:372–378

Chadha N, Mishra M, Prasad R, Varma A (2014) Root endophytic fungi: research update. J Biol Life Sci USA 5(2):135–158

Chadha N, Prasad R, Varma A (2015) Plant promoting activities of fungal endophytes associated with tomato roots from central Himalaya, India and their interaction with *Piriformospora indica*. Int J Pharma BioSci 6(1):333–343

Choi WY, Rim SO, Lee JH, Lee JM, Lee IJ, Cho KJ, Rhee IK, Kwon JB, Kim JG (2005) Isolation of gibberellins producing fungi from the root of several *Sesamum indicum* plants. J Microbiol Biotechnol 15:22–28

Clay K, Schardl C (2002) Evolutionary origins and ecological consequences of endophyte symbiosis with grasses. Am Nat 160:99–127

Contreras-Cornejo HA, Macias-Rodriguez L, Cortes-Penagos C, Lopez-Bucio J (2009) *Trichoderma virens*, a plant beneficial fungus, enhances biomass production and promotes lateral root growth through an auxin-dependent mechanism in Arabidopsis. Plant Physiol 149:1579–1592

Druege U, Baltruschat H, Franken P (2007) *Piriformospora indica* promotes adventitious root formation in cuttings. Sci Hortic 112:422–426

Duong LM, Jeewon R, Lumyong S, Hyde KD (2006) DGGE coupled with ribosomal DNA phylogenies reveal uncharacterized fungal phylotypes on living leaves of *Magnolia liliifera*. Fungal Divers 23:121–138

Faeth SH, Fagan WF (2002) Fungal endophytes: common host plant symbionts but uncommon mutualists. Integr Comp Biol 42:360–368

Fouda AH, Hassan SED, Eid AM, Ewais EED (2015) Biotechnological applications of fungal endophytes associated with medicinal plant *Asclepias sinaica* (Bioss.). Ann Agric Sci 260:95–104

Franken P (2012) The plant strengthening root endophyte *Piriformospora indica*: potential application and the biology behind. Appl Microbiol Biotechnol 96:1455–1464

Fu SF, Wei JY, Chen HW, Liu YY, Lu HY, Chou JY (2015) Indole-3-acetic acid: a widespread physiological code in interactions of fungi with other organisms. Plant Signal Behav 10: e1048052

Ghimire SR, Charlton ND, Craven KD (2009) The mycorrhizal fungus, *Sebacina vermifera*, enhances seed germination and biomass production in switchgrass (*Panicum virgatum* L). Bioenergy Res 2:51–58

Ghimire SR, Charlton ND, Bell JD, Krishnamurthy YL, Craven KD (2011) Biodiversity of fungal endophyte communities inhabiting switchgrass (*Panicum virgatum* L.) growing in the native tallgrass prairie of northern Oklahoma. Fungal Divers 47:19–27

Gill SS, Gill R, Trivedi DK, Anjum NA, Sharma KK, Ansari MW, Ansari AA, Johri AK, Prasad R, Pereira E, Varma A, Tuteja N (2016) *Piriformospora indica*: potential and significance in plant stress tolerance. Front Microbiol 7:332

Gond SK, Verma VC, Mishra A, Kumar A, Kharwar RN (2010) Role of fungal endophytes in plant protection. In: Arya A, Perelló AE (eds) Management of fungal plant pathogens. CAB International, Wallingford, pp 183–197

Hamayun M, Khan SA, Iqbal I, Na CI, Khan AL, Hwang YH, Lee BH, Lee IJ (2009) *Chrysosporium pseudomerdarium* produces gibberellins and promotes plant growth. J Microbiol 47:425–430

Harman GE (2000) Myths and dogmas of biocontrol—changes in perceptions derived from research on *Trichoderma harzianum* T-22. Plant Dis 84:377–393

Haselwandter K, Read DJ (1982) The significance of a root-fungus association in two Carex species of high-alpine plant communities. Oecologia 52:352–354

Hassan SE (2017) Plant growth-promoting activities for bacterial and fungal endophytes isolated from medicinal plant of *Teucrium polium* L. J Adv Res 8:687–695

Hersh MH, Vilgalys R, Clark JS (2011) Evaluating the impacts of multiple generalist fungal pathogens on temperate tree seedling survival. Ecology 93:511–520

Hilbert M, Voll LM, Ding Y, Hofmann J, Sharma M, Zuccaro A (2012) Indole derivative production by the root endophyte *Piriformospora indica* is not required for growth promotion but for biotrophic colonization of barley roots. New Phytol 196:520–534

Hiruma K, Gerlach N, Sacristán S, Nakano RT, Hacquard S, Kracher B, Neumann U, Ramírez D, Bucher M, O'Connell RJ, Schulze-Lefert P (2016) Root endophyte *Colletotrichum tofieldiae* confers plant fitness benefits that are phosphate status dependent. Cell 165:464–474

Hoffman MT, Arnold AE (2010) Diverse bacteria inhabit living hyphae of phylogenetically diverse fungal endophytes. Appl Environ Microbiol 76:4063–4075

Hu LY, Li D, Sun K, Cao W, Fu WQ, Zhang W, Dai CC (2018) Mutualistic fungus *Phomopsis liquidambari* increases root aerenchyma formation through auxin-mediated ethylene accumulation in rice (*Oryza sativa* L.). Plant Physiol Biochem 130:367–376

Huang CL, Jian FL, Huang HJ, Chang WC, Wu WL, Hwang CC, Lee RHT, Chiang TY (2014) Deciphering mycorrhizal fungi in cultivated Phalaenopsis microbiome with next-generation sequencing of multiple barcodes. Fungal Divers 66:77–88

Hyde KD, Doytong K (2008) The fungal endophyte dilemma. Fungal Divers 33:163–173

Johnson JM, Alex T, Oelmüller R (2014) *Piriformospora indica*: the versatile and multifunctional root endophytic fungus for enhanced yield and tolerance to biotic and abiotic stress in crop plants. J Trop Agr 52:103–122

Jumpponen A, Trappe JM (1998) Dark septate endophytes: a review of facultative biotrophic root-colonizing fungi. New Phytol 140:295–310

Kawaide H (2006) Biochemical and molecular analysis of gibberellins biosynthesis in fungi. Biosci Biotech Biochem 70:583–590

Khan SA, Hamayun M, Yoon HJ, Kim YH, Suh SJ, Hwang SK (2008) Plant growth promotion and *Penicillium citrinum*. BMC Microbiol 8:231

Khan AL, Hamayun M, Ahmad N, Hussain J, Kang SM, Kim YH, Muhammad A, Dong-Sheng T, Muhammad W, Ramalingan R, Young-Hyun H, In-Jung L (2011a) Salinity stress resistance offered by endophytic fungal interaction between *Penicillium minioluteum* LHL09 and *Glycine max* L. J Microbiol Biotechnol 21:893–902

Khan AL, Hamayun M, Kim YH, Kang SM, Lee IJ (2011b) Ameliorative symbiosis of endophyte (*Penicillium funiculosum* LHL06) under salt stress elevated plant growth of *Glycine max* L. Plant Physiol Biochem 49:852–862

Khan AL, Hamayun M, Kang SM, Kim YH, Jung HY, Lee JH, Lee IJ (2012) Endophytic fungal association via gibberellins and indole acetic acid can improve plant growth under abiotic stress: an example of *Paecilomyces formosus* LHL10. BMC Microbiol 12:3

Khan AL, Hussain J, Al-Harrasi A, Al-Rawahi A, Lee IJ (2015) Endophytic fungi: resource for gibberellins and crop abiotic stress resistance. Crit Rev Biotechnol 35:62–67

Khatabi B, Molitor A, Lindermayr C, Pfiffi S, Durner J, von Wettstein D, Kogel KH, Schäfer P (2012) Ethylene supports colonization of plant roots by the mutualistic fungus *Piriformospora indica*. PLoS One 7:e35502

Knapp DG, Pintye A, Kova'cs GM (2012) The dark side is not fastidious—dark septate endophytic fungi of native and invasive plants of semiarid sandy areas. PLoS One 7:e32570

Ko Ko TW, Stephenson SL, Bahkali AH, Hyde KD (2011) From morphology to molecular biology: can we use sequence data to identify fungal endophytes? Fungal Divers 50:113–120

Krings M, Taylor TN, Hass H, Kerp H, Dotzler N, Hermsen EJ (2007) Fungal endophytes in a 400-million-yr-old land plant: infection pathways, spatial distribution, and host responses. New Phytol 174:648–657

Kumar M, Prasad R, Kumar V, Tuteja N, Varma A (2017) Mycorrhizal fungi under biotic and abiotic stress. In: Varma A, Prasad R, Tuteja N (eds) Mycorrhiza. Springer International Publishing AG, Switzerland, pp 57–70

Lee YC, Johnson JM, Chien CT, Sun C, Cai D, Lou B, Oelmüller R, Yeh KW (2011) Growth promotion of Chinese cabbage and Arabidopsis by *Piriformospora indica* is not stimulated by mycelium-synthesized auxin. Mol Plant-Microbe Interact 24:421–431

Li HY, Shen M, Zhou ZP, Li T, Wei YL, Lin LB (2012) Diversity and cold adaptation of endophytic fungi from five dominant plant species collected from the Baima Snow Mountain, Southwest China. Fungal Divers 54:79–86

Malcolm GM, Kuldau GA, Gugino BK, Jimenez-Gasco Mdel M (2013) Hidden host plant associations of soilborne fungal pathogens: an ecological perspective. Phytopathology 103:538–544

Malla R, Prasad R, Kumari R, Giang PH, Pokharel U, Oelmueller R, Varma A (2004) Phosphorus solubilizing symbiotic fungus *Piriformospora indica*. Endocytobiosis Cell Res 15(2):579–600

Mandyam K, Loughin T, Jumpponen A (2010) Isolation and morphological and metabolic characterization of common endophytes in annually burned tallgrass prairie. Mycologia 102:813–821

Mandyam K, Fox C, Jumpponen A (2012) Septate endophyte colonization and host responses of grasses and forbs native to a tallgrass prairie. Mycorrhiza 22:109–119

Mishra M, Prasad R, Varma A (2015) Endophytic fungi: biodiversity and functions. Int J Pharm Bio Sci 6(1):18–36

Murphy BR, Batke SP, Doohan FM, Hodkinson TR (2015) Media manipulations and the culture of beneficial fungal root endophytes. Int J Biol 7:94

Murphy BR, Hodkinson TR, Doohan FM (2017) A fungal endophyte consortium counterbalances the negative effects of reduced nitrogen input on the yield of field-grown spring barley. J Agric Sci 155:1324–1331

Nassar AH, El-Tarabily KA, Sivasithamparam K (2005) Promotion of plant growth by an auxin-producing isolate of the yeast *Williopsis saturnus* endophytic in maize (*Zea mays* L.) roots. Biol Fertil Soil 42:97–108

Navarro L, Dunoyer P, Jay F, Arnold B, Dharmasiri N, Estelle M, Voinnet O, Jones JD (2006) A plant miRNA contributes to antibacterial resistance by repressing auxin signaling. Science 312:436–439

Nelissen H, Moloney M, Inzé D (2014) Translational research: from pot to plot. Plant Biotechnol J 12:277–285

Newsham KK (2011) A meta-analysis of plant responses to dark septate root endophytes. New Phytol 190:783–793

Pandey R, Mishra AK, Tiwari S, Singh HN, Kalra A (2011) Enhanced tolerance of Mentha arvensis against Meloidogyne incognita (Kofoid and White) Chitwood through mutualistic endophytes and PGPRs. J Plant Interact 6:247–253

Petrini O (1991) Fungal endophytes of tree leaves. In: Andrews J, Hirano S (eds) Microbial ecology of leaves. Springer, New York, pp 179–197

Porras-Alfaro A, Bayman P (2011) Hidden fungi, emergent properties: endophytes and microbiomes. Annu Rev Phytopathol 49:291–315

Porras-Alfaro A, Herrera J, Sinsabaugh RL, Odenbach KJ, Lowrey T, Natvig DO (2008) Novel root fungal consortium associated with a dominant desert grass. Appl Environ Microbiol 74:2805–2813

Prasad R (2008) Studies on interaction between a symbiotic fungus (*Piriformospora indica*), rhizobacteria and selected plants. Ph. D. thesis. CCS University, Meerut

Prasad R, Pham GH, Kumari R, Singh A, Yadav V, Sachdev M, Peskan T, Hehl S, Oelmuller R, Garg AP, Varma A (2005) Sebacinaceae: culturable mycorrhiza–like endosymbiotic fungi and their interaction with non-transformed and transformed roots. In: Declerck S, Strullu DG, Fortin JA In vitro culture of mycorrhizas, Springer, Berlin, 4: 291–312

Prasad R, Bagde US, Pushpangdan P, Varma A (2008) *Bacopa monniera* L.: pharmacological aspects and case study involving *Piriformospora indica*. Int J Integr Biol 3:100–110

Prasad R, Kamal S, Sharma PK, Oelmueller R, Varma A (2013) Root endophyte Piriformospora indica DSM 11827 alters plant morphology, enhances biomass and antioxidant activity of medicinal plant *Bacopa monnieri*. J Basic Microbiol 53(12):1016–1024

Prasad R, Chhabra S, Gill SS, Singh PK, Tuteja N (2020) The microbial symbionts: potential for the crop improvement in changing environments. In: Tuteja N, Tuteja R, Passricha N, Saifi SK (eds) Advancement in crop improvement techniques. Elsevier, Amsterdam, pp 233–240

Rademacher W (1994) Gibberellin formation in microorganisms. Plant Growth Reg 15:303–314

Rodriguez RJ, Redman RS (2008) More than 400 million years of evolution and some plants still can't make it on their own: plant stress tolerance via fungal symbiosis. J Exp Bot 59:1109–1114

Rodriguez RJ, White JF Jr, Arnold AE, Redman RS (2009) Fungal endophytes: diversity and functional roles. New Phytol 182:314–330

Rosier A, Bishnoi U, Lakshmanan V, Sherrier DJ, Bais HP (2016) A perspective on inter-kingdom signaling in plant-beneficial microbe interactions. Plant Mol Biol 90:537–548

Saikkonen K, Faeth SH, Helander M, Sullivan TJ (1998) Fungal endophytes: a continuum of interactions with host plants. Annu Rev Ecol Syst 29:319–343

Saikkonen K, Wali P, Helander M, Faeth SH (2004) Evolution of endophyte-plant symbioses. Trends Plant Sci 9:275–280

Schäfer P, Kogel KH (2009) The sebacinoid fungus *Piriformospora indica*: an orchid mycorrhiza which may increase host plant reproduction and fitness. In: Deising HB, Esser K (eds) The mycota, Plant relationships, vol 5. Springer, Heidelberg, pp 99–112

Schäfer P, Pfiffi S, Voll LM, Zajic D, Chandler PM, Waller F, Scholz U, Pons-Kühnemann J, Sonnewald S, Sonnewald U, Kogel KH (2009) Manipulation of plant innate immunity and gibberellin as factor of compatibility in the mutualistic association of barley roots with Piriformospora indica. Plant J 59:461–474

Schardl CL, Leuchtmann A, Spiering MJ (2004) Symbioses of grasses with seedborne fungal endophytes. Annu Rev Plant Biol 55:315–340

Schulz B, Boyle C (2005) The endophytic continuum. Mycol Res 109:661–686

Sherameti I, Shahollari B, Venus Y, Altschmied L, Varma A, Oelmüller R (2005) The endophytic fungus *Piriformospora indica* stimulates the expression of nitrate reductase and the starch degrading enzyme glucan-water dikinase in tobacco and Arabidopsis roots through a homeodomain transcription factor that binds to a conserved motif in their promoters. J Biol Chem 280:26241–26247

Shoresh M, Harman GE, Mastouri F (2010) Induced systemic resistance and plant responses to fungal biocontrol agents. Annu Rev Phytopathol 48:21–43

Siddikkee Zereen MI, Li CF, Dai CC (2016) Endophytic fungus *Phomopsis liquidambari* and different doses of N-fertilizer alter microbial community structure and function in rhizosphere of rice. Sci Rep 6:32270

Sirrenberg A, Goebel C, Grond S, Czempinski N, Ratzinger A, Karlovsky P, Santos P, Feussner I, Pawlowski K (2007) *Piriformospora indica* affects plant growth by auxin production. Physiol Plant 131:581–589

Spaepen S, Vanderleyden J, Remans R (2007) Indole-3-acetic acid in microbial and microorganism-plant signaling. FEMS Microbiol Rev 31:425–448

Stierle A, Strobel GA, Stierle D (1993) Taxol and taxane production by *Taxomyces andreanae*. Science 260:214–216

Stone J, Petrini O (1997) Endophytes of forest trees: a model for fungus-plant interactions. In: Carroll GC, Tudzynski P (eds) The Mycota V Part B. Springer, New York, pp 129–140

Stone JK, Bacon CW, White JF (2000) An overview of endophytic microbes: endophytism defined. In: Bacon CW, White JF (eds) Microbial endophytes. Dekker, New York, pp 3–30

Sun K, Cao W, Hu LY, Fu WQ, Gong JH, Kang N, Dai CC (2019) Symbiotic fungal endophyte *Phomopsis liquidambari*-rice system promotes nitrogen transformation by influencing below-ground straw decomposition in paddy soil. J Appl Microbiol 26(1):191–203

Taylor TN, Taylor EL (2000) The Rhynie chert ecosystem: a model for understanding fungal interactions. In: Bacon CW, White JF (eds) Microbial endophytes. Marcel Dekker, New York, pp 31–47

Teale WD, Paponov IA, Palme K (2006) Auxin in action: signalling, transport and the control of plant growth and development. Nat Rev Mol Cell Biol 7:847–859

Tudzynski B (1997) Fungal phytohormones in pathogenic and mutualistic associations. In: Carroll GC, Tudzynski P (eds) The mycota V, Part A, plant relationships. Springer, Berlin, pp 167–184

Upson R, Read DJ, Newsham KK (2009) Nitrogen form influences the response of *Deschampsia antarctica* to dark septate root endophytes. Mycorrhiza 20:1–11

Usuki F, Narisawa K (2007) A mutualistic symbiosis between a dark septate endophytic fungus, *Heteroconium chaetospira*, and a nonmycorrhizal plant, Chinese cabbage. Mycologia 99:175–184

Vandenkoornhuyse P, Baldauf SL, Leyval C, Straczek J, Young JP (2002) Extensive fungal diversity in plant roots. Science 295:2051

Verma S, Varma A, Rexer KH, Kost G, Sarbhoy A, Bisen P, Butehorn B, Franken P (1998) Piriformospora indica, gen. et sp. nov., a new root-colonizing fungus. Mycologia 95:896–903

Varma A, Sherameti I, Tripathi S, Prasad R, Das A, Sharma M et al (2012) The symbiotic fungus *Piriformospora indica*: review. In: Hock B (ed) Fungal associations, The mycota IX, 2nd edn. Springer, Berlin

Varma A, Sree KS, Arora M, Bajaj R, Prasad R, Kharkwal AC (2014) Functions of novel symbiotic fungus- *Piriformospora indica*. Proc Indian Natl Sci Acad 80(2):429–441

Viterbo A, Landau U, Kim S, Chernin L, Chet I (2010) Characterization of ACC deaminase from the biocontrol and plant growth-promoting agent *Trichoderma asperellum* T203. FEMS Microbiol Lett 305:42–48

Waller F, Achatz B, Baltruschat H, Fodor J, Becker K, Fischer M, Heier T, Hückelhoven R, Neumann C, von Wettstein D, Franken P, Kogel KH (2005) The endophytic fungus *Piriformospora indica* reprograms barley to salt-stress tolerance, disease resistance, and higher yield. Proc Natl Acad Sci 102:13386–13391

Waqas M, Khan AL, Kamran M, Hamayun M, Kang SM, Kim YH, Lee IJ (2012) Endophytic fungi produce gibberellins and indoleacetic acid and promotes host-plant growth during stress. Molecules 17:10754–10773

Wehner J, Powell JR, Muller LAH, Caruso T, Veresoglou SD, Hempel S, Rillig MC (2014) Determinants of root-associated fungal communities within Asteraceae in a semi-arid grassland. J Ecol 102:425–436

Weiss M, Selosse M-A, Rexer K-H, Urban A, Oberwinkler F (2004) Sebacinales: a hitherto overlooked cosm of heterobasidiomycetes with a broad mycorrhizal potential. Mycol Res 108:1003–1010

Weiß M, Waller F, Zuccaro A, Selosse MA (2016) Sebacinales – one thousand and one interactions with land plants. New Phytol 211:20–40

Xu L, Wu C, Oelmüller R, Zhang W (2018) Role of phytohormones in *Piriformospora indi*ca-induced growth promotion and stress tolerance in plants: more questions than answers. Front Microbiol 9:1646

Yadav V, Kumar M, Deep DK, Kumar H, Sharma R, Tripathi T, Tuteja N, Saxena AK, Johri AK (2010) A phosphate transporter from the root endophytic fungus *Piriformospora indica* plays a role in phosphate transport to the host plant. J Biol Chem 285:26532–26544

Chapter 7
Biological Control of Plant Diseases: Opportunities and Limitations

Akanksha Singh, Vipin Kumar Singh, Abhishek Kumar Dwivedy, Deepika, Shikha Tiwari, Awanindra Dwivedi, and Nawal Kishore Dubey

Abstract Plant diseases are important challenge to agriculture worldwide. Annually millions of tons of agricultural produce are lost due to the actions of plants pathogens. Past historical evidences are available showing the great mass migration and death of humans caused by the disease outbreak. Currently, several measures have been adopted to control the loss of crop productivity caused by fungal diseases. Physical and chemical approaches have gained huge success in managing the plant diseases, but being costly and toxic to natural environment in most of the cases, these are not preferred by the farmers. Moreover, the use of agrochemicals to control the plant pathogens has evoked the phenomenon of pest resistance and thus aggravating the seriousness of plant diseases and loss of crop productivity. To minimize the risks of synthetic chemicals, biological control measures have been introduced to control the fast multiplication of several plant diseases; however, under natural environmental conditions, their efficiency is very much affected. Plant systems have evolved several mechanisms to deal with the encountered pathogens. Enhancing the plant immunity against diseases caused by important plant pathogens by identifying and introducing the genes promoting the diseases resistance may serve as a good option in near future to control the plant disease for human welfare.

7.1 Introduction

Plant diseases have always been a challenge to plant growth and crop production in several parts of the world. Plant diseases can affect plants productivity by interfering with several processes such as the absorbance and translocation of water and nutrients, photosynthesis, flower and fruit development, and plant growth and

A. Singh · V. K. Singh · A. K. Dwivedy · Deepika · S. Tiwari · N. K. Dubey (✉)
Centre of Advanced Study in Botany, Institute of Science, Banaras Hindu University, Varanasi, India
e-mail: nkdubey@bhu.ac.in

A. Dwivedi
National Centre for Disease Control, New Delhi, India

A. Varma et al. (eds.), *Plant Microbiome Paradigm*,
https://doi.org/10.1007/978-3-030-50395-6_7

development. The infection of plants by pathogens can have serious consequences on plant health that further affects human health in several ways. Plant diseases are well-known to reduce the food availability to humans by ultimately interfering with crop yields. Some major outbreaks of plant diseases are the Irish Potato Famine (from 1845 to 1849) also known as the Great Hunger, caused by fungus *Phytophthora infestans* culminated into rapid disease spread throughout Ireland and ruined one-half of the potato crop during this period and about three-quarters of the crop over the next 7 years. Due to this famine, Ireland's 1845 population of eight million dropped to 5.5 million by 1860.

In India, Bengal famine (1945) occurred due to *Helminthosporium oryzae*, causing brown spot disease of rice, leading to severe yield loss of rice and caused death of two million people due to starvation (Fry 2012). There are several other plant diseases causing great losses to plant productivity but do not lead to severe human suffering. For instance, in 1970, in the USA, there was the occurrence of one of the most devastating epidemics "Southern corn leaf blight" caused by fungus *Helminthosporium maydis* which destroyed about 15% of the US corn crop causing loss of about $1 billion (Ullstrup 1972). In 1870, *Hemileia vastatrix* wiped out the coffee plantation of Sri Lanka, and coffee production was shifted toward the western hemisphere. From 1930 to 1970, there was induction of Dutch elm disease caused by *Ceratocystis ulmi* (*Ophiostoma ulmi*) which destroyed large number of elm trees in the USA causing great damage to the environment of that region (Sinclair and Campana 1978).

Plant diseases may cause both qualitative and quantitative losses of crops leading to reduced amount of crop ingredient, contamination of crops with toxic products of pests (mycotoxins), and diminished crop productivity. Fungal pathogens have been recorded to cause most devastating and universal crop diseases, causing largest crop losses per season. For example, rice blast disease caused by *Magnaporthe grisea* destroyed about 157 million tons of cultivated rice annually, which is enough quantity to feed 60 million people worldwide (Pennisi 2001). Out of US$1.3 trillion food production capacity worldwide, 31–40% (US$500 billion) losses occurred due to biotic stress caused by plant diseases, insects, and weeds. Crop losses due to pathogens are more severe in developing countries (cereals, 22%) in comparison with developed countries (cereals 6%) (Oerke et al. 1994). 6–20% (US$120 billion) crop losses also occur due to abiotic stresses like drought, flood, frost, nutrient deficiencies, etc.

Various methods have been be used to control plant diseases which can vary depending on the kind of pathogen, host, and the way of interaction between them as well as environmental conditions. While controlling plant disease, plants are generally treated as populations rather than as individuals because the damage or loss of one or a few plants is usually considered insignificant, although certain hosts (especially trees, ornamentals, and, sometimes, other virus-infected plants) may be treated individually (Kareem 2015). Therefore, control measures aim to save population rather than few individual of plants. Most serious diseases of crop plants appear on a few plants in an area year after year, spread rapidly, and are difficult to cure after their development. Therefore, almost all control methods are aimed to protect plants from becoming diseased rather than curing them after they become diseased. Plant disease control methods can be categorized in physical, chemical, and biological approaches.

The physical approaches most commonly used in controlling plant diseases are temperature (high or low), dry air, and various types of radiation. Similarly, plastic or net covering of row crops may protect the crop from infection by preventing pathogens or vectors from reaching the plants (Kareem 2015). Besides this, various types of chemicals or pesticides have also been in use to control plant diseases.

Although physical and chemical approaches help in controlling growth of plant pathogens, they also have some drawbacks. Excessively high or prolonged high temperatures treatment during soil sterilization may destroy some saprophytic flora and may also cause accumulation of toxic substances which inhibit further growth of plants (Kareem 2015). Chemical pesticides may cause some adverse effect on environment, on human health, as well as on other organisms. The ever-rising biological control methods are free from such limitations. In general, biological control refers to the use of living organisms also called as antagonistic microorganism (e.g., fungi or bacteria) to suppress the activities and population of one or more plant pathogen and thus reduces the occurrence of disease.

The present review throws lights on different type of biological control; plant-pathogen interaction; impact on physical and chemical method of plant protection; limitations of biological control; and future prospects.

7.2 Plant Immune System and Soil-Borne Plant Pathogens

During the course of evolution, plants have developed sophisticated machinery to deal with the multiple pathogen attack. The primary resistance to most of the pathogen attack is waxy cuticle deposition in different plant parts as well as the synthesis of biomolecules with antimicrobial nature.

Plants are endowed with a variety of intrinsic survival mechanisms. Most of the pathogenic bacteria and fungi get access to internal plant system through natural openings such as stomata and hydathodes. Alternatively, they may enter through lesions induced by mechanical injury in plants (Jones and Dangl 2006). However, the mechanisms adapted by soil-borne nematodes and aphids are quite different. They are equipped with so-called specially designed sharp penetration/piercing tool "stylet" in order to obtain the required nutrient materials. The interaction at this initial stage triggers the transfer of virulent factors into host system.

The immunity displayed by plant system is very much different to that of animal systems. Due to absence of locomotion, they cannot avoid attack of pathogens like the animals. The plant immune responses are the results of innate immunity harbored by the individual cells as well as biomolecular signals emerging from the point of pathogen attack (Chisholm et al. 2006). The interaction of pathogen with the plants results into development of defense mechanisms to neutralize the harmful impacts.

Basically, plant immunity can be categorized into two broad types. In one line of the defense system, gradually emanating microbe- or pathogen-associated molecular patterns (MAMPS or PAMPs) are perceived by the receptor system known as pattern recognition receptors (PRRs) situated in cell membrane and able to mark the patterns and respond accordingly. Such immunity has been designated as pattern-triggered immunity (PTI) (Boller and Felix 2009). Another line of plant immunity is relied upon the intracellular responses of NB-LRR (nucleotide-binding leucine-rich repeats) protein expressed under the strict control of number of plant resistance (R) genes. The NB-LRR proteins are effective in recognizing the multitude of effector molecules secreted by pathogens. Such kind of defense mechanisms has been reported to result into considerable protection from diseases resulting from the pathogens of biotrophic or semibiotrophic nature but not from pathogens exhibiting nectrotrophic nature (Glazebrook 2005). These effector-based responses have been referred as effector-triggered immunity (ETI). Under some conditions, effector molecules may weaken the immunity leading to survival and multiplication of pathogens followed by good chances of host susceptibility to a particular disease. Such responses mediated by effector molecules are termed as effector-triggered susceptibility (ETS).

Conformational changes happening in NB-LRR proteins during the different stages of plant-pathogen interactions are not known in much detail (Takken and Goverse 2012; Bonardi et al. 2012). Limited studies performed by self-assembling or oligomerization prior to and after the pathogen interface have contradictory results regarding their molecular arrangements (Eitas and Dangl 2010). The different changes observed in NB-LRR proteins during the pathogen interface can be described in terms of oligomerization, intramolecular reorganization, and dimer formation at N-terminal. Some studies have documented the role of one part of NB-LRR pair as “sensor” and another one as “helper” in effector-mediated host responses. The sensor part is essentially required to activate the NB-LRR in response to effectors delivered by pathogens, while helper section is known to contribute for proper functioning (Bonardi et al. 2011, 2012). However, the pair formed may also be of heteromeric nature and reveal the presence of even larger number of currently known list of such proteins of significance importance in plant immunity.

Rhizospheric region of plants is inhabited by a variety of soil microorganisms and their released substances (the microbiome) because of the presence of root exudates enriched with amino acids, carbohydrates, and organic compounds (Pieterse et al. 2016; Singh et al. 2019; Prasad et al. 2020). The composition and biological activities of microbiome are regulated significantly by soil physico-chemical and biological characteristics. The microbiome plays an important role in maintaining the root morphology, efficient acquisition of mineral nutrients, and prevention from entry of soil-borne pathogenic microbes.

Since the plant system identifies the rhizospheric microbes as non-self, they are eliminated from their vicinity as a result of innate immunity system. The association of advantageous microbe is facilitated by changes in plant immune responses and is the primary stair for mutualistic association. Lebeis et al. (2015) have recently demonstrated the potential of plant-produced compound salicylic acid in modifying the composition of endophytic microbes. The presence of specific microbial community inhabiting under the influence of root exudates is very much helpful to restrict the entry of several soil-borne pathogens to get access within plant. Such soil system supported with pathogen inhibitory characteristics attributed to microbiological activities generally emerging in singly growing crops under the conditions of intense disease occurrence has been designated as suppressive soil in recent literatures (Raaijmakers and Mazzola 2016; Schlatter et al. 2017). Suppressive soils represent the promising instance of root exudates supported diverse array of rhizospheric microbes against soil-borne pathogens as the very first line of plant protection strategy (Weller 2007). Presence of such rhizosphere dwelling antagonizing microbial community can have different outcomes. These microbial community may restrict the pathogen invasion at preliminary stage, suppresses pathogenicity after establishment with host and lead to little loss to agricultural productivity if able to cross the barrier or reduced potential ability to produce diseases at successive stages of cropping (Cook and Baker 1983; Weller et al. 2002). Recently, the potential inhibitory action of rhizospheric microbiome against the soil-borne pathogen, *Fusarium oxysporum*, has been well illustrated by Mendes et al. (2018). Analytical investigation of diverse microbes surviving in the rhizosphere of resistant plant revealed the presence of bacterial species belonging to families' Pseudomonadaceae and Bacillaceae. High-throughput analysis of metagenome in the rhizosphere vicinity demonstrated the abundance of genes involved in biosynthetic pathways governing the production of potent antimicrobials including phenazine and rhamnolipids. The soil suppression property can be classified as general and specific depending on the microbial types, effectiveness, and transferability from one field/soil to another field/soil. General suppression is the innate feature of numerous soil systems responsible for the inhibitory action over a broad range of soil-borne pathogens, non-transferrable, susceptible to abolition by soil heating and can be enhanced by the agricultural practices favoring the multiplication, biological activity, and diversity of soil microbiome (Cook 2014; Weller et al. 2002). The microbiome here in broader sense refers to the microbes along with their synthesized metabolic product such as sugars, amino acids, proteins, etc. present in soil environment. According to Cook (2014), general suppression can be considered as fire of microbes inflaming the root-derived metabolites and other nutrient substances essential to support the growth of multitude of soil-borne pathogens. The specific suppression is the gradually developing outcome of general suppression appearing over a certain period of regular cropping (Cook 2014). Unlike general suppression, specific suppression can be successfully developed in conducive soil through transfer of small quantity of soil or inoculation with individual microbial species or group of some preferred species. Interestingly, with increase in the amount of soil or inoculum density, there is no change in extent of specific suppressiveness indicating deviation from dose-dependent behavior.

7.3 Indirect and Direct Host Recognition of Pathogen Effectors

Innate immune systems of plants are able to recognize all external factors including different alliance and pathogens in order to defend themselves. Plant uses cell wall as an important primary barrier to limits the infection. It is one of the most active organelles of plant system with abundant sources of lignin, cellulose, and hemicelluloses that are resistant against enzymatic degradation (Dhingra et al. 2012) and may serve as a nutritional source for the microbiota (Willis 2016). When confrontation is established in between pathogen and host, pathogen must be able to modulate these defense systems by releasing a type of pathogen proteins called effectors (Toruño et al. 2016) in order to weaken the plant immunity. They release a number of enzymes as well as toxins to enter inside the host and make an etiological/parasitic relationship. Enzymes are also required during late stage of invasion of pathogen. For example, in consideration of liberation of monosaccharides and oligosaccharides which is basically required for proper growth and developmental phases of pathogen, they kill the plant tissue and cause pathogenicity. Some studies on *Neurospora crassa* and *Thermoascus aurantiacus* have been reported that they possess a lytic polysaccharide monooxygenases functioning in the presence of external electron source (Kubicek et al. 2014).

There are numbers of enzymes released by pathogens during course of infection; for example, cellulases which contain endoglucanases, exoglucanases, and β-glucosidases (Juturu and Wu 2014) play a potential role in hydrolysis of insoluble cellulose into simpler units. Cellulose is converted into an intermediate substances cellobiose and then into glucose by the key action of C1 and Cx enzyme subsequently (Wood 1960). Further, pectinase, a type of cell wall-degrading enzyme, is responsible for soft-rot diseases via tissue maceration and creating foul-smell (Charkowski 2018). Among various kind of pectinases, pectin lyases (also known as pectate transeliminases) are the most devastating one that degrade pectin polymers into 4,5-unsaturated oligogalacturonides through β-elimination mechanism (Yadav et al. 2009). It was first time reported in *Erwinia carotovora* and *Bacillus* culture medium. Recent report suggests that pectin lyases were not secreted by plant pathogen only but also in plant genome (*Arabidopsis*) for various developmental and physiological processes (Marín-Rodríguez et al. 2002). Besides that, cutinases are released under starvation condition responsible for degrading the cuticular surface of the host plant in order to make entrance inside. It is also reported that cutinases enhance the adhesion of the microbe's spores to the leaves (Epstein and Nicholson 2016).

The second way to induce pathogenesis is the production of phytotoxins. The phytotoxins are poisonous substances involved in development of diseases and are categorized as host-specific and non-host-specific. Host-specific toxin works at narrow range, i.e., target-selected plant cultivars, and causes physical and physiological changes including alteration in respiration, cell permeability, halted protein synthesis, and CO_2 synthesis; in contrast non-host-specific toxin works at broad spectrum, i.e., induce toxicity to many plants as well as animals (Yoder 1980; Pusztahelyi et al. 2015). Some of the important pathogen-derived toxins and diseases caused by them are listed in Table 8.1.

There is an essential link between the plants and environmental condition that favors the growth of plants. Upon the pathogen attack, numerous changes are displayed by the plants for their stability against adverse condition. Plants synthesize broad range of secondary metabolites as a natural protectant which not only helps in protection but also helps in accelerating the mechanism of primary metabolites facilitating in plant growth and development. In nature, diverse plant species are available producing toxigenic compounds having the ability to inhibit or halt the growth and proliferation of other organisms (Wittstock and Gershenzon 2002). The diagram representing effect of some plant-synthesized compound on important cellular processes is depicted in Fig. 8.1.

7.4 Signaling in Plant-Pathogen Interaction

The pathogens interacting with the host are recognized by well-orchestrated, coordinated, and developed plant receptor system responsible for the disease resistance. Generally, the association of host with pathogens triggers an array of strategic mechanisms designated as hypersensitivity reaction characterized by degeneration of infected cells, thus limiting the further propagation of pathogens (Heath 1998). One of the primary responses during hypersensitivity reactions is appeared in the form of generation of reactive oxygen species (ROS) and nitric oxide (NO) by the cellular enzymatic systems comprising of NADPH oxidase (Keller et al. 1998; Lamb and Dixon 1997) and nitric oxide synthase (Chandok et al. 2003), respectively. The synthesis and release of ROS and NO are not only essential for the degeneration of host cells but also required for the enhanced and coordinated actions of important immune programs mediated by salicylic acid production, induction of ion transport, modulation in extent of protein phosphorylation, maintenance of external pH, transmembrane potential, interaction between cell wall proteins, and changes in calcium ion concentrations (Shirasu et al. 1997; Kapoor et al. 2019). These processes induce the integrated signaling pathway culminating into the emergence of systemic acquired resistance (SAR) providing resistance against secondary infection arising due to either same or different kind of invading pathogens. Successful events of SAR are the important outcome of expression of different gene families collectively called as SAR genes, causing inhibition of pathogens multiplication and hence disease development (Cameron et al. 1994). Some of the important biological molecules mediated signaling pathways linked to plant defense response are discussed briefly.

Table 8.1 List of host-specific and non-specific toxins

S. No.	Pathogens	Related toxins	Chemical nature	Disease development	Targeted host	References
I. Host-specific/host-selective toxins						
1.	*Alternaria alternata*	AK toxin ACT toxin AF toxin	Epoxy-decatrienoic esters	Black spot Brown spot Black spot	Japanese pear Tangerine Strawberry	Wolpert et al. (2002)
2.	*Helminthosporium sacchari*	HS toxin	Heterodetic cyclic peptide	Leaf blight	Sugarcane	Walton (1996), Scheffer and Livingston (1984)
3.	*Cochiobolus heterostrophus*	T toxin	Polyketol	Southern leaf blight	Corn	Walton (1996)
4.	*Phyllosticta maydis*	PM toxin	Linear polyketols	Yellow leaf blight	Corn	Wolpert et al. (2002)
5.	*Periconia circinata*	Peritoxin/PC-toxin	Cyclized peptide and chlorinated polyketide	Root rot	Sorghum	Walton (1996), Wolpert et al. (2002)
6.	*Helminthosporium victoria*	Victorin	Chlorinated peptides	Blight diseases	Oats	Graniti (1991), Walton (1996)
II. Non host-specific/non-host-selective toxins						
7.	*Pseudomonas syringae* pv. *tabaci*	Tabtoxin	Dipeptide (threonine +tabtoxinine)	Wildfire diseases	Tobacco and others	Patil (1974), Graniti (1991)
8.	*Pseudomonas syringae* pv. *Phaseolicola*	Phaseolotoxin	Tripeptide toxin (ornithine +alanine+arginine)	Halo blight	Beans and legumes	Patil (1974)
9.	*Alternaria alternata*	Tentotoxin	Cyclic tetrapeptide	Chlorosis in seedling	Many plants	Gross (1985), Ballio (1991)
10.	*Cercospora* sp.	Cercosporin	–	Leaf spot and leaf blight	Many plants	Ballio (1991)

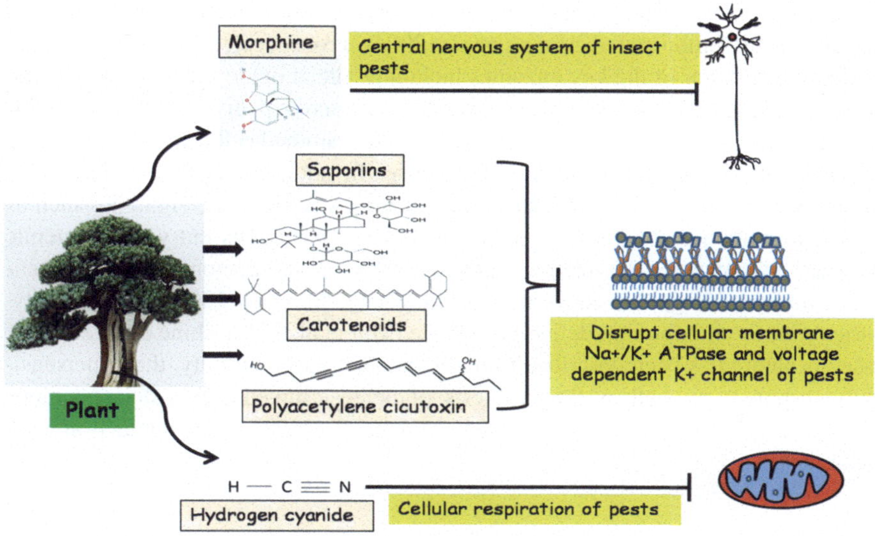

Fig. 8.1 Effect of plant-synthesized molecules on biological processes of pathogen

7.4.1 Nitric Oxide Mediated Signaling Pathway in Plant-Pathogen Interaction

Apart from hypersensitivity responses, experimental investigations have revealed the role of nitric oxide in inducing the signaling pathway corresponding to the synthesis of phytoalexin (Noritake et al. 1996). Furthermore, the induction of phenylalanine ammonia lyase (PAL) after treating the tobacco cells with the enzyme nitric oxide synthase (NOS) or substances releasing NO is also recognized (Durner et al. 1998). Similarly, enhancement of the biosynthesis of antimicrobials by enzymes equivalent to nitric oxide synthase has also been demonstrated. The inhibitor of NOS has been illustrated to considerably reduce the expression of enzyme chalcone synthase (Modolo et al. 2002) involved in biosynthesis of antibiotics derived from flavonoids and isoflavonoid (Dixon and Paiva 1995). It has been suggested that nitric oxide-dependent defense response works by involvement of cyclic GMP as the presence of guanylate cyclase partially retards the synthesis of PAL. Hence, the partial inhibition suggests the role of molecules other than nitric oxide in downstream signal transduction pathway. Studies on mitogen-activated protein kinases (MAPK) have indicated its contribution in plant defense response. Although, experimental evidences are available showing the nitric oxide-induced activation of MAPK (Clarke et al. 2000), its direct role in deciphering the association with defense response is still under preliminary investigation (Xu et al. 2018). Expression of defense-associated genes resulting from wound formation has been observed to diminish under the influence of nitric oxide (Orozco-Cárdenas and Ryan 2002). Furthermore, the inhibition of hydrogen peroxide formation by catalase and restricted synthesis of proteinase inhibitors working downstream of jasmonic acid has also been demonstrated. Huang et al. (2004) have revealed the effect of nitric oxide on activation of the key enzymes including allene oxide synthase (AOS) and lipoxygenase (LOX2) of jasmonic acid signal transduction pathway. Nitric oxide has been demonstrated to play an important role in programmed cell death (PCD) during pathogen attack via hypersensitivity response. In addition, evidences have also shown non-specific defense mechanism mediated by nitric oxide signaling such as papilla formation in plants in order to restrict the pathogens entry and systemic acquired resistance (SAR). Interestingly, studies have also demonstrated the potential of nitric oxide-guided signaling in pathogens (Mur et al. 2006). Conclusively, nitric oxide as an important defense molecule (Mur et al. 2013; Shine et al. 2018) of gaseous nature has been described to participate in plant immunity, the hypersensitivity response, as well as the synthesis of defense-linked cell wall appositions (Zeidler et al. 2004; Prats et al. 2005; Mur et al. 2008) upon interaction with pathogens.

7.4.2 Salicylic Acid Mediated Signaling Pathway in Plant-Pathogen Interaction

Resistance or susceptibility of a host toward a particular disease is the result of interaction between different molecules synthesized and released by plant and pathogens. Resistance in plants is achieved by induction of defense reactions inhibiting the entry and multiplication of pathogens within host system. In general, plant-pathogen interactions rely on the molecular communication between plant resistance (*R*) gene and avirulence (*Avr*) gene-encoded products. Absence of consanguineous genes in either of the interacting components results into failure or interrupted action of defense machinery. Salicylic acid has been identified as an important molecule of plant immunity (Huang et al. 2018) involved not only in signaling pathways conferring both systemic (Bhar et al. 2018) and local disease resistance but also in expression of pathogenesis-related (PR) proteins (Dempsey et al. 1999). Few evidences have been presented revealing the well-established and coordinated actions of salicylic acid with ethylene and jasmonic acid-derived molecule methyl jasmonate to induce the enhanced biosynthesis of pathogen-related protein upon challenge with pathogens (Xu et al. 1994; Lawton et al. 1994). However, experimental evidences on evolution of defense strategies not relying upon salicylic acid but depending on ethylene and jasmonic acid have also been reported (Pieterse et al. 1996; Penninckx et al. 1996). The signaling pathways based on salicylic acid is comprised of hydrogen peroxide scavenging enzymes such as catalase and ascorbate peroxidase, protein exhibiting strong affinity toward salicylic acid, proteins kinase synthesized under the influence of salicylic acid, NPR1 protein, and transcription factors of TGA/OBF family. Enhanced cytosolic content of calcium ions subsequent to pathogen interaction has been reported to induce the synthesis of phenylalanine ammonia lyase (PAL) which in turn promotes the formation of salicylic acid. SA-induced protein kinase (SIPK) transfers the signals to NPR1 (non-expresser of *PR* genes). The signals are finally perceived by PR-1 through the bZIP transcription factors of TGA/OBF family (Klessig et al. 2000). Apart from NPR1 system, NPR3/NPR4 receptor system has also been elucidated to be engaged in salicylic acid-mediated immune response following challenge with pathogen but in a different manner (Ding et al. 2018). While NPR1 functions as transcriptional inducer, NPR3/NPR4 serves as ligases and facilitates in degeneration of NPR1. Salicylic acid has been demonstrated to disrupt the functionality of NPR3/NPR4 receptor system and enhances the synthesis of downstream components of immune system. Mutation analysis in receptor system has revealed contrasting effects. Mutation in NPR1 receptor enhanced the salicylic acid directed immune response, but mutation in NPR3/NPR4 system hindered the immune response suggesting the opposite ways to regulate the expression of genes conferring resistance to pathogens.

7.4.3 *Jasmonic Acid Mediated Signaling Pathway in Plant-Pathogen Interaction*

Jasmonic acids and related biomolecules are lipid-derived signaling components produced by plants upon challenge with pathogens (Gfeller et al. 2010). Subsequent to biosynthesis, the important phytohormone jasmonic acid may be biologically transformed either to methylated form to produce methyl jasmonate form via the actions of jasmonic acid carboxyl methyltransferases (JMT) or linked with amino acid isoleucine with the action of JA conjugate synthase JAR1 (Staswick and Tiryaki 2004) to give rise to jasmonoyl-isoleucine (JA-Ile) (Fonseca et al. 2009). Two distinct pathways of jasmonic acid-mediated defense response are recognized in *Arabidopsis*. One of the signaling pathway designated as MYC is regulated by transcription factors including VSP2. Another signaling cascade involves the participation of transcription factors ERF1 and ORA59 and jasmonic acid-induced gene plant defensin (PDF1.2) to control the signaling reaction. Induction of ERF pathway upon challenge with pathogen is dependent on coordinated activity of jasmonic acid as well as ethylene; however, the molecular mechanisms deciphering the involvement of COI1/JAZ in directing this pathway are poorly recognized so far (Pieterse et al. 2012). The transcription factors EIN3 and EIL1 have been described to associate with JAZ proteins followed by binding with HDA6 serving as co-repressor to restrict the biological activity of EIN3/EIL1 in cells not exposed to pathogen attack. In contrast, the cells encountering the pathogens eliciting the production of jasmonic acid and ethylene exhibit the elevated transcriptional activity of EIN3/EIL1 subsequent to COI1 directed proteasomal degeneration of JAZ proteins. This cascade facilitates the induction of *ERF1* and probably *ORA59* genes too, along with other downstream genes including *PDF1.2*. Fundamentally, the ERF part of jasmonic acid signaling pathway is associated with defense against necrotrophs, while MYC part is activated upon lesions caused by insects and herbivores. The activated pathway initiated by insect-mediated lesion or herbivory is now triggered to another location of plant part to immunize the whole plant. The transcription factor MYC is reported to hinder the development without the involvement of jasmonic acid-mediated defense reaction. For instance, in root, MYC2 acts by suppressing the expression of AP2-domain PLETHORA (PLT), while in leaves, MYC inhibits the expression of genes associated with photosynthetic processes as well as retardation of cell multiplication and enlargement (Guo et al. 2018). The so-called chemical defense strategy against lesions involves the biochemical synthesis of hazardous biomolecules with repellent nature so that plant could avoid itself from upcoming further damages caused by insects and herbivores (Howe and Jander 2008). Recent studies have shown the impact of elevated CO_2 concentrations on plant defense mechanisms against loss resulting from insect feeding via alteration in jasmonic acid signal transduction pathway (Lu et al. 2018).

In addition to plant immunity against pathogens, growers around the world have adapted some strategies, viz., physical and chemical methods, since long in order to effectively control the pathogens growth and disease development.

7.5 Global Impact of Physical and Chemical Strategies of Plant Disease Control

Since agriculture began, cultivars have had to face harmful organism, viz., plant pathogens, animal pests, and weeds, imposing negative impacts on growth of plants. About 16% production loss of agriculture has been reported because of plant diseases caused by plant pathogens (viruses, bacteria, and fungi); they also alter the quality including reduction in crop production (Montesinos and Bardaji 2008). Thus, in order to control them, various physical (mechanical) and chemical (pesticides) strategies have been adapted by people over time.

Physical strategies adapted by growers are heat treatment which includes hot water treatment of plant propagules as well as hot air treatment, soil solarization, use of certain wavelength of light for controlling growth of plant pathogens, drying of stored grains and foods, as well as refrigeration (cold treatment/low temperature treatment). Soil-borne pathogens lead to heavy loss of agriculture crops. Repeated cultivation of same crop on same land increases the amount of pathogen inoculums; thus, soil solarization (solar heating of soil) reduces the inoculum density of associated pathogens (Katan 1981). It has been reported that temperature is necessary for killing of pathogens. For example, *Pythium* and *Phytophthora* inoculums can be easily eliminated from soil by heating at 45 °C for 30 min. However, some sclerotia-forming fungi, viz., *Sclerotium rolfsii*, need some higher temperature up to 54 °C for same period. However, some heat-tolerant pathogens have been reported which remains unaffected by heat treatment, viz., *Monosporascus antipodes* and *Macrophomina phaseolina* (Tjamos et al. 2013).

Upper layer soil sterilization either by aerated steam or by hot water treatment is done by passing steam through perforated pipes at a depth of 15 cm (Singh and Pandey 2012). At 50 °C, nematodes, some oomycetes, and other water molds get killed, whereas most plant pathogenic fungi and bacteria, along with some worms, slugs, and centipedes usually get killed at temperature 60–72 °C. At about 82 °C, most weeds, plant pathogenic bacteria, most plant viruses in plant debris, and most insects get killed. Some plant viruses, like tobacco mosaic virus (TMV), get killed at or near the boiling point, i.e., 95–100 °C. Heat sterilization of soil can also be achieved by heat produced electrically rather than supplied by steam or hot water. In process of controlling pathogen's growth, sometimes beneficial saprophytic flora may also be destroyed by excessive or prolonged high temperature heating leading to negative impact on plant growth (Kareem 2015). Further, Bhardwaj and Raj (2004) have reported that soil solarization for 40 days effectively controls the collar and root rot disease of strawberry caused by *Sclerotium rolfsii.* Bacterial canker disease of tomato (*Clavibacter michiganensis* subsp. *michiganensis*) has also been successfully managed by soil solarization for 1–2 months (Akhtar et al. 2008).

Additionally, hot water treatment of certain seeds, bulbs, and nursery stocks kills pathogens which may be present inside seed coats, bulb scales, or on external plant surfaces or in wounds. Hot water treatment of seeds helps in controlling loose smut disease of cereals. Similarly, treatment of bulbs and nursery stocks with hot water protects them from nematodes that may be present inside host plant. The temperature and duration of the hot water used for treatment may vary with the different host-pathogen combinations; for instance, to control the loose smut of wheat, the seed is kept in hot water at 52 °C for 11 min, whereas bulbs are treated for 3 h at 43 °C to control *D. dipsaci* (Kareem 2015). Further, controlled burning may also alter the environment: high temperature destructs plant pathogens and helps in controlling plant disease by heating effect (Zentmyer and Bald 1977). It was initiated to control the blind seed disease of perennial rye grass caused by *Gloeotinia temulenta*. It also effectively controlled *Claviceps purpurea* (ergot of rye) and *Anguina agrostis* (seed nematode) (Singh and Pandey 2012).

Chemical strategies include use of synthetic pesticides, viz., inorganic compounds such as copper, sulfur, and organic compounds. However, the first use of sulfur compounds as pesticides had been started in 2500–1500 BC. Chemical fungicides used for controlling pathogen's growth have been categorized in two groups on the basis of their nature. These fungicides have been classified as protectant fungicides and systemic fungicides. Protectant fungicide (inorganic and organic fungicides) protects plants from infection only at applied regions, whereas systemic fungicides penetrate and move from their site of application to other parts of the plants for their protection against pathogens. Bordeaux mixture (copper compound), the first generation fungicides, had been used in 1885 for the control of powdery and downy mildew. Moreover, in 1942, DDT and lindane discovery as insecticide gave rise to a new era in management of agriculture loss because of insects. These chlorinated hydrocarbon compounds are still in use in some areas; however, their use has been restricted as they are persistent in nature and have adverse effect on public health (Oerke 2006) and natural ecosystem. The total consumption of pesticides has been increased about 3.5 times in period of 1980–2011. In 2011, about 55% of synthetic pesticides have been sold in North America and Europe, while 45% of synthetic pesticides have been sold in Asia and rest of the world including South America. Because of great demand of synthetic pesticides, their global markets have been expected to reach 3.2 million tons in 2019 (Hajek and Eilenberg 2018).

Further, among organic chemicals, chloranil (quinone) fungicide was the first organic fungicide introduced in 1940 (Horsfall 1956). The quinone fungicide dithianon (a protectant fungicide) has been widely sprayed to control foliage diseases of fruits and ornamentals with less toxicity to plants and animals. Dithiocarbamates, the most important and versatile group of organic fungicides (McCallan 1967), have displayed toxicity toward fungi. Kareem (2015) have reported application of some dithiocarbamic acid derivatives, viz., Thiram, Ferbam, Nabam, Maneb, Zineb, and Mancozeb. Thiram, Ferbam, and Nabam have been used for controlling fungi like *Pythium*, *Rhizoctonia*, *Fusarium*, and *Protomyces*, whereas Maneb has been used for controlling anthracnose caused by *Colletotrichum lindemuthianum*, downy mildew, and fruit rot of chillies. Zineb, an excellent, safe, multipurpose foliar and soil fungicide, has been used to control leaf spots, blights, and fruit rots of vegetables.

Dicarboximide fungicide group (heterocyclic nitrogen compound) contains various fungicides like iprodione, procymidone, vinclozolin, and captan; basically, these fungicides have been used to control some sclerotia producing ascomycetes like *Botrytis*, *Sclerotinia*, and *Monilinia.* These fungicides are evidently known to disrupt mitotic processes and also destruction of fungal cell membrane integrity leading to inhibited conidial germination and mycelia growth (Ogle 2016).

Besides protectant fungicides, systemic fungicides show better efficacy in controlling pathogen's growth as they penetrate inside the host tissue and thus are less affected by weather (Ogle 2016). Carboxin and oxycarboxin are systemic fungicides derivatives primarily used in controlling basidiomycete group of fungi (Mathre 1970). Carboxin acts by blocking the transfer of electrons from succinic dehydrogenase (SDS) to coenzyme Q in the complex II region of the electron transport pathway (Mathre 1971). The specificity of carboxin for various organisms is usually determined by the sensitivity to SDS systems. Other systemic fungicides which have been used are benzimidazoles group. This group includes benomyl, carbendazim, thiabendazole, and thiophanate. They are effective against numerous types of diseases caused by a wide varieties of fungi. Benzimidazoles are effective against various fungal diseases as these compounds are converted into methyl benzimidazole carbamate (MBC, carbendazim) after their application which interferes with nuclear division of sensitive fungi; although these are ineffective against oomycetes and zygomycetes. According to Kareem (2015), oxanthiins was the first reported fungicide exhibiting systemic activity. Carboxin and oxycarboxin are effective against some smuts, rust fungi, and *Rhizoctonia.* Oxanthiins concentrated in cells of fungi are observed to inhibit succinic dehydrogenase enzyme and kill the target fungi by inhibiting mitochondrial respiration. Metalaxyl and furalaxyl are classified under Acyl alanine class of fungicides. These compounds possess excellent

Fig. 8.2 Summary of negative impacts of use of synthetic pesticides on human health and environment

controlling efficacy against various peronosporales (Sijpesteijn 1982) but are unable to control fungal pathogens belonging to ascomycetes, basidiomycetes, and deuteromycetes. Metalaxyl specifically inhibits RNA synthesis in *Pythium splendens* Braun (Kerkenaar 1981) and *Phytophthora megasperma* Drechs (Davidse et al. 1981). Although they serve as one of the important way to control the plant diseases, the huge applications of synthetic pesticides have several detrimental consequences. The negative impacts of pesticide application are shown in Fig. 8.2.

These synthetic pesticides showed significant crop protection against plant diseases even though they are not much acceptable because of residual toxicity and environmental pollution. According to a report of World Health Organization, about 25 million populations have suffered from pesticides poisoning including 20,000 unintentional deaths per year Jeyaratnam (1990). Now, growers have oriented toward a balanced use of synthetic pesticides with lower mammalian toxicity and reduced negative impact on environment. A number of synthetic pesticides have been banned as they promote the resistance development in the targeted pathogens and increase the environmental toxicity problems. For example, methyl bromide which was used for controlling soil-borne pathogens has been banned in most of countries in early 2000, because of its ozone depletion behavior (Hajek and Eilenberg 2018). Nowadays, there are strict rules on use of synthetic pesticides and high pressure to eliminate the hazardous synthetic pesticides from the markets. Considering the lesser effectiveness of physical approaches toward pathogen's growth and control as well as negative impact of chemical approaches over consumers' health and environment, present-day growers are showing their interests toward safe and effective control of pathogen's growth through biological control.

7.6 Biological Control

Biological control methods have long been accepted as attractive alternatives to physical and chemical control strategies for pest management, as it offers an eco-friendly and powerful alternative to the synthetic pesticides for controlling plant diseases (Isman 2006).

In plant pathology, the term "biological control" or "biocontrol" has oftenly been used for the use of antagonistic microorganisms to suppress plant disease; however, it does also include the use of host-specific pathogens in order to control weed populations. In short "biological control" consists of the involvement of living organisms to suppress the activity or survival of pathogens. Many workers have defined the term "biological control," but a commonly accepted definition is "The use of living organisms to suppress the population of a specific pest organism, making it less abundant or less damaging than it would otherwise be" (Eilenberg et al. 2001). The organisms that suppress the growth of pathogens are referred to as the biological control agents (BCAs). This unfriendly relation between BCAs and pathogens comes under antagonism. Further, antagonisms have been classified in mainly three categories, (a) amensalism, (b) competition, and (c) hyperparasitisim or exploitation (Pal and Gardener 2006). The antagonistic relation between living organisms and pathogen may occur, in either direct way or indirect way. Direct antagonism is outcome of physical contact which involves a high degree of specificity by the mechanisms expressed by the BCAs toward the pathogens. For example, hyperparasitism by obligate parasites of a plant pathogen is considered to be the most direct type of antagonism. In contrast to this, indirect antagonisms result from stimulation of defense pathways of host plant by nonpathogenic BCAs. Further, Iavicoli et al. (2003) have reported an antibiotic 2,4-diacetylphloroglucinol (DAPG) produced by *Pseudomonas* to be capable in induction of host defenses. In addition to this, DAPG producers can further help in suppression of pathogen's activity by making competition for organic nutrients in the rhizosphere of wheat, as they are capable to colonize roots (Raaijmakers et al. 2002). The figure showing the mechanisms and advantages of biological control is presented in Fig. 8.3.

Moreover, various strategies have been developed by BCAs for suppressing pathogens, among which antibiotic-mediated suppression of pathogens including hyperparasitism and predation is the major strategies. Most microorganisms produce some toxins which are inhibitory to the other microorganisms and have been classified as antibiotics. Some of the important antibiotics secreted by microbes and their targeted pathogens are summarized in Table 8.2.

Hyperparasitisms (parasitoids) are among the other strategies developed by the BCAs against growth of pathogens and have been widely used (Tougeron and Tena 2018). Trutmann et al. (1982) reported hyperparasitic activity of *Coniothyrium minitans* over *Sclerotinia sclerotiorum*, a sclerotia-forming plant pathogen. Other hyperparasitic relations have been seen in case of *Pythium oligandrum* over *Pythium ultimum* and *Sporidesmium sclerotivorum* over *Sclerotinia minor*, *Spharellopsis*

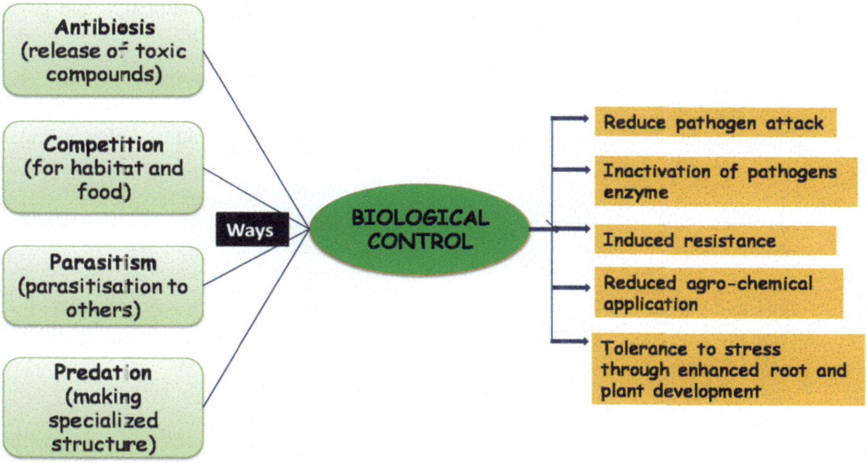

Fig. 8.3 Mechanisms and advantages of biological control

filum, and *Ampelomyces quisqualis* parasitizing on rust and mildew fungi (Rosenheim et al. 1995).

Microorganisms secrete various types of metabolites inhibiting growth as well as biological activities of plant pathogens, for example, microorganisms produce several lytic enzymes hydrolyzing a wide variety of polymeric compounds, including chitin, proteins, cellulose, hemicellulose, and DNA suppressing activities of plant pathogens directly. For example, *Trichoderma harzianum* controls *R. solani* and *S. rolfsii* by recognizing and attaching to the pathogenic fungus *R. solani* and *S. rolfsii* (Barak and Chet 1990) due to production of lectins (carbohydrate-binding proteins) and catalyzes the extracellular secretion of lytic enzymes such as β-1,3-glucanase, chitinase, protease, and lipase causing suppressed growth of pathogens. *Sclerotium rolfsii* is reported to be controlled by *Serratia marcescens* due to chitinase production (Ordentlich et al. 1988). Besides lytic enzymes, there are also various other microbial byproducts suppressing pathogen's growth; for example, *Pseudomonas fluorescens* CHA0 suppresses black rot disease of tobacco caused by *Thielaviopsis basicola*, primarily due to HCN production but also due to substantial production of antibiotics and siderophores (Voisard et al. 1989). *Enterobacter cloacae* produces ammonia which suppresses growth of *Pythium ultimum*, the causal organism of damping off of cotton (Howell et al. 1988).

Abundant nonpathogenic plant-associated microbes are known to protect the plants from various diseases by rapid colonization and thereby exhausting the limited available substrates so that none are available for pathogens to grow. Iron is extremely limited in the rhizosphere; however, its bioavailability is largely dependent on soil pH. The survival in iron-deficient environment is facilitated by secretion of iron-binding ligands called siderophores by numerous microorganisms (Das et al. 2007). Almost all microorganisms produce siderophores, which may be catechol type or hydroxamate type (Neilands 1981). Kloepper et al. (1980) have

Table 8.2 List of microorganisms used in biological control, associated antibiotics, and target pathogens

S. no.	Microorganisms	Antibiotic compounds	Targeted pathogens and associated disease	References
1	*P. fluorescens* Pf-5	Pyoluteorin Pyrrolnitrin	*Pythium ultimum* and *R. solani* Damping off	Howell and Stipanovic (1980)
2	*Agrobacterium radiobacter*	Agrocin 84	*Agrobacterium tumefaciens* Crown gall	Kerr (1980)
3	*Burkholderia cepacia*	Pyrrolnitrin Pseudane	*R. solani* and *Pyricularia oryzae* Damping off and rice blast	Homma et al. (1989)
4	*Pseudomonas cepacia* strain RB425	Pyrrolnitrin	*R. solani* and *F. oxysporum* Damping off	Yoshihisa et al. (1989)
5	*Bacillus cereus* UW85	Zwittermicin A	*Phytophthora medicaginis* and *P. aphanidermatum* Damping off	Smith et al. (1993)
6	*Bacillus subtilis* AU195	Bacillomycin D	*Aspergillus flavus* Aflatoxin Contaminations	Moyne et al. (2001)
7	*Trichoderma virens*	Gliotoxin	*Rhizoctonia solani* Root rot	Wilhite et al. (2001)
8	*Bacillus amyloliquefaciens* FZB42	Bacillomycin Fengycin	*Fusarium oxysporum* Wilt	Koumoutsi et al. (2004)
9	*Bacillus subtilis* BBG100	Mycosubtilin	*Pythium aphanidermatum* Damping off	Leclère et al. (2005)
10	*Lysobacter* sp. Strain SB-K88	Xanthobaccin A	*Aphanomyces cochlioides* Damping off	Islam et al. (2005)
11	*Streptomyces rimosus*	Oxytetracycline	*Xanthomonas arboricola* pv. *pruni* Bacterial spot	Stockwell and Duffy (2012)
12	*Streptomyces* sp. KNF2047	Neopeptin A and B	*Sphaerotheca fusca* Powdery mildew of cucumber	Kim et al. (2007)
13	*Serratia plymuthica* strain C-1 *Chromobacterium* sp. strain C-61 *Lysobacter enzymogenes* strain C-3	–	*Phytophthora capsici* Phytophthora blight of pepper	Kim et al. (2008)

(continued)

Table 8.2 (continued)

S. no.	Microorganisms	Antibiotic compounds	Targeted pathogens and associated disease	References
14	*Trichoderma harzianum*	Pyrone	*Gaeumannomyces graminis var. tritici*	Vinale et al. (2008)
15	*Lysobacter enzymogenes* C3 strain	–	*Fusarium graminearum* Fusarium head blight	Li et al. (2008a)
16	*S. malaysiensis*	Malayamycin	*Stagonospora nodorum* Blotch of wheat	Li et al. (2008b)
17	*Bacillus subtilis* CMB32 strain	–	*Colletotrichum gloeosporioides* Anthracnose disease of pepper, blueberry, and mango	Kim et al. (2010)
18	*S. cavourensis* subsp. *cavourensis* SY224	2-Furancarboxaldehyde	*Colletotrichum gloeosporioides* Anthracnose of pepper	Lee et al. (2012)
19	*P. fluorescens* and *P. aureofaciens* strain	Phenazine-1-carboxylate (Phz) 2,4-diacetylphloroglucinol (Phi) Phenazine-1-carboxylate	*Gaeumannomyces graminis* var. *tritici* Take-all disease of wheat	Hill et al. (2018)

demonstrated the importance of siderophore production as an effective mechanism of biological control of *Erwinia carotovora* by several plant growth promoting *Pseudomonas fluorescens* strains A1, BK1, TL3B1, and B10.

7.7 Limitations of Biological Control

Although biological controls of pathogens have shown significant role in management of plant diseases and emerged as a strong and rising alternate solution to the synthetic pesticides, there are some limitations with it. Since biocontrol involves introduction of non-native living organisms, serious ecological impacts may be associated with them. For example, non-native species may become invasive and may cause negative impacts over environment, as they may spread beyond the region of introduction (Jennings et al. 2017). Moreover, it is applicable at small scale, but the feasibility at large scale is still warranted. Although BCAs are genetically stable, nevertheless their uses have not received very much success because of continuous climate change. There are some BCAs showing their predatory behavior only in nutrient-limited conditions and not in normal growing conditions. For example, *Trichoderma* sp. does not attack directly on the *Rhizoctonia solani* when fresh bark compost is added. The reason behind this is the availability of cellulose, as at lower concentrations of cellulose, genes governing the synthesis of chitinase in *Trichoderma* sp. are activated and produce the enzyme for parasitic activity (Pal and Gardener 2006).

7.8 Opportunities for the Future

Potential microbial pest control agents (MPCAs) must be used for disease control after their laboratory, greenhouse, and field testing. MPCAs can be applied to the soil, seed, foliage, and harvested products to protect them from pathogens. Research must be done to enhance efficacy of biocontrol agents in terms of their survival ability in natural environment; ability to compete in plant rhizosphere; tolerance toward adverse environmental conditions like extreme moisture, temperature, and pH; and tolerance toward commonly used pesticides. Techniques like protoplast fusion, transformation, and gene cloning maybe considered to develop new strains of bacteria and industrial fungi for improved performance of biocontrol agents.

7.9 Conclusion

Plant-pathogen interaction and disease development have always been a big challenge. To protect themselves, plants have evolved several natural immune systems. In parallel with the course of evolution of plant's immune system, modes of pathogen attack and disease development have also evolved; hence plants health and crop production remain a concern globally. With time farmers have applied different strategies to control the pathogen's growth. Physical and chemical control strategies are important techniques for controlling devastating plant diseases, but they have several limitations too. Biological control strategies are much important as compared to physico-chemical methodologies, from environment protection point of view. Biological control methods employing antagonistic living organisms for effective control of pathogens have superiority in terms of absence of resistance development. Improving plant immunity by enhancing the cellular concentrations of important molecules offering resistance to diseases is another important way to manage the plant diseases. Most importantly, the suppression of plant diseases can also be practiced by identifying and introducing the genes conferring resistance to plant diseases. Thorough elucidation of complex signaling pathways during plant-pathogen interaction would be very much helpful in development of plant varieties resistant to number of plant diseases. Improving soil microbiome is also another important way to effectively encounter the challenges of serious plant diseases.

Acknowledgments Authors are thankful to University Grants Commission (UGC) New Delhi, Council of Scientific and Industrial Research (CSIR) New Delhi, Head, CAS in Botany, Banaras

Hindu University, Varanasi, DST-PURSE, and ISLS, Banaras Hindu University, Varanasi, India for providing necessary facilities.

References

Akhtar J, Khalid A, Kumar B (2008) Soil solarization: a non-chemical tool for plant protection. Green Farming 1:50–53

Ballio A (1991) Non-host-selective fungal phytotoxins: biochemical aspects of their mode of action. Experientia 47:783–790

Barak R, Chet I (1990) Lectin of *Sclerotium rolfsii*: its purification and possible function in fungal–fungal interaction. J Appl Microbiol 69:101–112

Bhar A, Chatterjee M, Gupta S, Das S (2018) Salicylic acid regulates systemic defense signaling in chickpea during *Fusarium oxysporum* f. sp. ciceri race 1 infection. Plant Mol Biol Report 36:162–175

Bhardwaj U, Raj H (2004) Mulching with transparent polyethylene and root dip in fungicides for the management of collar and root rot of strawberry. Indian Phytopathol 57(1):48–52

Boller T, Felix G (2009) A renaissance of elicitors: perception of microbe-associated molecular patterns and danger signals by pattern recognition receptors. Annu Rev Plant Biol 60:379–406

Bonardi V, Tang S, Stallmann A, Roberts M, Cherkis K, Dangl JL (2011) Expanded functions for a family of plant intracellular immune receptors beyond specific recognition of pathogen effectors. Proc Natl Acad Sci 108:16463–16468

Bonardi V, Cherkis K, Nishimura MT, Dangl JL (2012) A new eye on NLR proteins: focused on clarity or diffused by complexity? Curr Opin Immunol 24:41–50

Cameron RK, Dixon RA, Lamb CJ (1994) Biologically induced systemic acquired resistance in *Arabidopsis thaliana*. Plant J 5:715–725

Chandok MR, Ytterberg AJ, van Wijk KJ, Klessig DF (2003) The pathogen-inducible nitric oxide synthase (iNOS) in plants is a variant of the P protein of the glycine decarboxilase complex. Cell 113:469–482

Charkowski AO (2018) The changing face of bacterial soft-rot diseases. Annu Rev Phytopathol 56:269–288

Chisholm ST, Coaker G, Day B, Staskawicz BJ (2006) Host–microbe interactions: shaping the evolution of the plant immune response. Cell 124:803–814

Clarke A, Desikan R, Hurst RD, Hancock JT, Neill SJ (2000) NO way back: nitric oxide and programmed cell death in *Arabidopsis thaliana* suspension cultures. Plant J 24:667–677

Cook RJ (2014) Plant health management: pathogen suppressive soils. In: Encyclopedia of agriculture and food systems. Elsevier, pp 441–455

Cook RJ, Baker KF (1983) The nature and practice of biological control of plant pathogens. APS, St. Paul, p 539

Das A, Prasad R, Srivastava A, Giang PH, Bhatnagar K, Varma A (2007) Fungal siderophores: structure, functions and regulations. In: Varma A, Chincholkar SB (eds) Microbial siderophores, vol 12. Springer-Verlag, Berlin-Heidelberg, pp 1–42

Davidse LC, Gerritsma OCM, Hofman AE (1981) Mode d'action du metalaxyl. Phytiatrie-Phytofarmacie 30:235–244

Dempsey DMA, Shah J, Klessig DF (1999) Salicylic acid and disease resistance in plants. Crit Rev Plant Sci 18:547–575

Dhingra D, Michael M, Rajput H, Patil RT (2012) Dietary fibre in foods: a review. J Food Sci Technol 49(3):255–266

Ding Y, Sun T, Ao K, Peng Y, Zhang Y, Li X, Zhang Y (2018) Opposite roles of salicylic acid receptors NPR1 and NPR3/NPR4 in transcriptional regulation of plant immunity. Cell 173:1454–1467

Dixon RA, Paiva N (1995) Stress-induced phenylpropanoid metabolism. Plant Cell 7:1085–1097
Durner J, Wendehenne D, Klessig DF (1998) Defense gene induction in tobacco by nitric oxide, cyclic GMP, and cyclic ADPribose. Proc Natl Acad Sci USA 95:10328–10333
Eilenberg J, Hajek A, Lomer C (2001) Suggestions for unifying the terminology in biological control. BioControl 46:387–400
Eitas TK, Dangl JL (2010) NB-LRR proteins: pairs, pieces, perception, partners, and pathways. Curr Opin Plant Biol 13:472–477
Epstein L, Nicholson R (2016) Adhesion and adhesives of fungi and oomycetes. In: Smith AM (ed) Biological adhesives. Springer, Cham, pp 25–55
Fonseca S, Chini A, Hamberg M, Adie B, Porzel A, Kramell R, Miersch O, Wasternack C, Solano R (2009) (+)-7-*iso*-Jasmonoyl-L-isoleucine is the endogenous bioactive jasmonate. Nat Chem Biol 5:344–350
Fry WE (2012) Principles of plant disease management. Academic, New York, pp 1–11
Gfeller A, Dubugnon L, Liechti R, Farmer EE (2010) Jasmonate biochemical pathway. Sci Signal 3:cm3
Glazebrook J (2005) Contrasting mechanisms of defense against biotrophic and necrotrophic pathogens. Annu Rev Phytopathol 43:205–227
Graniti A (1991) Phytotoxins and their involvement in plant diseases. Introduction Exp 47:751–755
Gross GG (1985) Biosynthesis and metabolism of phenolic acids and monolignols. In: Higuchi T (ed) Biosynthesis and biodegradation of Wood components. Academic, New York, pp 229–271
Guo Q, Major IT, Howe GA (2018) Resolution of growth–defense conflict: mechanistic insights from jasmonate signaling. Curr Opin Plant Biol 44:72–81
Hajek AE, Eilenberg J (2018) Natural enemies: an introduction to biological control, 2nd edn. Cambridge University Press, pp 1–21
Heath MC (1998) Apoptosis, programmed cell death and the hypersensitive response. Eur J Plant Pathol 104:117–124
Hill S, Hammer PE, Ligon J (2018) The role of antifungal metabolites in biological control of plant disease. In: Gresshoff PM (ed) Technology transfer of plant biotechnology. CRC Press, Boca Raton, FL, pp 41–48
Homma Y, Kato Z, Hirayama F, Konno K, Shirahama H, Suzui T (1989) Production of antibiotics by *Pseudomonas cepacia* as an agent for biological control of soilborne plant pathogens. Soil Biol Biochem 21:723–728
Horsfall JG (1956) Principles of fungicidal action, vol 30. Chronica Botanica, Waltham, pp 2–40
Howe GA, Jander G (2008) Plant immunity to insect herbivores. Annu Rev Plant Biol 59:41–66
Howell CR, Stipanovic RD (1980) Suppression of *Pythium ultimum*-induced damping-off of cotton seedlings by Pseudomonas fluorescens and its antibiotic, pyoluteorin. Phytopathology 70:712–715
Howell CR, Beier RC, Stipanovic RD (1988) Production of ammonia by Enterobacter cloacae and its possible role in the biological control of *Pythium* preemergence damping-off by the bacterium. Phytopathology 78:1075–1078
Huang X, Stettmaier K, Michel C, Hutzler P, Mueller MJ, Durner J (2004) Nitric oxide is induced by wounding and influences jasmonic acid signaling in *Arabidopsis thaliana*. Planta 218:938–946
Huang X, Zhu GQ, Liu Q, Chen L, Li YJ, Hou BK (2018) Modulation of plant salicylic acid-associated immune responses via glycosylation of dihydroxybenzoic acids. Plant Physiol 176:01530
Iavicoli A, Boutet E, Buchala A, Métraux JP (2003) Induced systemic resistance in *Arabidopsis thaliana* in response to root inoculation with *Pseudomonas fluorescens* CHA0. Mol Plant-Microbe Interact 16:851–858
Islam MT, Hashidoko Y, Deora A, Ito T, Tahara S (2005) Suppression of damping-off disease in host plants by the rhizoplane bacterium Lysobacter sp. strain SB-K88 is linked to plant colonization and antibiosis against soilborne Peronosporomycetes. Appl Environ Microbiol 71:3786–3796

Isman MB (2006) Botanical insecticides, deterrents, and repellents in modern agriculture and an increasingly regulated world. Annu Rev Entomol 51:45–66

Jennings DE, Duan JJ, Follett PA (2017) Environmental impacts of arthropod biological control: an ecological perspective. In: Coll M, Wajnberg E (eds) Environmental pest management: challenges for agronomists, ecologists, economists and policymakers. Wiley, Chichester, p 105

Jeyaratnam J (1990) Acute pesticide poisoning: a major global health problem. World Health Stat Q 43(3):139–144

Jones JD, Dangl JL (2006) The plant immune system. Nature 444:323

Juturu V, Wu JC (2014) Microbial cellulases: engineering, production and applications. Renew Sust Energ Rev 33:188–203

Kapoor D, Singh S, Kumar V, Romero R, Prasad R, Singh J (2019) Antioxidant enzymes regulation in plants in reference to reactive oxygen species (ROS) and reactive nitrogen species (RNS). Plant Gene. https://doi.org/10.1016/j.plgene.2019.100182

Kareem A (2015) Management of plant diseases. Research gate, pp 4–42

Katan J (1981) Solar heating (solarization) of soil for control of soilborne pests. Annu Rev Phytopathol 19:211–236

Keller T, Damude HG, Werner D, Doerner P, Dixon RA, Lamb C (1998) A plant homolog of the neutrophil NADPH oxidase gp91phox subunit gene encodes a plasma membrane protein with Ca2+ binding motifs. Plant Cell 10:255–266

Kerkenaar A (1981) On the antifungal mode of action of metalaxyl, an inhibitor of nucleic acid synthesis in *Pythium splendens*. Pestic Biochem Physiol 16:1–13

Kerr A (1980) Biological control of crown gall through production of agrocin 84. Plant Dis 64:25–30

Kim YS, Kim HM, Chang C, Hwang IC, Oh H, Ahn JS, Kim KD, Hwang BK, Kim BS (2007) Biological evaluation of neopeptins isolated from a Streptomyces strain. Pest Manag Sci 63 (12):1208–1214

Kim YC, Jung H, Kim KY, Park SK (2008) An effective biocontrol bioformulation against Phytophthora blight of pepper using growth mixtures of combined chitinolytic bacteria under different field conditions. Eur J Plant Pathol 20(4):373–382

Kim PI, Ryu J, Kim YH, Chi YT (2010) Production of biosurfactant lipopeptides Iturin A, fengycin and surfactin A from Bacillus subtilis CMB32 for control of Colletotrichum gloeosporioides. J Microbiol Biotechnol 20(1):138–145

Klessig DF, Durner J, Noad R, Navarre DA, Wendehenne D, Kumar D, Zhou JM, Shah J, Zhang S, Kachroo P, Trifa Y (2000) Nitric oxide and salicylic acid signaling in plant defense. Proc Natl Acad Sci 97:8849–8855

Kloepper JW, Leong J, Teintze M, Schroth MN (1980) Pseudomonas siderophores: a mechanism explaining disease-suppressive soils. Curr Microbiol 4:317–320

Koumoutsi A, Chen XH, Henne A, Liesegang H, Hitzeroth G, Franke P, Vater J, Borriss R (2004) Structural and functional characterization of gene clusters directing nonribosomal synthesis of bioactive cyclic lipopeptides in *Bacillus amyloliquefaciens* strain FZB42. J Bacteriol 186:1084–1096

Kubicek CP, Starr TL, Glass NL (2014) Plant cell wall–degrading enzymes and their secretion in plant-pathogenic fungi. Annu Rev Phytopathol 52:427–451

Lamb C, Dixon RA (1997) The oxidative burst in plant disease resistance. Ann Rev Plant Physiol Plant Mol Biol 48:251–275

Lawton KA, Potter SL, Uknes S, Ryals J (1994) Acquired resistance signal transduction in Arabidopsis is ethylene independent. Plant Cell 6:581–588

Lebeis SL, Paredes SH, Lundberg DS, Breakfield N, Gehring J, McDonald M, Malfatti S, Del Rio TG, Jones CD, Tringe SG, Dangl JL (2015) Salicylic acid modulates colonization of the root microbiome by specific bacterial taxa. Science 349:860–864

Leclère V, Béchet M, Adam A, Guez JS, Wathelet B, Ongena M, Thonart P, Gancel F, Chollet-Imbert M, Jacques P (2005) Mycosubtilin overproduction by *Bacillus subtilis* BBG100

enhances the organism's antagonistic and biocontrol activities. Appl Environ Microbiol 71:4577–4584
Lee SY, Tindwa H, Lee YS, Naing KW, Hong SH, Nam Y, Kim KY (2012) Biocontrol of anthracnose in pepper using chitinase, beta-1, 3 glucanase, and 2-furancarboxaldehyde produced by Streptomyces cavourensis SY224. J Microbiol Biotechnol 22(10):1359–1366
Li S, Jochum CC, Yu F, Zaleta-Rivera K, Du L, Harris SD, Yuen GY (2008a) An antibiotic complex from Lysobacter enzymogenes strain C3: antimicrobial activity and role in plant disease control. *Phytopathology, 98*(6), 695–701.
Li W, Csukai M, Corran A, Crowley P, Solomon PS, Oliver RP (2008b) Malayamycin, a new streptomycete antifungal compound, specifically inhibits sporulation of Stagonospora nodorum (Berk) Castell and Germano, the cause of wheat glume blotch disease. *Pest Management Science: formerly Pesticide Science, 64*(12), 1294–1302.
Lu C, Qi J, Hettenhausen C, Lei Y, Zhang J, Zhang M, Zhang C, Song J, Li J, Cao G, Malook SU (2018) Elevated CO2 differentially affects tobacco and rice defense against lepidopteran larvae via the jasmonic acid signaling pathway. J Integr Plant Biol 60:412–431
Marín-Rodríguez MC, Orchard J, Seymour GB (2002) Pectate lyases, cell wall degradation and fruit softening. J Exp Bot 53:2115–2119
Mathre DE (1970) Mode of action of oxathiin systemic fungicides. I. Effect of carboxin and oxycarboxin on the general metabolism of several basidiomycetes. Phytopathol 60(4):671–676
Mathre DE (1971) Mode of action of oxathiin systemic fungicides: III. Effect on mitochondrial activities. Pestic Biochem Physiol 1:216–224
McCallan SEA (1967) History of fungicides. In: Torgeson DC (ed) Fungicides: an advanced treatise, vol 1. Academic, New York, pp 1–37
Mendes LW, Raaijmakers JM, de Hollander M, Mendes R, Tsai SM (2018) Influence of resistance breeding in common bean on rhizosphere microbiome composition and function. ISME J 12:212–224
Modolo LV, Cunha FQ, Braga MR, Salgado L (2002) Nitric oxide synthase-mediated phytoalexin accumulation in soybean cotyledons in response to the *Diaporthe phaseolorum* f. sp. meridionalis elicitor. Plant Physiol 130:1288–1297
Montesinos E, Bardaji E (2008) Synthetic antimicrobial peptides as agricultural pesticides for plant-disease control. Chem Biodivers 5:1225–1237
Moyne AL, Shelby R, Cleveland TE, Tuzun S (2001) Bacillomycin D: an iturin with antifungal activity against *Aspergillus flavus*. J Appl Microbiol 90:622–629
Mur LA, Carver TL, Prats E (2006) NO way to live; the various roles of nitric oxide in plant–pathogen interactions. J Exp Bot 57:489–505
Mur LA, Kenton P, Lloyd AJ, Ougham H, Prats E (2008) The hypersensitive response; the centenary is upon us but how much do we know? J Exp Bot 59:501–520
Mur LA, Prats E, Pierre S, Hall MA, Hebelstrup KH (2013) Integrating nitric oxideinto salicylic acid and jasmonic acid/ethylene plant defense pathways. Front Plant Sci 4:215
Neilands JB (1981) Microbial iron compounds. Annu Rev Biochem 50(1):715–731
Noritake T, Kawakita K, Doke N (1996) Nitric oxide induces phytoalexin accumulation in potato tuber tissues. Plant Cell Physiol 37:113–116
Oerke EC (2006) Crop losses to pests. J Agric Sci 144:31–43
Oerke EC, Dehne HW, Schönbeck F, Weber A (1994) Crop production and crop protection: estimated losses in major food and cash crops. Elsevier, Amsterdam
Ogle H (2016) Disease management: chemicals, pp 373–389
Ordentlich A, Elad Y, Chet I (1988) The role of chitinase of *Serratia marcescens* in biocontrol of *Sclerotium rolfsii*. Phytopathology 78:84–88
Orozco-Cárdenas ML, Ryan C (2002) Nitric oxide negatively modulates wound signaling in tomato plants. Plant Physiol 130:487–493
Pal KK, Gardener BM (2006) Biological control of plant pathogens. Plant Health Instruct 2:1117–1142
Patil SS (1974) Toxins produced by phytopathogenic bacteria. Annu Rev Phytopathol 12:259–279

Penninckx IA, Eggermont K, Terras FR, Thomma BP, De Samblanx GW, Buchala A, Métraux JP, Manners JM, Broekaert WF (1996) Pathogen-induced systemic activation of a plant defensin gene in Arabidopsis follows a salicylic acid-independent pathway. Plant Cell 8:2309–2323

Pennisi E (2001) The push to pit genomics against fungal pathogens. Science 292:2273–2274

Pieterse CM, Van Wees SC, Hoffland E, Van Pelt JA, Van Loon LC (1996) Systemic resistance in Arabidopsis induced by biocontrol bacteria is independent of salicylic acid accumulation and pathogenesis-related gene expression. Plant Cell 8:1225–1237

Pieterse CM, Van der Does D, Zamioudis C, Leon-Reyes A, Van Wees SC (2012) Hormonal modulation of plant immunity. Annu Rev Cell Dev Biol 28:489

Pieterse CM, de Jonge R, Berendsen RL (2016) The soil-borne supremacy. Trends Plant Sci 21:171–173

Prasad R, Chhabra S, Gill SS, Singh PK, Tuteja N (2020) The microbial symbionts: potential for the crop improvement in changing environments. In: Tuteja N, Tuteja R, Passricha N, Saifi SK (eds) Advancement in crop improvement techniques. Elsevier, Amsterdam, Netherlands, pp 233–240

Prats E, Mur LA, Sanderson R, Carver TL (2005) Nitric oxide contributes both to papilla-based resistance and the hypersensitive response in barley attacked by *Blumeria graminis* f. sp. hordei. Mol Plant Pathol 6:65–78

Pusztahelyi T, Holb IJ, Pócsi I (2015) Secondary metabolites in fungus-plant interactions. Front Plant Sci 6:573

Raaijmakers JM, Mazzola M (2016) Ecology. Soil immune responses. Science 352:1392–1393

Raaijmakers JM, Vlami M, De Souza JT (2002) Antibiotic production by bacterial biocontrol agents. Antonie Van Leeuwenhoek 81:537

Rosenheim JA, Kaya HK, Ehler LE, Marois JJ, Jaffee BA (1995) Intraguild predation among biological-control agents: theory and evidence. Biol Control 5:303–335

Scheffer RP, Livingston RS (1984) Host-selective toxins and their role in plant diseases. Science 223:17–21

Schlatter D, Kinkel L, Thomashow L, Weller D, Paulitz T (2017) Disease suppressive soils: new insights from the soil microbiome. Phytopathology 107:1284–1297

Shine MB, Xiao X, Kachroo P, Kachroo A (2018) Signaling mechanisms underlying systemic acquired resistance to microbial pathogens. Plant Sci 279:81. https://doi.org/10.1016/j.plantsci.2018.01.001

Shirasu K, Nakajima H, Rajasekhar VK, Dixon RA, Lamb C (1997) Salicylic acid potentiates an agonist-dependent gain control that amplifies pathogen signals in the activation of defense mechanisms. Plant Cell 9:261–270

Sijpesteijn AK (1982) Mechanism of action of fungicides. In: Fungicide resistance in crop protection. Centre for Agricultural Publishing and Documentation Wageningen, pp 32–45

Sinclair WA, Campana RJ (1978) Dutch elm disease: perspectives after 60 years. In: Sinclair WA, Campana RJ (eds) Search-agriculture. Cornell University, Ithaca

Singh VK, Pandey P (2012) Physical methods in management of plant diseases. In: Singh VK, Singh Y, Singh A (eds) Eco-friendly innovative approaches in plant disease management. International Book, New Delhi, pp 21–30

Singh D, Raina TK, Kumar A, Singh J, Prasad R (2019) Plant microbiome: a reservoir of novel genes and metabolites. Plant Gene. https://doi.org/10.1016/j.plgene.2019.100177

Smith KP, Havey MJ, Handelsman J (1993) Suppression of cottony leak of cucumber with *Bacillus cereus* strain UW85. Plant Dis 77:139–142

Staswick PE, Tiryaki I (2004) The oxylipin signal jasmonic acid is activated by an enzyme that conjugates it to isoleucine in Arabidopsis. Plant Cell 16:2117–2127

Stockwell VO, Duffy B (2012) Use of antibiotics in plant agriculture. Rev Sci Tech 31(1):199–210

Takken FL, Goverse A (2012) How to build a pathogen detector: structural basis of NB-LRR function. Curr Opin Plant Biol 15:375–384

Tjamos EC, Papavizas GC, Cook RJ (eds) (2013) Biological control of plant diseases: progress and challenges for the future, vol 230. Springer Science & Business Media, New York, pp 1–20

Toruño TY, Stergiopoulos I, Coaker G (2016) Plant-pathogen effectors: cellular probes interfering with plant defenses in spatial and temporal manners. Annu Rev Phytopathol 54:419–441

Tougeron K, Tena A (2018) Hyperparasitoids as new targets in biological control in a global change context. Biol Control 130:164. https://doi.org/10.1016/j.biocontrol.2018.09.003

Trutmann P, Keane PJ, Merriman PR (1982) Biological control of *Sclerotinia sclerotiorum* on aerial parts of plants by the hyperparasite *Coniothyrium minitans*. Trans Br Mycol Soc 78:521–529

Ullstrup AJ (1972) The impacts of the southern corn leaf blight epidemics of 1970-1971. Annu Rev Phytopathol 10:37–50

Vinale, F., Sivasithamparam, K., Ghisalberti, E. L., Marra, R., Woo, S. L., & Lorito, M. (2008). Trichoderma–plant–pathogen interactions. Soil Biology and Biochemistry, 40(1), 1–10

Voisard C, Keel C, Haas D, Dèfago G (1989) Cyanide production by *Pseudomonas fluorescens* helps suppress black root rot of tobacco under gnotobiotic conditions. EMBO J 8:351–358

Walton JD (1996) Host-selective toxins: agents of compatibility. Plant Cell 8:1723

Weller DM (2007) Pseudomonas biocontrol agents of soilborne pathogens: looking back over 30 years. Phytopathology 97:250–256

Weller DM, Raaijmakers JM, McSpadden Gardener BB, Thomashow LS (2002) Microbial populations responsible for specific suppressiveness to plant pathogens. Annu Rev Phytopathol 40:309–348

Wilhite SE, Lumsden RD, Straney DC (2001) Peptide synthetase gene in *Trichoderma virens*. Appl Environ Microbiol 67:5055–5062

Willis JD (2016) Modification of carbohydrate active enzymes in switchgrass (*Panicum virgatum* L.) to improve saccharification and biomass yields for biofuels

Wittstock U, Gershenzon J (2002) Constitutive plant toxins and their role in defense against herbivores and pathogens. Curr Opin Plant Biol 5:300–307

Wolpert TJ, Dunkle LD, Ciuffetti LM (2002) Host-selective toxins and avirulence determinants: what's in a name? Annu Rev Phytopathol 40:251–285

Wood RKS (1960) Pectic and cellulolytic enzymes in plant disease. Annu Rev Plant Physiol 11:299–322

Xu YI, Chang PFL, Liu D, Narasimhan ML, Raghothama KG, Hasegawa PM, Bressan RA (1994) Plant defense genes are synergistically induced by ethylene and methyl jasmonate. Plant Cell 6:1077–1085

Xu HY, Zhang C, Li ZC, Wang ZR, Jiang XX, Shi YF, Tian SN, Braun E, Mei Y, Qiu WL, Li S (2018) The MAPK kinase kinase GmMEKK1 regulates cell death and defense responses. Plant Physiol 178:907–922

Yadav S, Yadav PK, Yadav D, Yadav KDS (2009) Pectin lyase: a review. Process Biochem 44:1–10

Yoder OC (1980) Toxins in pathogenesis. Annu Rev Phytopathol 18:103–129

Yoshihisa H, Zenji S, Fukushi H, Katsuhiro K, Haruhisa S, Takahito S (1989) Production of antibiotics by *Pseudomonas cepacia* as an agent for biological control of soilborne plant pathogens. Soil Biol Biochem 21:723–728

Zeidler D, Zahringer U, Gerber I, Dubery I, Hartung T, Bors W (2004) Innate immunity in *Arabidopsis thaliana*: lipopolysaccharides activate nitric oxide synthase (NOS) and induce defense genes. Proc Natl Acad Sci USA 101:15811–15816

Zentmyer GA, Bald JG (1977) Management of the environment. Plant Dis 1:121–144

Chapter 8
Circadian Redox Rhythms Play an Important Role in Plant-Pathogen Interaction

Snigdha Tiwari, Siddarth N. Rahul, Alka Sehrawat, and Beena Rawat

Abstract Plants have improved robust mechanisms to recognize and respond to various environmental stimuli such as stress like biotic and abiotic stress. Biotic stress includes fungal, bacterial, and viral pathogens invasion, progression, and development of diseases. The plants have to encounter these microorganisms during 24-h a day. So as to protect themselves, plants have evolved numerous passive and active defence mechanisms to disrupt the progression of pathogens. The plant circadian clock stimulates daily rhythm in the activity of countless processes. These rhythms are harmonized to the diurnal day/night cycle by light as well as temperature. They control defence responses in plants against biotrophic and necrotrophic pathogens by modulating the expression of various genes of defence component pathways such as opening and closing of stomata (*CIRCADIAN CLOCK ASSOCIATED 1* (*CCA1*) and *LATE ELONGATED HYPOCOTYL* (*LHY*)), ROS and RNS, Myb factor, E3 ubiquitin protein ligase and ubiquitin-like protein, arachidonic acid, Isochorismate Synthase1 (ICS1), EDS1, COI, JAZ genes, etc. Additionally, redox state of a cell regulates defence components and circadian rhythm. Altogether these studies have revealed that circadian clock helps in regulating the pathways of defence essential for the plant survival.

S. Tiwari (✉) · B. Rawat
Department of Agriculture and Agri-Informatics, School of Biological Engineering and Life Sciences, Shobhit Institute of Engineering and Technology (Deemed to be University), Meerut, Uttar Pradesh, India
e-mail: snigdha.tiwari@shobhituniversity.ac.in

S. N. Rahul
Department of Plant Pathology, College of Agriculture NDUA&T, Ayodhya (Faizabad), Uttar Pradesh, India

A. Sehrawat
Department of Biotechnology, School of Biological Engineering and Life Sciences, Shobhit Institute of Engineering and Technology (Deemed to be University), Meerut, Uttar Pradesh, India

A. Varma et al. (eds.), *Plant Microbiome Paradigm*,
https://doi.org/10.1007/978-3-030-50395-6_8

8.1 Introduction

The term circadian (diurnal) was given by Franz Halberg in 1959, from two Latin words, i.e. "circa" (about) and "dies" (day). These rhythms are the subclass of biological rhythms which includes a period, comprise of whole one cycle of 24 h period (Dunlap et al. 2004). Another defining attribute of these rhythms is they are generated from within and self-sustaining, so circadian rhythm can endure under constant environmental conditions, i.e. constant light (or dark) and constant temperature. Under mentioned situations, the organism is dispossessed of exterior time cues and was observed for 24 h free-running period. Temperature compensation, a third attribute of entire circadian rhythms; the period vestiges relatively constant over a wide range of ambient temperatures (Pittendrigh 1954). This was thought to be one aspect of an overall circadian mechanism that shields the clock against any modifications in cellular metabolism.

The scientific literature was thought to be initiated when a French astronomer, de Mairan, observed leaf movements (on daily basis) in sensitive heliotropic plant (probably *Mimosa pudica*) sustained in persistent darkness, indicating their endogenous origin (de Mairan 1729).

It's almost took a century to accurately measured these leaf movements and was realized that almost an era passed before length period of these leaf movements was precisely stated and was observed to be of 24 h, which makes these rhythms on 24-h basis or so called circadian and suggested that these were due to internal causes and not just reactions to environmental time cues. In addition, studies conducted by de Candolle observed that in *M. pudica*, the free-running period was from 22 h to 23 h slightly shorter than 24 h and also discovered that by turning around the alternation of light and dark period, the circadian rhythm could be revert back. In plants, normal circadian rhythms are essential for biomass production and the capability to grow in varied climatic conditions (Böhlenius et al. 2006; Dodd et al. 2005; Kloosterman et al. 2013). However, disturbance in circadian clock rhythmic activity affects cellular health and the responses to various environmental stress factors.

Circadian rhythms and redox state of cell influence one another uninterruptedly, and various environmental stressors will probably influence both attributes. Healthy and completely functional plant cells reflect that cellular homeostasis was actively maintained and plants and environment were in a *dynamic equilibrium* (Strasser 1988). Under this equilibrium, relation between circadian rhythms and the redox state was operative, with a functional clock resonating with the environmental cycles (Guadagno et al. 2018). Clock mutant's use was observed to be crucial in detecting the reactive oxygen species (ROS) homeostasis regulated by circadian rhythm under various environmental conditions (Baxter et al. 2014; Greenham and McClung 2015). ROS production and the enzymatic scavenger's activity had revealed that both synchronously peak at certain daytimes (Lai et al. 2012). The redox state was directly under regulation of circadian clock at the cellular level thereby synchronizing with the time-based activity of many scavengers. Any fluctuations in environmental factors (one/many) might affect the coordination between the clock and plant redox state eventually manipulating various plant activities like growth, phenological stage, metabolism, etc. (Ahmad et al. 2008; Das and RoyChoudhury 2014; Gyöngyösi and Káldi 2014).

8.2 Circadian Clock and Plant Immunity

As plants are sessile in nature, they have to manage with many types of biotic and abiotic stresses in their environment and thus possess intricate, dynamic mechanisms to regulate their growth and development (Van Loon 2016). Abiotic stresses are several and include heat or cold, drought or submergence, lack of nutrients, salinity, osmotic stress, etc. Abiotic stress occurrence is temporally flexible over the annual or circadian cycle in most of the cases and, due to their immobile nature, has to face these stresses under different conditions (Suzuki et al. 2012). Biotic stress includes bacterial, fungal, viral, protozoan, nematode, etc. diseases. On contrary to animals, definite immune systems in plants are lacking. So they rely on the innate immunity of every cell as well as the occurrence of systemic signals at the site of infection (Schulze-Lefert and Panstruga 2011). However, with the plant specific characteristics of defence system, the molecular mechanisms adopted by these organisms are very much alike with the animals. Evidence regarding the association of circadian clocks with plant immunity in a timed upregulation has been studied (Bhardwaj et al. 2011; Goodspeed et al. 2012). In case of many abiotic and biotic stresses in plants, redox-based signalling was found to be involved (Spoel and Loake 2011). With the increase in world's population, there will be drastically rise in food demand globally, and this can be reduced due to stress conditions faced by plants. Taking this into consideration, research studies on plant health are critical to our improvement as a species, and the circadian clock is progressively found to regulate to a greater extent a healthy surroundings for growth and development of plants.

Plants are continually subjected to biotic stresses, and incidence of these stresses fluctuates over 24-h cycle. It was revealed that in *Arabidopsis*, maximum oomycetes activity (Hyaloperonospora) occurred during dusk time, i.e. 12 noon (ZT12) and greatest susceptibility to bacterial infection of *Pseudomonas* was noticed to be in the evening time around 4:00 pm (ZT16) (Spoel and Van Oijen 2014). Invasion of *Pseudomonas syringae* pv. tomato (Pst) DC3000 in leaf lamina is through stomatal opening. Activation of pattern-triggered immunity (PTI) prompts rapid closure of stomatal to prevent further progression of bacteria into the intercellular space of leaf. PTI-prompted stomatal closure involves components of both the salicylic acid (SA) and abscisic acid (ABA) signalling pathways, as these are shown to reveal circadian regulation, in association with ROS and RNS. In order to increase the progression and disease development in *Arabidopsis*, *P. syringae* pv. tomato (Pst) DC3000 embraced a virulence approach by producing coronatine (COR) which opens stomata, to enable bacterial progression at night. It was also observed that the Pst DC3118, a COR-defective mutant, was less active in the dark as compared to light conditions for infecting *Arabidopsis* (Panchal et al. 2016). In case of biotrophic bacterial pathogen infection, plants showed resistance controlled by the circadian clock (Bhardwaj et al. 2011). In *Arabidopsis*, CIRCADIAN CLOCK ASSOCIATED 1 (CCA1), the central oscillator component, was revealed to act as a positive integrator between defence pathways and the clock showing resistance against a pathogen (oomycete) (Wang et al. 2011b), but variation in host susceptibility to phytopathogens was a result of endogenously driven circadian rhythms which have not been revealed (Wang et al. 2011a; Griebel and Zeier 2008). The circadian clock provides crucial timing information. Light play a major time-setting mechanism (zeitgeber) in clock synchronization (Oakenfull and Davis 2017; Roden and Ingle 2009). The time of inoculation and leaf age were found to play an important role in non-host resistance. The causal agent of blast disease, *Pyricularia oryzae* rate of entry of into old leaves of *Arabidopsis* pen2-1 (Table 8.1), was observed and noted to be significantly higher as compared to young leaves after inoculation at dusk time, but there was hardly any difference at the time of dawn. Therefore, light and/or the circadian clock may play important roles in non-host resistance in *Arabidopsis* against *P. oryzae*.

During infection of *Phytopthora infestans* and related oomycetes in plant, germinating spores released arachidonate (Ricker and Bostock 1992), is the potent inducer of systemic resistance against pathogens (Bostock et al. 1981, 1986; Cohen et al. 1991; Coquoz et al. 1995; Fidantsef et al. 1999) also found to regulate the genes DEA1, a circadian clock-regulated protein of unknown function (Choi et al. 1992; Weyman et al. 2006) (Table 8.1). In case of tomato plants, the DEA1 gene get induced upon *Phytophthora infestans* infection and was found to be steadily expressed under long days but constitutively expressed under short days (Weyman et al. 2006). Research studies also revealed that few eukaryotic microbes, such as *Botrytis cinerea* and *Hyaloperonospora arabidopsidis*, are able to deploy the *Arabidopsis* circadian clock (Zhang et al. 2013).

Presence of stomata on the plant surface opens in the day and closes at night at regular intervals, a course which is controlled by the circadian rhythm in anticipation of humidity and light changes (Greenham and McClung 2015). Although this process is critical for water exchange and photosynthesis, some pathogens can use open stomata as gateways in order to access nutrients, and space within the plant tissue and pathogen invasion is restricted by close stomata. Further than frontline physical barriers, plants have progressive complex surveillance systems to sense intruders like pathogens and pests. When receptors on cell surface identify a pathogen, the plant instantaneously closes stomata at the site of invasion. Unregulated circadian rhythm damages closure of stomata, leading towards more intense disease (Zhang et al. 2013). A plant rapidly closes its stomata as soon as flagellins (pathogen-associated molecular patterns, PAMPs) are detected as a part of pattern-triggered immunity (PTI) (Bigeard et al. 2015; Melotto et al. 2006). Core morning clock genes like *LATE ELONGATED HYPOCOTYL* (*LHY*) and *CIRCADIAN CLOCK ASSOCIATED 1* (*CCA1*) (Table 8.1) had found to regulate stomatal immunity through their target *GLYCINE-RICH RNA-BINDING PROTEIN 7* (*GRP7*), also known as *COLD AND CIRCADIAN REGULATED 2* (*CCR2*) (Zhang

Table 8.1 List of genes regulated by circadian clock and involved in plant defence

S. no.	Name of gene	Time of expression (gene regulation)	Function	References
1.	LATE ELONGATED HYPOCOTYL (LHY) And CIRCADIAN CLOCK ASSOCIATED 1 (CCA1)	Expressed in morning	Provide resistance against *Pseudomonas syringae* and *Hyaloperonospora* sp.	Wang et al. (2011a)
2.	pen2–1	Expressed during dusk time	*P. oryzae*	
3.	*GLYCINE-RICH RNA-BINDING PROTEIN 7* (*GRP7*)		Enhances resistance against a necrotrophic bacterium *Pectobacterium carotovorum* SCC1 or a biotrophic virus tobacco mosaic virus.	Lee et al. (2012)
4.	*TIME FOR COFFEE* (*TIC-2*)	Expressed at night	*Pseudomonas syringae* (*Pst* DC3000)	Shin et al. (2012)
5.	Isochorismate Synthase1 (ICS1)	Expressed at night	Biotrophic pathogens like *Magnaporthe grisea*; powdery mildew and rust fungus	Bhardwaj et al. (2011), Miller et al. (2015), Wang et al. (2011b)
6.	CCI1, MYC2, and the JAZ genes	Expressed in morning	*Botrytis cinerea*	Hevia et al. (2015), Ingle et al. (2015)
7.	DEA1	Steadily expressed under long days but constitutively expressed under short days	*Phytophthora infestans*	Weyman et al. (2006)

et al. 2013). GRP7 is component of an external loop of the circadian rhythm (Table 8.1) (Heintzen et al. 1997), which in addition to being part of stomatal defence also attached to few transcripts of PAMP receptor and during infection increase the translation of any one of them (Nicaise et al. 2013). It also enhances resistance against *Pectobacterium carotovorum* SCC1, a necrotrophic bacterium or tobacco mosaic virus (biotrophic virus) (Lee et al. 2012). In addition, *TIME FOR COFFEE* (*TIC*), a night-expressed clock gene, is likewise important for stomatal opening through circadian oscillation and effective stomatal defence (Table 8.1) (Korneli et al. 2014). Several genes of *Arabidopsis thaliana* are involved in PTI, and at dawn their basal defence peaks and/or is induced more strongly (Zhou et al. 2015; Korneli et al. 2014; Melotto et al. 2006), when conditions favourable for pathogen invasion are present such as high humidity and opened stomata (Korneli et al. 2014;

Wang et al. 2011a, b). Therefore, plants showed highest resistance when infiltrated directly with bacteria, bypassing the stomata (Korneli et al. 2014; Melotto et al. 2006).

There are certain pathogens which have developed precise proteins, known as effectors, which are transported into plant cells to overcome PTI effects and thus enhance virulence. In order to counteract effectors, plants have intracellular nucleotide-binding and leucine-rich repeat (NB-LRR) immune receptors to identify these or their associated activities to prompt effector-triggered immunity (ETI) (Spoel and Dong 2012). ETI is additional extreme defence responses as compared to PTI, as it generally involved programmed cell death (PCD) of the diseased tissues. In case of infection of an oomycete pathogen, Emwa1, an effector produced by *Hyaloperonospora arabidopsidis* (Hpa), is recognized by RECOGNITION OF PERONOSPORA PARASITICA 4 (RPP4), the immune receptor gene. This gene is target of the CCA1 core clock component, and its levels are high during morning, when the probability of *H. arabidopsidis* infection is the maximum (Wang et al. 2011b). This provides a direct genetic relationship between the circadian clock and defence.

8.3 Components of Plant Defence Regulated by Circadian Rhythm

8.3.1 Transcription Factor

In plants, circadian regulation would benefit in modulating pertinent defence compounds when they encounter pathogens or herbivores. *Arabidopsis* transcriptome associated with defence-associated transcripts was also regulated by the circadian and diurnal regulation (Graf et al. 2010; Dong et al. 2011; Ryals et al. 1996; Ni et al. 2009). CIRCADIAN CLOCK ASSOCIATED1 (CCA1), a Myb domain transcription factor that had both transcript and protein morning-phased expression, has been revealed to regulate the expression of a many defence genes (Table 8.1) (Knight et al. 2008; Ryals et al. 1996) and binds to promoter gene sequences known as evening elements (EE) (McClung 2008) to control their expression (Harmer 2009). The reason behind the rhythmic transcription of defence genes may be due to co-localization in the genome for efficient gene regulation as suggested for immunity genes in *Drosophila* (Lu et al. 2009; Mizoguchi et al. 2002), or it may possibly due to functional coordination, to prime defence responses at certain times of day when infections are most likely. As per the studies conducted on Arabidopsis, where out of 3975 only 63 circadian controlled genes appeared as co-localization clusters within the genome (Graf et al. 2010), it appears that periodic transcription of defence genes is having more probability for functional co-ordination as compared because of co-localization (Zhang et al. 2013). The transcription factor can also regulate the expression of ROS-producing genes which might be also regulated by the circadian rhythm during plant pathogen interaction. In case of gene isolated from oilseed rape, *BnaNAC87* TF acts like a transcriptional activator along with a unique regulator of reactive oxygen species level and plant cell death. As this transcription factor (TF) can affect the expression of ROS-producing marker genes and enzymes responsible for ROS-scavenging as well as proteins involved in mediating defence in plants or leaf senescence (Yan et al. 2018).

Resistance against the downy mildew disease caused by biotrophic pathogen was found to largely depend on a gene clusters such as R gene RPP4, which deliberates full immunity to this strain of pathogen that initiates defence and ultimately programme cell death. The regulation of this gene was regulated by CCA1, the circadian regulator revealing increase expression at early morning and night, overlapping with the pathogen sporulation and spores dispersal. Therefore, artificial infection at dusk compared to dawn leads to CCA1-dependent increase of susceptibility to downy mildew infection (Wang et al. 2011a). Thus, it seems that plants time the expression of defence genes and R genes involved in programmed cell death to overlap with daytime at which pathogen threat is impending (Spoel and Van Oijen 2014)

8.3.2 Arachidonic Acid

In Solanaceae, this polyunsaturated fatty acid elicits programmed cell death and systemic defence responses via an α-linolenate/jasmonate-independent route presumably involving salicylic acid (Coquoz et al. 1995; Yu et al. 1997; Knight et al. 2001). Arachidonate induced DEA1, genes encoding a circadian rhythm-regulated protein of unidentified function, 3-hydroxy-3-methylglutaryl coenzyme A, specific forms of reductases and family 1 pathogenesis-related (PR) proteins while remaining uninduced by jasmonate or wounding (Choi et al. 1992; Fidantsef and Bostock 1998; Fidantsef et al. 1999; Rivard et al. 2004; Weyman et al. 2006).

Notably, leaves of tomato (*Solanum lycopersicum*) treated with AA showed reduced susceptibility to *Botrytis cinerea* infection, confirming AA as a potent elicitor in plants (Savchenko et al. 2010). For optimal elicitor activity, free carboxyl group is necessary (Preisig and Kuc 1985). Potato lipoxygenases (LOXs) improve the activity of AA. LOX catalyses the peroxidation of AA and plays roles in the rapid hypersensitive cell death elicited by AA (Vaughn and Lulai 1992).

8.3.3 Ubiquitin Protein

During stress conditions, production of ROS may cause oxidation of proteins. The protein goes through several types of direct or indirect alterations. Direct modifications involve various chemical alterations such as carboxylation, nitrosylation, disulphide bond formation, and glutathionylation. Protein carbonylation is commonly used as a marker for estimating protein oxidation (Moller et al. 2007). Indirect alterations of proteins can occur as a result of interaction with the products of LPO. When the ROS concentration crosses its threshold value, it leads to the site-specific modification of amino acids like Lys, Arg, Thr, Pro, and Trp and increased chances to undergo proteolytic degradation (Moller et al. 2007). The oxidized protein therefore becomes enhanced targets for digestion by proteolytic enzymes by getting preconditioned for ubiquitination-mediated proteasomal degradation (Das and RoyChoudhury 2014). E3 ubiquitin protein ligase and ubiquitin-like protein are ubiquitin pathway proteins which are one of the major protein turnover mechanisms that play important roles in controlling various cellular functions such as cell cycle, transcription development, circadian clocks, nutrient sensing, and signal transduction (Jonkers and Rep 2009).

8.3.4 Reactive Oxygen Species (ROS)/Reactive Nitrogen Species (RNS)

ROS and RNS are involved in programmed cell death on activation of ETI (Delledonne et al. 1998, 2001; Torres et al. 2002; Kangasjärvi and Kangasjärvi 2014), and circadian control of their production, scavenging, and gene targets was recently shown (Lai et al. 2012). During ETI, NADPH oxidase is responsible for ROS production, few of which depict circadian expression (Bhardwaj et al. 2011) and are associated with mammalian oxidases which are involved in respiratory burst in pathogen-activated phagocytes (Torres et al. 2002). Notably, NADPH oxidase AtRBOHD of *Arabidopsis* induced by ETI is itself regulated by RNS. RNS species generated on infection precisely *S*-nitrosylate AtRBOHD at Cys890 hinders binding of flavin adenine dinucleotide (FAD) and damages its capability to produce ROS. In accordance, if mutation of Cys890 occurs it rendered RBOHD unresponsive to RNS, resultant enhanced pathogen-prompted ROS production and programmed cell death (PCD) (Feechan et al. 2005). Moreover GSNOR1, a denitrosylase gene, governed pathogen-induced *S*-nitrosylation of AtRBOHD. Plants having mutant gene, gsnor1, accumulate increased levels of SNO proteins, display low activity of NADPH oxidase, and are extremely susceptible to pathogen infection (Feechan et al. 2005; Yun et al. 2011). The role of circadian rhythm was also found in regulating the expression of the Arabidopsis catalase (CAT) genes. When plants were placed in continuous light, oscillations in abundance mRNA of CAT2 and CAT3 were continued for several cycles. That robust oscillations in abundance mRNA continue with the circadian (Ç24 h) period for several cycles in plants depressed of external time cues and revealed that both CAT2 and CAT3 are regulated by an endogenous circadian clock (McClung 1997). Supposed H_2O_2 behaves like a signal transducer that is transmitting information related to external environment to the circadian pacemaker. Disruption of circadian oscillator may lead to susceptibility of oxidative stress, another suggested mechanism which links ROS and the circadian clock expression (Qian et al. 2010).

Forthcoming research should have emphasis on determining if ROS levels after a stress event may retune the periodicity of scavenger activity and affect clock gene expression. Moreover, it remains uncertain how the clock performs in mainly stressed plants which are near to mortality (Sanchez et al. 2011; Zhang et al. 2013; Resco de Dios and Gessler 2018). Taking this into consideration, a functional characterization of the effects of environmental noise on the primary oscillator is key to assimilating metabolic information, such as ROS dynamics, into current clock models (Einset et al. 2007; Miller et al. 2010; Zhang and Kay 2010; Obata and Fernie 2012; Haydon et al. 2013).

8.3.5 Phytohormones

Plant hormones are tuners of responses to biotic and abiotic stresses. They are involved in several complicated networks, through which they moderate responses to different stimuli. The hormones which primarily regulate plant defence against pathogens are jasmonic acid (JA), ethylene (ET), salicylic acid (SA), and abscisic acid (ABA) (Prasad et al. 2017). In case of biotrophic pathogen, effective defence against biotrophic pathogens is mainly due to PCD and activation of defence related to salicylic acid-dependent defence pathways which trigger SAR. On the contrary, in necrotrophs, a different set of defence responses are activated via jasmonic acid and ethylene signalling which activates induced systemic resistance (ISR) (Glazebrook 2005).

Circadian regulation of plant growth hormones has been revealed in daily rhythmic hormone accumulation and expression of genes related to hormone biosynthesis, signalling, and response. Some of these hormone genes are directly controlled by core clock proteins (Atamian and Harmer 2016). The circadian clock also gates plant responses to some growth hormones. It is still not that much clarified whether the circadian clock controls defence through regulating the rhythmicity of phytohormones. Recent studies, however, showed that the role of the circadian clock in defence is at least moderately coordinated through the regulation of the two defence hormones salicylic acid (SA) and jasmonic acid (JA) (Lu et al. 2017).

The regulation of SA pathway by circadian clock has been supported by different studies such as the basal SA levels that oscillate daily, with a peak at night (Goodspeed et al. 2012). In relation to this, expression of main genes affecting salicylic acid levels, including Isochorismate Synthase1 (ICS1), EDS1, EDS, and ACCELERATED CELL DEATH 6 (ACD6), also displays circadian oscillations (Bhardwaj et al. 2011; Miller et al. 2015; Wang et al. 2011a, b). While the transcript level of NPR1 (SA receptors) remains constant, NPR1 monomer accumulates rhythmically, with a peak at night (Miller et al. 2015; Zhou et al. 2015). Biosynthesis of salicylic acid was also shown to be directly under the control of circadian rhythm; the clock protein CHE binds to the promoter region of ICS1 (a major gene involved in SA biosynthesis) and affects the basal oscillation of ICS1 transcript and SA

(Table 8.1) (Zheng et al. 2015). In the case of PTI and ETI defence responses against pathogens infection, it has been widely reported that increased in SA levels are led by apoplastic H_2O_2 bursts mediated by NADPH oxidases and extracellular peroxidases (PRXs; Torres et al. 2002; Joo et al. 2005; Tsuda et al. 2008; O'Brien et al. 2012; Mammarella et al. 2015). Activation of these defence responses during plant-pathogen interaction by recognition of different pathogen molecules results in activating several signals which includes ROS and SA. Alterations in the timing and levels at which these signals are produced both in PTI and ETI regulate differences in the speed and strength at which these immune reactions are recognized to be operative in counteracting potential pathogens with little cost on fitness (Tsuda et al. 2008; Katagiri and Tsuda 2010). The associations of ROS with SA were proposed to mediate the establishment of SAR (Durrant and Dong 2004). ROS metabolism might also affect the expression of NPR1, a marker gene for SA as well as an important facilitator of these systemic responses, by adjusting redox state of NPR1 (Mou et al. 2003).

Jasmonic acids (JAs) are a group of lipid-derived molecules that play key roles in plant defence. Jasmonic acid and salicylic acid work antagonistically, i.e. commonly higher JA levels inhibit accumulation of SA, and its signalling thus favours resistance against necrotrophic microorganisms and most insect herbivores; on the contrary, increased SA levels also inhibit accumulation and signalling of JA and promote resistance against most biotrophic pathogens. It was demonstrated in several studies that the JA pathway is circadian regulated. The JA level oscillates during a day with a peak at midday (Goodspeed et al. 2013). Expression of few key JA biosynthetic genes is circadian regulated (Covington and Harmer 2007), including those directly targeted by CCA1 (Nagel et al. 2015). Expression of some core JA signalling genes, e.g. COI1, MYC2, and the JAZ genes (Table 8.1), also displayed circadian cycling that is reliant on the clock protein TIC (Shin et al. 2012). *Arabidopsis* showed resistance against *Botrytis cinerea*; the necrotrophic pathogen is also time-of-day dependent and involves several main clock genes and intact JA signalling (Hevia et al. 2015; Ingle et al. 2015). Moreover, *Arabidopsis* activates a sequence of various defence responses over the time course of invasion of *B. cinerea* (Windram et al. 2012). Together, these studies provide sustenance about the importance of the circadian clock and the JA pathway in plant defence against both herbivores and necrotrophic fungal pathogens. However, clock genes other than core clock genes (TIC) that activate plant defence through a direct control of either JA accumulation or JA signalling need to be discovered. It was stated that ROS produced by RbohF and RbohD enzymes are essential for jasmonic acid-induced gene expression controlled by MYC2 transcription factor basically involved in JA-mediated response where mutant plants of *RbohD* and *RbohF* treated with methyl jasmonate (MeJA) found to be unsuccessful in increasing the various expression levels of MYC2 downstream genes (Maruta et al. 2011). A vibrant collaboration between JA and ROS was known to control biosynthesis of lignin in response to damage of cell wall where ROS generated by RbohD and JA-isoleucine generated by JASMONIC ACID RESISTANT1 were observed to form a feedback loop having a negative nature that influences lignin accumulation. It was discovered that intracellular production of ROS in *cat2* mutant *Arabidopsis* plants leads to activate the JA pathway and its associated genes with accumulation of glutathione as an intermediate (Han et al. 2013).

PHOSPHATE TRANSPORTER 4;1 (*PHT4;1*) (negative regulator of salicylic acid-related plant defence) expression is controlled by light and the circadian clock. CCA1-binding site, two copies of have been located in the gene promoter sequence, which proposes a possible role of CCA1 both in controlling of *PHT4;1* expression as well as in plant defence responses (Wang et al. 2011a).

A study has also discovered the relation of the circadian clock with the gibberellin (GA) hormone (Arana et al. 2011). The study showed that regulation of GA receptor expression was done through the circadian clock, which leads to more stability of DELLA proteins during daytime and increased sensitivity of GA at night. This type of regulation was critical for periodic diurnal growth and for the rhythmic expression of clock-associated genes associated with abiotic and biotic stresses (McClung 2011). The gated GA sensitivity by the circadian rhythm incorporates a novel layer of regulation linking the circadian rhythm (clock) with the additional environmental and endogenous signals.

8.4 Conclusion and Future Prospects

The circadian rhythm regulates plant defence against a broad range of pathogens. The circadian clock affects both preformed and induced defence. This chapter focused on circadian regulation of stomatal defence, PTI, ETI, defence gene expression, and signalling pathways of defence facilitated by SA, JA, and ROS as well as how the redox state in a cell will going to affect the expression of the defence components which are playing vital role during plant-pathogen interaction. The regulation of ROS/RNS by the central oscillator/circadian rhythm will either directly or indirectly affect the plant response, i.e. resistance, tolerance, and susceptible during pathogen interaction. Further research in this area will bring insight into the rhythmic oscillations in ROS production during plant-pathogen interaction and its effect in regulating the components of signalling pathways both at the biochemical and molecular level. The modulation of defence components by both circadian rhythm and redox state of a cell will be an interesting area to induce resistance in plants during pathogen attack.

References

Ahmad P, Sarwat M, Sharma S (2008) Reactive oxygen species, antioxidants and signaling in plants. J Plant Biol 51:167–173

Arana M, Marín-de la Rosa N, Maloof J, Blázquez M, Alabadí D (2011) Circadian oscillation of gibberellin signaling in *Arabidopsis*. Proc Natl Acad Sci USA 108:9292–9297

Atamian HS, Harmer SL (2016) Circadian regulation of hormone signaling and plant physiology. Plant Mol Biol 91:691–702

Baxter A, Mittler R, Suzuki N (2014) ROS as key players in plant stress signalling. J Exp Bot 65:1229–1240

Bhardwaj V, Meier S, Petersen LN, Ingle RA, Roden LC (2011) Defence responses of Arabidopsis thaliana to infection by Pseudomonas syringae are regulated by the circadian clock. PLoS One 6:1–8

Bigeard J, Colcombet J, Hirt H, Zhang D (2015) Signaling mechanisms in pattern-triggered immunity (PTI) Mol. Plant 8(4):521–539

Böhlenius H, Huang T, Charbonnel-Campaa L, Brunner AM, Jansson S, Strauss SH, Nilsson O (2006) CO/FT regulatory module controls timing of flowering and seasonal growth cessation in trees. Science 312:1040–1043

Bostock RM, Kuć JA, Laine RA (1981) Eicosapentaenoic and arachidonic acids from Phytophthora infestans elicit fungitoxic sesquiterpenes in the potato. Science 212:67–69

Bostock RM, Schaeffer DA, Hammerschmidt R (1986) Comparison of elicitor activities of arachidonic-acid, fatty-acids and glucans from Phytophthora infestans in hypersensitivity expression in potato tuber. Physiol Mol Plant Pathol 29:349–360

Cohen Y, Gisi U, Mosinger E (1991) Systemic resistance of potato plants against Phytophthora infestans induced by unsaturated fatty acids. Physiol Mol Plant Pathol 38:255–263

Choi D, Ward BL, Bostock RM (1992) Differential induction and suppression of potato 3-hydroxy-3-methylglutaryl coenzyme A reductase genes in response to *Phytophthora infestans* and to its elicitor arachidonic acid. Plant Cell 4:1333–1344

Coquoz JL, Buchala AJ, Meuwly P, Métraux JP (1995) Arachidonic acid induces local but not systemic synthesis of salicylic acid and confers systemic resistance in potato plants to *Phytophthora infestans* and *Alternaria solani*. Phytopathology 85:1219–1224

Covington MF, Harmer SL (2007) The circadian clock regulates auxin signaling and responses in Arabidopsis. PLoS Biol 5:e222

Das K, RoyChoudhury A (2014) Reactive oxygen species (ROS) and response of antioxidants as ROS-scavengers during environmental stress in plants. Front Environ Sci 1–13

de Mairan J (1729) Observation botanique. Hist Acad Roy Sci:35–36

Delledonne M, Xia Y, Dixon RA, Lamb C (1998) Nitric oxide functions as a signal in plant disease resistance. Nature 394:585–588

Delledonne M, Zeier J, Marocco A, Lamb C (2001) Signal interactions between nitric oxide and reactive oxygen intermediates in the plant hypersensitive disease resistance response. Proc Natl Acad Sci USA 98:13454–13459

Dodd AN, Salathia N, Hall A, Kevei E, Toth R, Nagy F, Hibberd JM, Millar AJ, Webb AAR (2005) Plant circadian clocks increase photosynthesis, growth, survival, and competitive advantage. Science 309:630–633

Dong MA, Farre EM, Thomashow MF (2011) Circadian clock-associated 1 and late elongated hypocotyl regulate expression of the C-repeat binding factor (CBF) pathway in Arabidopsis. Proc Natl Acad Sci USA 108:7241–7246

Dunlap JC, Loros JJ, DeCoursey P (2004) Chronobiology: biological timekeeping. Sinauer Associates, Sunderland, MA

Durrant WE, Dong X (2004) Systemic acquired resistance. Annu Rev Phytopathol 42:185–209

Einset J, Winge P, Bones A (2007) ROS signaling pathways in chilling stress. Plant Signal Behav 2:365–367

Feechan A, Kwon E, Yun BW, Wang Y, Pallas JA, Loake GJ (2005) A central role for S-nitrosothiols in plant disease resistance. Proc Natl Acad Sci USA 102:8054–8059

Fidantsef AL, Bostock RM (1998) Characterization of potato tuber lipoxygenase cDNAs and lipoxygenase expression in potato tubers and leaves. Physiol Plant 102(2):257–271

Fidantsef AL, Stout MJ, Thaler JS, Duffey SS, Bostock RM (1999) Signal interactions in pathogen and insect attack: expression of lipoxygenase, proteinase inhibitor II, and pathogenesis-related protein P4 in the tomato, *Lycopersicon esculentum*. Physiol Mol Plant Pathol 54:97–114

Glazebrook J (2005) Contrasting mechanisms of defence against biotrophic and necrotrophic pathogens. Annu Rev Phytopathol 43:205–227
Goodspeed D, Chehab EW, Min-Venditti A, Braam J, Covington MF (2012) Arabidopsis synchronizes jasmonate-mediated defence with insect circadian behavior. PNAS 109:4674–4677
Goodspeed D, Liu JD, Chehab EW, Sheng Z, Francisco M et al (2013) Postharvest circadian entrainment enhances crop pest resistance and phytochemical cycling. Curr Biol 23:1235–1241
Graf A, Schlereth A, Stitt M, Smith AM (2010) Circadian control of carbohydrate availability for growth in Arabidopsis plants at night. Proc Natl Acad Sci USA 107:9458–9463
Greenham K, McClung CR (2015) Integrating circadian dynamics with physiological processes in plants. Nat Rev Genet 16:598–610
Griebel T, Zeier J (2008) Light regulation and daytime dependency of inducible plant defences in Arabidopsis: phytochrome signalling controls systemic acquired resistance rather than local defence. Plant Physiol 147:790–801
Guadagno CR, Ewers BE, Weinig C (2018) Circadian rhythms and redox state in plants: till stress do us Part. Front Plant Sci 9:1–9
Gyöngyösi N, Káldi K (2014) Interconnections of reactive oxygen species homeostasis and circadian rhythm in Neurospora crassa. Antioxid Redox Signal 20:3007–3023
Han Y, Chaouch S, Mhamdi A, Queval G, Zechmann B, Noctor G (2013) Functional analysis of Arabidopsis mutants points to novel roles for glutathione in coupling H2O2 to activation of salicylic acid accumulation and signaling. Antioxid Redox Signal 18:2106–2121
Harmer S (2009) The circadian system in higher plants. Annu Rev Plant Biol 60:357–377
Haydon MJ, Mielczarek O, Robertson FC, Hubbard KE, Webb AAR (2013) Photosynthetic entrainment of the *Arabidopsis thaliana* circadian clock. Nature 502:689–692
Heintzen C, Nater M, Apel K, Staiger D (1997) AtGRP7, a nuclear RNA-binding protein as a component of a circadian-regulated negative feedback loop in *Arabidopsis thaliana.* Proc Natl Acad Sci USA 94(16):8515–8520
Hevia MA, Canessa P, Muller-Esparza H, Larrondo LF (2015) A circadian oscillator in the fungus *Botrytis cinerea* regulates virulence when infecting *Arabidopsis thaliana*. PNAS 112:8744–8749
Ingle RA, Stoker C, Stone W, Adams N, Smith R et al (2015) Jasmonate signalling drives time-of-day differences in susceptibility of Arabidopsis to the fungal pathogen *Botrytis cinerea.* Plant J 84:937–948
Jonkers W, Rep M (2009) Lessons from fungal F-box proteins. Eukaryot Cell 8:677–695
Joo JH, Wang S, Chen JG, Jones AM, Fedoroff NV (2005) Different signaling and cell death roles of heterotrimeric G protein alpha and beta subunits in the Arabidopsis oxidative stress response to ozone. Plant Cell 17:957–970
Kangasjärvi S, Kangasjärvi J (2014) Towards understanding extracellular ROS sensory and signaling systems in plants. Adv Bot 538946
Katagiri F, Tsuda K (2010) Understanding the plant immune system. Mol Plant Microbe Interact 23:1531–1536
Kloosterman B, Abelenda JA, Carretero-Gomez M, Oortwijn M, De Boer JM, Kowitwanich K et al (2013) Naturally occurring allele diversity allows potato cultivation in northern latitudes. Nature 495:246–250
Knight H, Thomson AJW, McWatters HG (2008) Sensitive to freezing6 integrates cellular and environmental inputs to the plant circadian clock. Plant Physiol 148:293–303
Knight VI, Wang H, Lincoln JE, Lulai EC, Gilchrist DG, Bostock RM (2001) Hydroperoxides of fatty acids induce programmed cell death in tomato protoplasts. Physiol Mol Plant Pathol 59:277–286
Korneli C, Danisman S, Staiger D (2014) Differential control of pre-invasive and post-invasive antibacterial defence by the Arabidopsis circadian clock. Plant Cell Physiol 55(9):1613–1622
Lai AG, Doherty CJ, Mueller-Roeber B, Kay SA, Schippers JH, Dijkwel PP (2012) Circadian clock-associated 1 regulates ROS homeostasis and oxidative stress responses. Proc Natl Acad Sci USA 109:17129–17134

Lee HJ, Kim JS, Yoo SJ, Kang EY (2012) Different roles of glycine-rich RNA-binding protein7 in plant defence against *Pectobacterium carotovorum*, *Botrytis cinerea*, and tobacco mosaic viruses. Plant Physiol Biochem 60:46–52

Lu SX, Knowles SM, Andronis C, Ong MS, Tobin EM (2009) Circadian clock associated1 and late elongated hypocotyl function synergistically in the circadian clock of Arabidopsis. Plant Physiol 150:834–843

Lu H, McClung CR, Zhang C (2017) Tick tock: circadian regulation of plant innate immunity. Annu Rev Phytopathol 55:287–311

Mammarella ND, Cheng Z, Fu ZQ, Daudi A, Bolwell GP, Dong X, Ausubel FM (2015) Apoplastic peroxidases are required for salicylic acid-mediated defence against *Pseudomonas syringae*. Phytochemistry 112:110–121

Maruta T, Inoue T, Tamoi M, Yabuta Y, Yoshimura K, Ishikawa T et al (2011) Arabidopsis NADPH oxidases, Atrboh D and Atrboh F, are essential for jasmonic acid-induced expression of genes regulated by MYC2 transcription factor. Plant Sci 180:655–660

McClung CR (1997) Regulation of catalases in Arabidopsis. Free Radic Biol Med 23:489–496

McClung CR (2008) Comes a time. Curr Opin Plant Biol 11:514–520

McClung C (2011) The genetics of plant clocks. Adv Genet 74:105–113

Melotto M, Underwood W, Koczan J, Nomura K, He SY (2006) Plant stomata function in innate immunity against bacterial invasion. Cell 126(5):969–980

Miller G, Suzuki N, Ciftci-Yilmaz S, Mittler R (2010) Reactive oxygen species homeostasis and signaling during drought and salinity stresses. Plant Cell Environ 33:453–467

Miller M, Song Q, Shi X, Juenger TE, Chen ZJ (2015) Natural variation in timing of stress-responsive gene expression predicts heterosis in intraspecific hybrids of Arabidopsis. Nat Commun 6:7453

Mizoguchi T, Wheatley K, Hanzawa Y, Wright L, Mizoguchi M et al (2002) LHY and CCA1 are partially redundant genes required to maintain circadian rhythms in Arabidopsis. Dev Cell 2:629–641

Moller IM, Jensen PE, Hansson A (2007) Oxidative modifications to cellular components in plants. Annu Rev Plant Biol 58:459–481

Mou Z, Fan W, Dong X (2003) Inducers of plant systemic acquired resistance regulate NPR1 function through redox changes. Cell 113(7):935–944

Nagel DH, Doherty CJ, Pruneda-Paz JL, Schmitz RJ, Ecker JR, Kay SA (2015) Genome-wide identification of CCA1 targets uncovers an expanded clock network in Arabidopsis. PNAS 112: E4802–E4810

Ni Z, Kim ED, Ha M, Lackey E, Liu J et al (2009) Altered circadian rhythms regulate growth vigour in hybrids and allopolyploids. Nature 457:327–331

Nicaise V, Joe A, Jeong BR, Korneli C, Boutrot F, Westedt I, Staiger D, Alfano JR, Zipfel C (2013) Pseudomonas HopU1 modulates plant immune receptor levels by blocking the interaction of their mRNAs with GRP7. EMBO J 32(5):701–712

Oakenfull RJ, Davis SJ (2017) Shining a light on the Arabidopsis circadian clock. Plant Cell Environ 40(11):2571–2585

O'Brien JA, Daudi A, Finch P, Butt VS, Whitelegge JP, Souda P, Ausubel FM, Bolwell GP (2012) A peroxidase-dependent apoplastic oxidative burst in cultured Arabidopsis cells functions in MAMP-elicited defence. Plant Physiol 158(4):2013–2027

Obata T, Fernie AR (2012) The use of metabolomics to dissect plant responses to abiotic stress. Cell Mol Life Sci 69:3225–3243

Panchal S, Roy D, Chitrakar R, Price L, Breitbach ZS, Armstrong DW, Melotto M (2016) Coronatine facilitates *Pseudomonas syringae* infection of Arabidopsis leaves at night. Front Plant Sci 7:880

Pittendrigh CS (1954) On the temperature independence in the clock system controlling emergence time in Drosophila. Proc Natl Acad Sci USA 40:1018–1029

Prasad R, Gupta N, Satti U, Wang S, Ahmed ISA, Varma A (2017) Management of fungal pathogens by mycorrhiza. In: Varma A, Prasad R, Tuteja N (eds) Mycorrhiza. Springer International Publishing AG, Switzerland, pp 179–194

Preisig CL, Kuc JA (1985) Arachidonic acid-related elicitors of the hypersensitive response in potato and enhancement of their activities by glucans from *Phytophthora infestans* (Mont) de Bary. Arch Biochem Biophys 236:379–389

Qian H, Yu S, Sun Z, Xie X, Liu W, Fu Z (2010) Effects of copper sulfate, hydrogen peroxide and N-phenyl-2-naphthylamine on oxidative stress and the expression of genes involved photosynthesis and microcystin disposition in *Microcystis aeruginosa*. Aquat Toxicol 99:405–412

Resco de Dios V, Gessler A (2018) Circadian regulation of photosynthesis and transpiration from genes to ecosystems. Environ Exp Bot 152:37–48

Ricker KE, Bostock RM (1992) Evidence for release of the elicitor arachidonic acid and its metabolites from sporangia of *Phytophthora infestans* during infection of potato. Physiol Mol Plant Pathol 41:61–72

Rivard D, Cloutier C, Michaud D (2004) Colorado potato beetles show differential digestive compensatory responses to host plants expressing distinct sets of defence proteins. Arch Insect Biochem Physiol 55:114–123

Roden LC, Ingle RA (2009) Lights, rhythms, infection: the role of light and the circadian clock in determining the outcome of plant-pathogen interactions. Plant Cell 21:2546–2552

Ryals JA, Neuenschwander UH, Willits MG, Molina A, Steiner HY et al (1996) Systemic acquired resistance. Plant Cell 8:1809–1819

Sanchez A, Shin J, Davis SJ (2011) Abiotic stress and plant circadian clock. Plant Signal Behav 6:223–231

Savchenko T, Walley JW, Chehab EW, Xiao Y, Kaspi R, Pye MF, Mohamed ME, Lazarus CM, Bostock RM, Dehesh K (2010) Arachidonic acid: an evolutionarily conserved signaling molecule modulates plant stress signaling networks. Plant Cell 22:3193–3205

Schulze-Lefert P, Panstruga R (2011) A molecular evolutionary concept connecting nonhost resistance, pathogen host range, and pathogen speciation. Trends Plant Sci 16(3):117–125

Shin J, Heidrich K, Sanchez-Villarreal A, Parker JE, Davis SJ (2012) TIME FOR COFFEE represses accumulation of the MYC2 transcription factor to provide time-of-day regulation of jasmonate signaling in Arabidopsis. Plant Cell 24:2470–2482

Spoel SH, Dong X (2012) How do plants achieve immunity? Defence without specialized immune cells. Nat Rev Immunol 12(2):89–100

Spoel SH, Loake GJ (2011) Redox-based protein modifications: the missing link in plant immune signalling. Curr Opin Plant Biol 14:358–364

Spoel SH, Van Oijen G (2014) Circadian redox signaling in plant immunity and abiotic stress. Antioxid Redox Signal 20:3024–3039

Strasser RJ (1988) Applications of chlorophyll Fluorescence. In: Lichtenthaler HK (ed) A concept for stress and its application in remote sensing. Kluwer Academic Publishers, Dordrecht, pp 333–337

Suzuki N, Koussevitzky S, Mittler R, Miller G (2012) ROS and redox signaling in the response of plants to abiotic stress. Plant Cell Environ 35:259–270

Torres MA, Dangl JL, Jones JD (2002) Arabidopsis gp91phox homologues AtrbohD and AtrbohF are required for accumulation of reactive oxygen intermediates in the plant defence response. Proc Natl Acad Sci USA 99:517–522

Tsuda K, Sato M, Glazebrook J, Cohen JD, Katagiri F (2008) Interplay between MAMP-triggered and SA-mediated defence responses. Plant J 53:763–775

Van Loon LC (2016) The intelligent behavior of plants. Trends Plant Sci 21(4):289–294

Vaughn SF, Lulai EC (1992) Further evidence that lipoxygenase activity is required for arachidonic acid-elicited hypersensitivity in potato callus cultures. Plant Sci 84:91–98

Wang GY, Shi JL, Ng G, Battle SL, Zhang C, Lu H (2011a) Circadian clock-regulated phosphate transporter PHT4;1 plays an important role in Arabidopsis defence. Mol Plant 4:516–526

Wang W, Barnaby JY, Tada Y, Li H, Tor M, Caldelari D, Lee DU, Fu XD, Dong X (2011b) Timing of plant immune responses by a central circadian regulator. Nature 470:110–114

Weyman PD, Pan Z, Feng Q, Gilchrist DG, Bostock RM (2006) A circadian rhythm-regulated tomato genes is induced by arachidonic acid and *Phytophthora infestans* infection. Plant Physiol 140:235–248

Windram O, Madhou P, McHattie S, Hill C, Hickman R et al (2012) Arabidopsis defence against *Botrytis cinerea*: chronology and regulation deciphered by high-resolution temporal transcriptomic analysis. Plant Cell 24:3530–3557

Yan J, Tong T, Li X, Chen Q, Dai M, Niu F, Yang M, Yang MK, Jiang Y (2018) A novel NAC-type transcription factor, NAC87, from oilseed rape modulates reactive oxygen species accumulation and cell death. Plant Cell Physiol 59(2):290–303

Yu D, Liu Y, Fan B, Klessig DF, Chen Z (1997) Is the high basal level of salicylic acid important for disease resistance in potato? Plant Physiol 115:343–349

Yun BW, Feechan A, Yin M, Saidi NB, Le Bihan T, Yu M, Moore JW, Kang JG, Kwon E, Spoel SH, Pallas JA, Loake GJ (2011) S-nitrosylation of NADPH oxidase regulates cell death in plant immunity. Nature 478:264–268

Zhang EE, Kay SA (2010) Clocks not winding down: unravelling circadian networks. Nat Rev Mol Cell Biol 11:764–776

Zhang C, Xie Q, Anderson RG, Ng G, Seitz NC, Peterson T et al (2013) Crosstalk between the circadian clock and innate immunity in Arabidopsis. PLoS Pathog 9(6):e1003370

Zheng XY, Zhou M, Yoo H, Pruneda-Paz JL, Spivey NW, Kay SA, Dong X (2015) Spatial and temporal regulation of biosynthesis of the plant immune signal salicylic acid. Proc Natl Acad Sci 112:9166–9173

Zhou M, Wang W, Karapetyan S, Mwimba M, Marques J, Buchler NE, Dong X (2015) Redox rhythm reinforces the circadian clock to gate immune response. Nature 523(7561):472–476

Chapter 9
Rhizospheric Microorganisms for the Remediation of Contaminants for Ecological Restoration

Ashita Rai, Jyoti Fulekar, and M. H. Fulekar

Abstract Rhizosphere defined by Hiltner (Gesellschaft 98:59–78, 1904) is the volume of soil that is influenced by the roots of plants, and according to Lynch, this term can be defined as the three units interacting: the plant, the soil and the microorganisms. The composition of rhizosphere structure is highly orientated by the type of plant, quantity and composition of root exudates and different root zones. Mycorrhiza is a Greek word for fungus and root. Mycorrhizal fungi form ample network of hyphae in the soil and provide microorganisms (bacteria, fungi and actinomycetes) to the surrounding root up to a depth of 1 mm and facilitates symbiotic association that develops rhizospheric environment. The present chapter deals with rhizosphere understanding, soil-microbial-plant interaction, significance and benefits for plant growth. The plant enzymes and microbial enzymes identified provide rhizodeposition that increases fertility of the soil and promotes plant growth, biomass as well as root-microbial exudates which develops rhizospheric ecology for the ecological restoration.

9.1 Introduction

Mutualistic association of bacteria, fungi and actinomycetes provides beneficial rhizosphere for plant growth by the enzymes secreted by microorganisms and host plant along 10 mm root area. Mycorrhizal association is identified by nutrient uptake and is mainly determined by the movement of carbon from plants to fungal

A. Rai
School of Environment and Sustainable Development, Central University of Gujarat, Gandhinagar, Gujarat, India

J. Fulekar
Center of Research for Development, Parul University, Vadodara, Gujarat, India

M. H. Fulekar (✉)
School of Environment and Sustainable Development, Central University of Gujarat, Gandhinagar, Gujarat, India

Center of Research for Development, Parul University, Vadodara, Gujarat, India

A. Varma et al. (eds.), *Plant Microbiome Paradigm*,
https://doi.org/10.1007/978-3-030-50395-6_9

counterpart and minerals, viz. phosphorus and/or nitrogen from fungal root to the plant (Singh and Fulekar 2017; Prasad et al. 2017). Mycorrhizal symbiosis triggers root bifurcation and increases absorptive surface of the roots thereby linking rhizospheric soil and host plant (Varma et al. 2020). Rhizospheric interactions of plant root with microorganisms are acutely complex yet significant in plant growth and soil health (Varma et al. 2017). Current advancement in microbiology has developed an understanding about the signals, pathways and mechanism of rhizospheric interaction.

Rhizosphere is a biologically functional zone of the soil where plant and microbial exudates provide supportive environment for plant and microbial growth (Prasad et al. 2020). Molecular biology developments also highlight rhizospheric microbial diversity. The research data suggests that rhizodeposits in the rhizosphere regulate microbial diversity nearby 1 mm zone of the root ensuring significant and defensive symbiosis for the uptake of essential nutrients and changes physicochemical characteristic of the rhizosphere. Thus, the studies on interaction between plant and microbial community of the rhizospheric soil are significant for understanding ecological processes, viz. cycling of nutrients and ecosystem functioning.

9.2 Mycorrhizal Morphology

Mycorrhizas are of two types: endomycorrhiza and ectomycorrhiza.

9.2.1 Endomycorrhiza (EM)

The fungus develops structures inside the cortex cells (Marschner 2012) as fungus-plant interface, the membranes of the fungus and the plant are in direct contact with each other. The EM is best known as arbuscular mycorrhiza (AM) and was formerly known as vesicular arbuscular mycorrhiza (VAM). Smith and Read (2008) reported AM fungi are the most abundant of EMs. The AM fungi are now classified in a separate phylum *Glomeromycota* which has precisely four genera (*Acaulospora*, *Gigaspora*, *Glomus*, *Sclerocystis*) and determined by the following features (Marschner 2012): arbuscules or hyphae within cortical cells and mycelium protruding the neighbouring soil. Arbuscular mycorrhiza either develops arbuscules or hyphae in the host plant (Dickson 2004).

9.2.2 Ectomycorrhiza

Ectomycorrhiza (ECM) are reported in northern hemisphere, especially in Pinaceae, Betulaceae, Fagaceae and Salicaceae (Marschner 2012). According to Smith and Read (2008), ECM occurs mainly on roots of woody plants and only occasionally on herbaceous and graminaceous perennial plants.

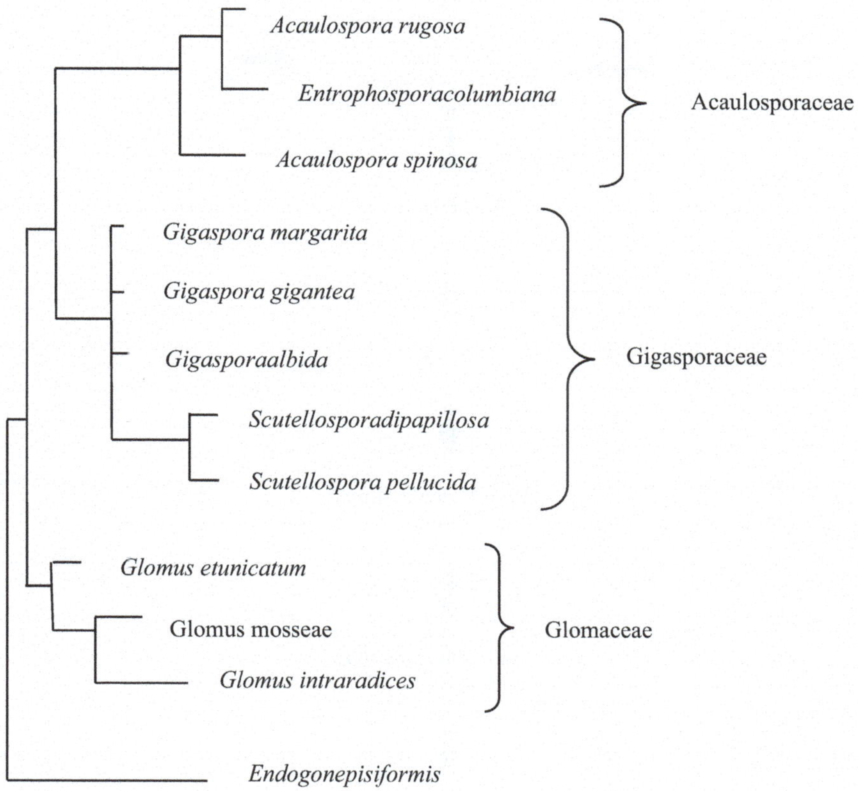

Fig. 9.1 Phylogenetic tree of arbuscular mycorrhiza 18S rDNA sequence (Source: Simon et al., 1993, reported in Letters in Nature)

ECM is characterised by (Marschner 2012) intertwined hyphae sheaths nearby roots; hyphal network that penetrates cortical cell and forms mycelial network (Hartig net) that surrounds the cortical cells thereby increasing root-fungus interface; and extensive hyphal (external) network. ECM fungi are Basidiomycetes and Ascomycetes (Marschner 2012) (Fig. 9.1).

9.3 Development of Mycorrhizal Soil (Method Developed by Fulekar et al.)

Pot culture techniques are being used for the development of mycorrhizal soil at R&D scale (Kharkwal et al. 2007). Arbuscular mycorrhiza is used as inoculum and grasses with fibrous roots, viz. sorghum as a host plant. Alluvial soil is mixed with in 3:1 ratio in a pot having perforations for proper infiltration and a provision for drainage. AM inoculum is properly into the soil, and sterilised seeds of the host plant are sown at a

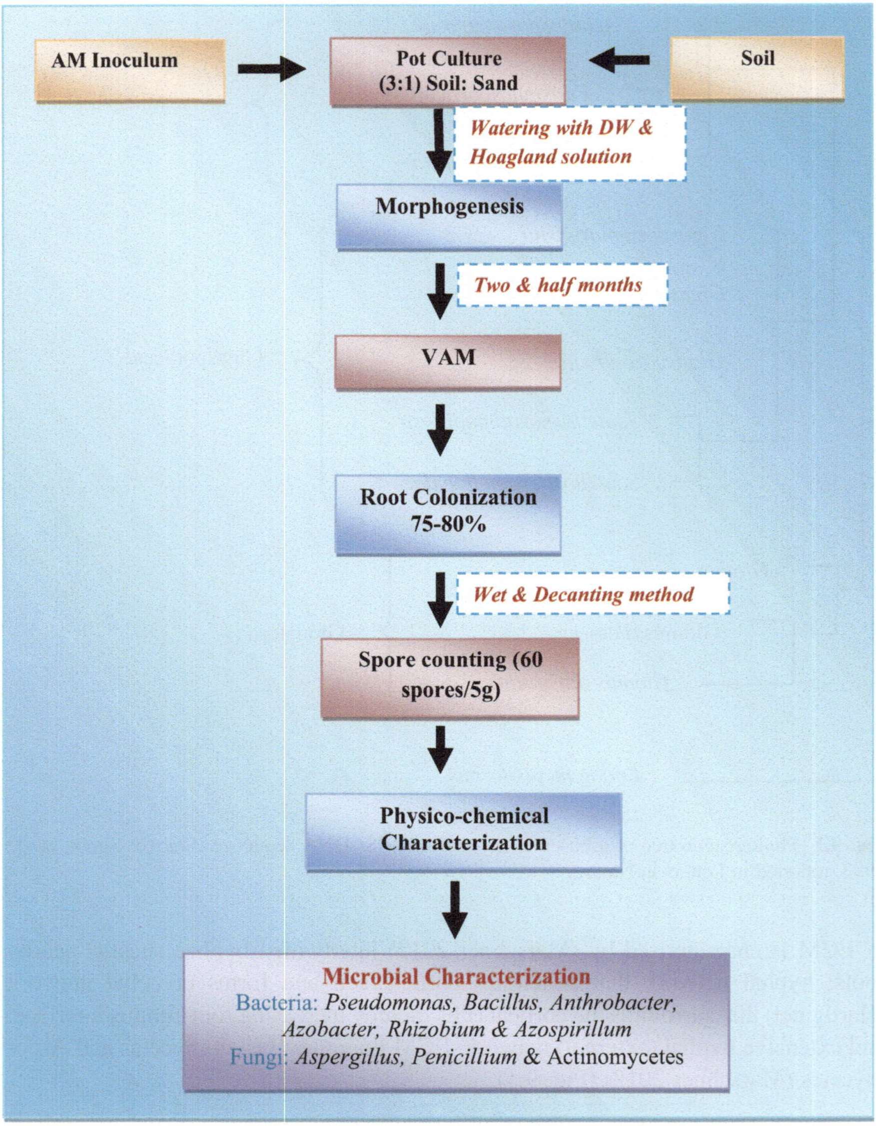

Fig. 9.2 Schematic diagram: method for the development of mycorrhizal soil

depth of 0.5 cm. The experiment is performed for 75 days in triplicate including control. The pots are kept at 27–28 °C in greenhouse and watered regularly to balance moisture level. After 15 days, AM are developed, and soil is assessed for physicochemical and microbial status at an interval of every 15 days up to 75 day. Root colonisation by AM is ensured using Trypan Blue method (Phillips and Hayman 1970), and spore counting is checked by wet sieving and decanting method (Gerdemann and Nicholson 1963). Colony-forming units of microorganism are counted using serial dilution method of developed soil suspension on the selective media (Fig. 9.2).

9.4 Estimation of Germination of Seed and Length of Root/Shoot

Seed germination is checked by counting the number of seedlings developed for a week at every 20 h interval. Host plant seedlings are considered germinated when the radical protrudes by at least 2 mm from the seed coat.

Seed germination percentage can be calculated using:

$$\text{Seed germination\%} = \frac{\text{Seedlings germinated}}{\text{Total number of seeds sown}} \times 100.$$

The experiment is performed for a period of two and half months. At an interval of every 15 days, host plant samples are randomly selected to measure the root/shoot length. Root length (root apex to root crown) and shoot length (crown of the host plant to shoot apex) are checked using Vernier callipers.

9.4.1 *Statistical Analysis*

Experiments are performed in triplicates. Mean and standard deviation (X ± S.D.) are calculated using standard statistical methods.

9.4.2 *Physicochemical Comparison Between Developed Mycorrhizal Soil and Alluvial*

S. no.	Parameters	Developed mycorrhizal soil	Alluvial soil	Method of determination
1.	pH	7.3	7.2	APHA, 1998
2.	Electrical conductivity (mMchs)	0.34	0.2	APHA,1998
3.	Moisture content (%)	42.2	35	APHA, 1998
4.	Water holding capacity (WHC) (%)	67	65	APHA,1998
5.	Organic carbon (gm/kg)	259	72	Walkley-Black method
6.	Nitrogen (gm/kg)	8.4	5.8	APHA, 1998
7.	Phosphorus (gm/kg)	0.81	0.72	APHA, 1998
8.	Sodium (mg/kg)	32	23	APHA, 1998
9.	Potassium (mg/kg)	22	21	APHA, 1998
10.	Heavy metals (ppm) Zinc Cadmium Lead	NA	10.5 BDL BDL	APHA, 1998

9.4.3 Mycorrhizal Soil Generally Consists of the Following Microbiota

Bacteria	Fungi	Actinomycetes
Alcaligenes sp.	*Aspergillus flavus*	*Micromonospora* sp.
Bacillus sp.	*Aspergillus fumigatus*	*Nocardia* sp.
Pseudomonas sp.	*Aspergillus niger*	
Sarcina sp.	*Penicillium* sp.	
Serratia sp.	*Rhizopus* sp.	
Streptococcus sp.	*Mucor* sp.	

9.5 Rhizosphere: Microhabitat for Microorganisms

Soil ecosystem is an important habitat for microorganisms associated along the root zones of the plant rhizosphere. In mycorrhizosphere, microbial community inhabits in the form of colonies together with the organic matter of the rhizosphere (Giri et al. 2005).

9.6 Soil-Plant-Microorganism Interaction

9.6.1 Plant-Bacteria Interaction

Bacteria interacts with the plant either mutualistically by facilitating the growth of the host or affects antagonistically by inhibiting growth using several mechanisms (Lazeano-Amora et al. 2010). Depending upon the soil microbe and host plant interactions, many bacteria support and facilitate the growth of the plant. Rhizospheric microorganisms interact with each other through simple surface attachment or obligate symbiosis plant, which is significant in plant growth, health and rhizospheric ecology (Perotto and Bonfante 1997) (Fig. 9.3).

9.6.2 Plant-Fungi Interaction

Arbuscular mycorrhiza is a diverse group of fungi linked symbiotically along the root zone of more than 90% of plant species (Bonfante and Genre 2010). Mycorrhiza is a Greek word for fungus and roots. Arbuscular mycorrhiza forms a huge network of hyphae in the rhizosphere and associated to the plant through symbiotic interfaces (Bonfante 2001; Parniske 2008). Hyphal network develops intercellularly, whereas arbuscules grow on the branches protruding the neighbouring cells. Plants

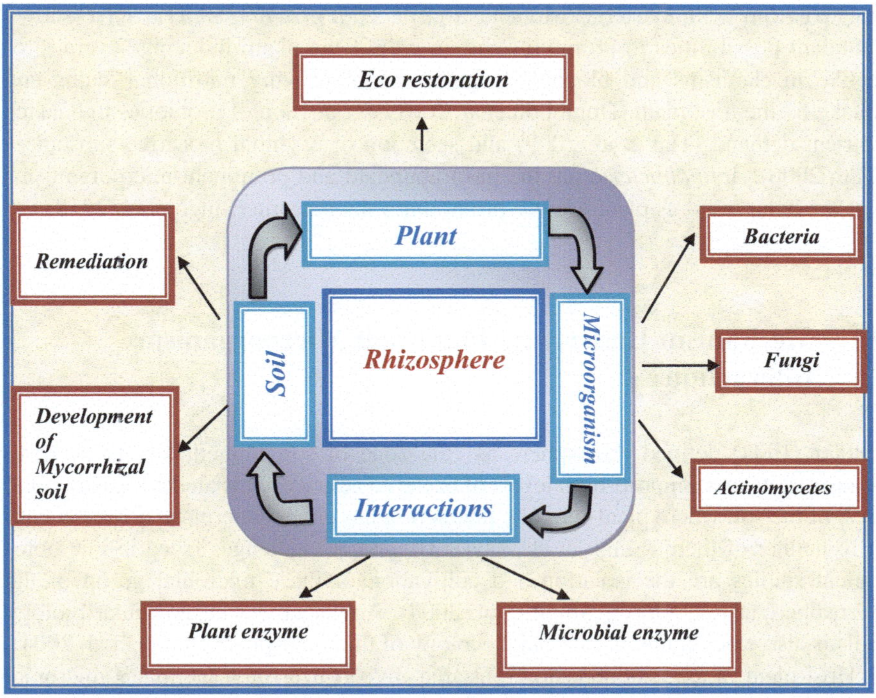

Fig. 9.3 Rhizosphere ecology for eco-restoration

symbiotically associated with AM show greater nutrient uptake, increased biomass of the plant and resistance to stress and pathogens (Smith and Read 2008). Arbuscular mycorrhiza plays an important role in improving the overall health of the host plant (Bonfante and Genre 2010).

9.6.3 Plant-Actinomycetes Interaction

Actinomycetes are the spore-forming and Gram-positive bacteria belonging to the order *Actinomycetales* (Bhatti et al. 2017). They intimately resemble fungi due to their branched and hyphae-type cells (Singh et al. 2018). The most common and prolific genus of actinobacteria is *Streptomyces.* The actinobacteria shows greater morphological diversity that is reflected in their habitat and cellular secretions. These prokaryotic microorganisms have characteristically higher guanine and cytosine constituent in their genomes with significant metabolic functionality. They cover approximately hundred genera with nearly thousand species inhabiting different categories of soil and are capable of recycling natural minerals and organic matter. Approximately more than 60% of essential bioactive microbial metabolites are isolated from actinobacteria (Moncheva et al. 2002). *Actinobacteria* have shown abundant possibilities in promoting the growth of the plant, inducing several protective mechanisms and biocontrol agents and improving nutritional status and uptake by the host plant (Doumbou et al. 2001), production of probiotics to balance nutrient demand (Tan et al. 2009) and secretion of essential bioactive substances (Lam 2006). *Actinobacteria* are the most abundant and prominent microorganisms that are being used as plant growth-promoting rhizobacteria (Singh et al. 2018).

9.7 Mechanism Underlying Plant-Soil-Microorganism Interaction

Hiltner (1904) defined rhizosphere as "the zone of soil immediately adjacent to legume roots that supports high levels of bacterial activity." It is an ecologically vital zone of the soil where plant-microbe interaction has either synergistic effect on both or to neither of them (Singh et al. 2004). The major challenges faced in microbiological studies are the isolation and cultivation of these microbial group of the rhizosphere at R&D level (Singh et al. 2004). Advancement in molecular biology tools is also elucidating on microbial ecology of the rhizosphere (Singh et al. 2004).

Host plant in association with arbuscular mycorrhiza fungi facilitates microbial diversity of the rhizosphere in the degradation of organic contaminants in the mycorrhizal soil for the ecological restoration (Korade and Fulekar 2009). This process is assisted by the secretion from the host plant like short-chain organic acids, phenolic group and small concentration of high molecular weight enzymes and proteins to facilitate bacterial enzyme induction, by building up organic carbon content to enhance microbial mineralisation rates or by providing habitat for the proliferation of microbial diversity (Korade and Fulekar 2009). Molecular biologists have identified five major plant enzymes secreted by the plant and associated rhizospheric microbial diversity. Dehalogenase plays a significant role in the dechlorination reactions of chlorinated hydrocarbons, nitro reductase is important in the first step of degradation of nitro aromatics, laccase serves to break down aromatic ring of organic pollutants, whereas peroxidise and nitrilase are important in oxidation reactions. These exudates and enzymes stimulate microbial activity and biochemical degradation of contaminants and increase the mineralisation rate.

Plant growth-promoting rhizobacteria (PGPR) are rhizospheric microorganisms inhabiting the soil which improves plant growth and its nutrient uptake capacity (NUC) by utilising wide array of processes, viz. organic matter mineralisation, nitrogen fixation and nutrient (potassium, phosphorus and zinc) solubilisation (Prasad et al. 2005, 2015).

Exponentially increasing population and increased food demand have been inclined towards the traditional use of chemical fertiliser for higher productivity, and its exhaustive use and avoidance in the agricultural practice have detrimentally degraded the efficiency of rhizospheric roots for the mobilisation and uptake of nutrients from the rhizosphere (Meena et al. 2017). PGPRs can improve nutrient uptake and availability or NUC by approximately 20–40%. Many PGPR species are already reported in the fixation of nitrogen and solubilisation of minerals like phosphorus and potassium, whereas other species have potential to increase the solubility of micronutrients like iron (Meena et al. 2017). However, the underlying molecular mechanisms and their applications in biotechnological perspective still need to be discussed.

Actinomycetes are actively involved in the degradation of organic contaminants and inhibit the proliferation of many plant pathogens in the rhizosphere. They break down complex polymers of the dead plant resulting in the production of many enzymes that are important in the crop production (Bhatti et al. 2017). *Actinobacteria* help in nitrogen fixation, buffering of the soil, biological control of soil environment by nitrogen fixation and decomposition of high molecular weight compounds (Bhatti et al. 2017) as well as increase the availability of mineral nutrient and enhance the production of metabolites and plant growth regulators. Further they improve soil health by forming compost and humus as well as by breaking down of complex plant and animal residues. Recent biochemical and molecular tool advancements for isolation of unculturable bacterial strains of the soil have facilitated the researchers to develop acute knowledge pertaining to the rhizospheric microbial diversity. In rhizosphere, the literature available on study of bacteria is much higher than the literature fond on the diversity study of fungal communities as molecular methods for isolation and characterisation of bacteria have been developed earlier.

Rhizodeposition in the rhizosphere has led to the accumulation of huge quantity of degradable carbon sources. Inflated rate of microbial ecology of the rhizosphere is 50 times more than that of the bulk soil. Complex food networks develop in the rhizosphere that links micro- as well as macroorganisms of the rhizosphere (Jeffery et al. 2010). In rhizosphere, a root surface occupies 15% of variety of bacterial strains (van Loon 2007). Bacteria are the most prominent microorganism in the rhizosphere. The studies on rhizosphere carried out by Fulekar and team reported bacterial species of *Pseudomonas*, *Bacillus*, *Arthrobacter*, *Azotobacter*, *Rhizobium*, *Azospirillum*, etc. in the rhizosphere.

Researchers have reported that Gram-positive bacteria precisely *Bacillus* is more predominant than the Gram-negative bacteria in several crops like potato (Smalla et al. 2001), wheat (Joshi and Bhatt 2010), etc. In the rhizosphere, microbial diversity is studied in terms of root colonisation and microbial proliferation along the root (Kloepper et al. 1991). The rhizosphere is thus considered as ecological richest zone of the soil in terms of bacterial diversity.

9.8 Rhizosphere: Plant-Microbe Interaction

Rhizosphere: Zone of the soil which is about 1 mm wide but has no distinct edge, surrounded by plant roots wherein biology and chemistry are influenced by the roots. In rhizosphere, biological and chemical activities are influenced by plant enzymes and microbial enzymes and their combined action (Shrivastava et al. 2014).

Plant microbial interactions have been observed in ecological environment of the soil. Depending upon the conditions like nutrient availability and uptake, rhizospheric soil environment, plant protection mechanism and the proliferating microorganism (Parmar and Dufresne 2011), the growing roots and microbial colonisation in and around the rhizospheric environment develop, and various kinds of relationship like associative, symbiotic, naturalistic or parasitic may develop. Interaction between plant and associated microbial diversity is facilitated by root enzymes and microbial secretions. Rhizospheric microflora develops an environment wherein plant and soil act as a link between them.

9.9 Rhizosphere Enzyme

The activities of plant roots, rhizospheric microorganisms and root microorganism interaction and enzymes are recognised as main factors for all activities occurring in rhizosphere environment. The enzymes commonly reported are dehalogenase, denitrogenase, nitrilase, laccase and peroxidase production, and activity of these enzymes is controlled by several factors depending upon the interaction of soil-plant microorganisms. The greater functional diversity of microbial community is in general interpreted proportional to higher activities of rhizosphere enzymes and involved in the degradation of the organic as well as inorganic contaminants. Therefore, role and function of each of the enzymes secreted by the interaction between soil, plant, microorganisms and the mechanism occurring in remediation of pollutant need to be studied in detail.

9.10 Conclusion

Rhizosphere is an ecological zone of the soil having varieties of nutrients produced from plant roots and microbial via. Rhizo-deposition. Rhizosphere environment has a potential source of bacteria with versatile capabilities which influence the favourable plant growth environment. In rhizosphere, bacteria are the most abundant organisms, and even the new organisms developed that also influence plant growth by varieties of direct and indirect mechanism in a wide range of agricultural crops. With the advancement of techniques in molecular biology and testing methods, the new breakthrough will increase the study area as well as our understanding on the interaction among three groups for the improvement of plant health. Thus, the rhizosphere replaces chemical fertilisers and pesticides and provides eco-friendly sustainable environment.

Acknowledgement The authors acknowledge the facilities provided by the School of Environment and Sustainable Development, Central University of Gujarat, and University Grants Commission (Non- Net Fellowship).

References

Bhatti AA, Haq S, Bhat RA (2017) Actinomycetes benefaction role in soil and plant health. Microb Pathog 111:458–467

Bonfante P (2001) At the interface between mycorrhizal fungi and plants: the structural organization of cell wall. In: Hock B (ed) Plasma membrane and cytoskeleton fungal associations. Springer, Berlin, pp 45–61

Bonfante P, Genre A (2010) Mechanisms underlying beneficial plant–fungus interactions in mycorrhizal symbiosis. Nat Commun 1(4):1–11

Dickson S (2004) The Arum-Paris continuum of mycorrhizal symbioses. New Phytol 163:187–200

Doumbou C, Salove MH, Crawford D, Beaulieu C (2001) Actinomycetes, promising tools to control plant diseases and to promote plant growth. Phytoprotection 82(3):85

Gerdemann JW, Nicholson TH (1963) Spores of mycorrhizal endogone species extracted from soil by wet sieving and decanting. Trans Br Mycol Soc 46:235–244

Giri B, Giang PH, Kumari R, Prasad R, Sachdev M, Garg AP, Oelmuller R, Varma A (2005) Mycorrhizosphere: strategies and functions. In: Buscot F, Varma A (eds) Microorganisms in soils: roles in genesis and functions, vol 3. Springer-Verlag, Berlin-Heidelberg, pp 213–252

Hiltner L (1904) Uber neuere Erfahrungen und Probleme auf dem Gebiete der Bodenbakteriologie unter besonderden berucksichtigung und Brache. Arb. Dtsch. Landwirtsch. Gesellschaft 98:59–78

Jeffery S, Garci C, Jones A, Montanarella L, Marmo L, Miko L, Ritz K, Peres G, Roombke J, van der Putten WH (2010) The soil environment. In: European atlas of soil biodiversity, European Commission. Publications office of the European Union, Luxembourg, pp 17–48

Joshi P, Bhatt AB (2010) Diversity and function of plant growth promoting rhizobacteria associated with wheat rhizosphere in north Himalayan region. Int J Environ Sci 1:1135–1144

Kharkwal H, Kharkwal AC, Prasad R, Varma A (2007) Medicinal plants for human life. In: Chauhan AK, Kharkwal H, Varma A (eds) Microbes for human life, vol 4. IK International, India, pp 19–32

Kloepper JW, Zablotowick RM, Tipping EM, Lifshitz R (1991) Plant growth promotion mediated by bacterial rhizosphere colonizers. In: Keister DL, Cregan PB (eds) The rhizosphere and plant growth. Kluwer Academic, Dordrecht, pp 315–326

Korade DL, Fulekar MH (2009) Rhizosphere remediation of chlorpyrifos in mycorhizospheric soil using ryegrass. In: Fulekar MH (ed) Environment science. Segment Books, New Delhi, pp 297–309

Lam KS (2006) Discovery of novel metabolites from marine actinomycetes. Curr Opin Microbiol 9 (3):245–251

Lazeano-Amora E, Guerrero-Zuniga, Rodriguez-Tovar A, Rodriguez-Dorantes, Vasquez-Murrieta MS (2010) Rhizospheric plant-microbe interactions that enhance the remediation of contaminated soil. In: Mendez-Vilas A (ed) Current research, technology and education topics in applied microbiology and microbial biotechnology. FORMATEX, Badajoz, pp 251–256

Marschner P (2012) Rhizosphere biology. In: Marschner P (ed) Marschner's mineral nutrition of higher plants, 3rd edn. Elsevier, Amsterdam, pp 369–388

Meena VS, Meena SK, Verma JP, Kumar A, Aeron A, Mishra PK, Bisht JK, Pattanayak A, Naveed M, Dotaniya M (2017) Plant beneficial rhizospheric microorganism (PBRM) strategies to improve nutrients use efficiency: a review. Ecol Eng 107:8–32

Moncheva P, Tishkov S, Dimitrova N, Chipeva V, Antonova-Nikolova S, Bogatzevska N (2002) Characteristics of soil Actinomycetes from Antarctica. J Cult Collect 3:3–14

Parmar N, Dufresne J (2011) Beneficial interactions of plant growth promoting rhizosphere microorganisms. In: Singh A et al (eds) Soil biology bioaugmentation, biostimulation and biocontrol. Springer, Berlin, pp 27–42

Parniske M (2008) Arbuscular mycorrhiza: the mother of plant root endosymbioses. Nat Rev Microbiol 6(10):763–775

Perotto S, Bonfante P (1997) Bacterial associations with mycorrhizal fungi: close and distant friends in the rhizosphere. Trends Microbiol 5(12):496–501

Phillips JM, Hayman DS (1970) Improved procedure for clearing roots and staining parasitic and vesicular arbuscular mycorrhizal fungus for rapid assessment of infection. Trans Br Mycol Soc 55:158–161

Prasad R, Garg AP, Varma A (2005) Interaction of medicinal plants with plant growth promoting rhizobacteria and symbiotic fungi. In: Podila GK, Varma A (eds) Basic research and applications of Mycorrhizae, vol 1. IK International, Delhi, pp 363–407

Prasad R, Kumar M, Varma A (2015) Role of PGPR in soil fertility and plant health. In: Egamberdieva D, Shrivastava S, Varma A (eds) Plant growth-promoting rhizobacteria (PGPR) and medicinal plants. Springer International, Cham, pp 247–260

Prasad R, Bhola D, Akdi K, Cruz C, Sairam KVSS, Tuteja N, Varma A (2017) Introduction to mycorrhiza: historical development. In: Varma A, Prasad R, Tuteja N (eds) Mycorrhiza. Springer International Publishing AG, Switzerland, pp 1–7

Prasad R, Chhabra S, Gill SS, Singh PK, Tuteja N (2020) The microbial symbionts: potential for the crop improvement in changing environments. In: Tuteja N, Tuteja R, Passricha N, Saifi SK (eds) Advancement in crop improvement techniques. Elsevier, Amsterdam, Netherlands, pp 233–240

Shrivastava S, Prasad R, Varma A (2014) Anatomy of root from eyes of a microbiologist. In: Morte A, Varma A (eds) Root engineering, vol 40. Springer, Cham, pp 3–22

Singh A, Fulekar MH (2017) Impact of heavy metals in mycorrhizosphere: strategy for phytoremediation. In: Fulekar MH (ed) Environment science. Segment Books, New Delhi, pp 257–274

Singh BK, Millard P, Whiteley AS, Murrell J (2004) Unravelling rhizosphere–microbial interactions: opportunities and limitations. Trends Microbiol 12(8):386–393

Singh DP, Patil HJ, Prabha R, Yandigeri MS, Prasad MS (2018) Crop improvement through microbial biotechnology. Elsevier, Amsterdam, pp 27–38

Smalla K, Wieland G, Buchner A, Zock A, Parzy J, Kaiser S, Roskot N, Heuer H, Berg G (2001) Bulk and rhizosphere soil bacterial communities studied by denaturing gradient gel electrophoresis: plant-dependent enrichment and seasonal shifts revealed. Appl Environ Microbiol 67 (10):4742–4751

Smith SE, Read DJ (2008) Mycorrhizal *symbiosis*. Academic, New York

Tan H, Deng Z, Cao L (2009) Isolation and characterization of actinomycetes from healthy goat faeces. Lett Appl Microbiol 49(2):248–253

van Loon LC (2007) Plant responses to plant growth promoting bacteria. Eur J Plant Pathol 119:243–254

Varma A, Prasad R, Tuteja N (2017) Mycorrhiza: function, diversity and state-of-art. Springer International Publishing, Switzerland. ISBN 978-3-319-53064-2. http://www.springer.com/us/book/9783319530635

Varma A, Swati T, Prasad R (2020) Plant microbe symbiosis. Springer International Publishing, Switzerland. ISBN 978-3-030-36247-8. https://www.springer.com/gp/book/9783030362478

Chapter 10
The Rhizosphere Microbiome: Microbial Communities and Plant Health

Sandeep Jain, Jyoti Jain, and Jayesh Singh

Abstract Exploration of rhizosphere and rhizosphere microbiome has been the research focus for last many decades. The rhizosphere is a junction for intercommunication among plants, insects, and microorganisms. It serves as diverse habitat with a nutrient-rich niche by providing a platform interaction among plants-soil-microorganism trio along with energy and matter trade-off. The rhizosphere microbiome also influences plant vigour, health, and defence against stresses by interfering with nutrient uptake, chemical signalling, and enzyme activity. Interaction among the microbiome, the environment, and the genetic makeup of host is well-known to contribute towards host health. The present chapter summarises the major effects of microbial communities present in rhizosphere on plant health and diseases.

10.1 Introduction

10.1.1 Rhizosphere: Hotspot of Root–Microbe Interaction

Rhizosphere acts as hotspot of root–microbe interaction. The etymology of rhizosphere can be expressed as rhiza (Greek), meaning root, and "sphere," meaning region influenced by roots. Lorenz Hiltner in the year 1904 was the first to use this term to describe the effects of root exudates on the proliferation of advantageous and deleterious microbes around roots (Hartmann et al. 2008). Rhizosphere is regarded as one of the most diverse ecosystems in the universe with huge energy flow (Barriuso et al. 2008). It is the region where intense biogeochemical activities take place depending on root architecture, exudates, and mucilage which influence the

S. Jain (✉)
Department of Plant Pathology, Punjab Agricultural University, Ludhiana, India
e-mail: sandeepjain@pau.edu

J. Jain · J. Singh
Department of Plant Breeding and Genetics, Punjab Agricultural University, Ludhiana, India
e-mail: jyotijain-pbg@pau.edu

A. Varma et al. (eds.), *Plant Microbiome Paradigm*,
https://doi.org/10.1007/978-3-030-50395-6_10

microorganisms feeding on these compounds within the soil matrix. Besides this, the pH and redox gradients of the rhizosphere are also assumed to influence rhizospheric microbial communities (Schmidt et al. 2011; Shrivastava et al. 2014; Prasad et al. 2015, 2020). The roots exudation effectively widens the functional boundary of the belowground plant–microbe interface as it may control the soil microbial community in their locale, prevent or defend against herbivores, invigorate beneficial symbioses, alter the physiochemical properties of the soil, or slow down the growth of competing plant species resulting in entirely different environment at the root interface compared to bulk soil (Bais et al. 2004, 2006; Hinsinger et al. 2005). This narrow region bordering and affected by plant roots is a junction for microbial colonisation and activity, termed as rhizodeposition. During rhizodeposition, various major processes take place like depletion of border cells and root cap, soluble root exudates, insoluble mucilage, organic carbon, carbon flow to rhizosphere symbionts, and lysis of root cells (Jones et al. 2009).

Rhizosphere is categorised into three zones based on microorganisms existing in close proximity to roots, and the influence of root exudates on them as endorhizosphere, rhizoplane, and ectorhizosphere. Endorhizosphere is made up root endodermis and cortex. Rhizoplane refers to the root facet where soil particles and microorganisms cohere and involves epidermis, cortical layers, and mucilaginous polysaccharides. The third zone is ectorhizosphere consisting of soil immediately adjoining the root (Prashar et al. 2013).

10.2 Rhizosphere Microbiome: Diversity of Microbial Communities

The complex soil communities of microorganisms associated with plant roots are regarded as the plant microbiome which is also considered as the plants' other genome. The root microbiome is dynamic and plays a key role in various nutrient cycling like nitrogen and phosphorus in addition to provide protection against various biotic and abiotic stresses, thus contributing towards plant health and increased productivity (Bender et al. 2016; Lladó et al. 2017; Berendsen et al. 2012; Kumar et al. 2020; Nath et al. 2018). Root–microbe interactions may be beneficial to the plant or to the microbes or to neither of them. Both plants and microbes can be benefitted directly through mutualistic relationship or indirectly via decomposition, nutrient cycling and solubilisation (Glick 1995), emanation of growth hormones (Narula et al. 2006), induction of the resistance (Pieterse et al. 2001), and antagonism of pathogens (Kloepper et al. 2004). According to two contrary schools of thoughts, the root exudation is an "active" or "passive" process. Cook et al. (1995) hypothesised that plants may actively moderate the root microbiome for their welfare by specifically influencing microbe in a targeted manner. Other school of thought is that root exudation is a passive process releasing waste products (Jones et al. 2009; Dennis et al. 2010).

By releasing nearly one half of their total photosynthetic fixed carbon, rhizosphere microbiome provides unique locale for a diverse conglomeration of microorganisms which are an important part of food chain that consumes the bulk of nutrients exuded by plant roots (Lynch and Whipps 1990; Bonkowski et al. 2009). It has been estimated that rhizosphere microbiome can harbour approximately hundred billion microbial cells per gram of root (Egamberdieva et al. 2008) which is many times more than (10^8) in bulk soil (Foster 1988). The rhizospheric microbial communities are different from those present within the root and are reduced in diversity compared with the microbial communities of bulk soil. Organisms like nitrogen-fixing bacteria, arbuscular mycorrhizae, plant growth-promoting rhizobacteria (PGPR), biological control agents, and mycoparasitic saprobes have been well documented for their beneficial impact on plant vigour (Singh et al. 2019). Rhizospheric microorganisms which are detrimental to plant health are various plant pathogens particularly oomycetes fungi, bacteria, and nematodes. Apart from these, rhizosphere also serves as a stockpile for certain human pathogens such as *Pseudomonas*, *Staphylococcus*, and *Stenotrophomonas* (Berg et al. 2005).

10.3 Rhizosphere Microbiome: Vigour and Well-Being

The diverse rhizospheric microbial population plays a key role towards plant vigour and yield as it protects against pathogens and produce phytohormones (Lu et al. 2018; Singh et al. 2019). Approximately nearly 5–20% of total carbon fixed through photosynthesis is exuded as root exudates and functions either as substrate, chemoattractant, or signalling molecules (Mendes et al. 2013). These chemicals allure advantageous microbes that elicit pest resistance, water holding, and the synthesis phytohormones like auxins and may influence plant phenotype. Interactions among microbes-root exudates-plant physiology aggressively influence rhizospheric microbial population and alters the plants phenotypic traits. At molecular level, cross-talk differs depending upon propinquity to other microbes and management practices, etc.

Naturally occurring plant communities are influenced by rhizosphere microorganisms either directly or indirectly (van der Heijden et al. 1998, 2006, 2008; Schnitzer et al. 2011). Microbial species diversity below ground has been found to be an indicator of aboveground plant diversity and productivity (Hooper et al. 2005; Lau and Lennon 2011). The underground microbial richness ensures plant productivity under diverse climatic conditions (Wagg et al. 2011). Thus, microorganisms present both in rhizosphere and bulk soil may be utilised as a bench mark of soil quality. The plant beneficial and plant pathogenic microorganisms are discussed below:

10.3.1 The Beneficial Rhizo-Microorganisms

The plant growth-promoting microbes present in rhizosphere including bacteria, fungi, actinomycetes, protozoa, and algae act through a variety of important mechanisms like biofertilisation, root growth stimulation, rhizoremediation, abiotic stress control, and direct disease control. The beneficial effects of different rhizosphere microorganisms like *Proteobacteria* (*Pseudomonas* and *Burkholderia*) and *Firmicutes* (*Bacillus* sp.) and for fungi imperfecti (deuteromycetes) genera like *Trichoderma*, *Gliocladium*, and *Piriformospora indica* along with non-pathogenic *Fusarium oxysporum* are well documented (Kogel et al. 2006; Prasad 2008; Qiang et al. 2012; Deshmukh and Shinde 2016; Prasad et al. 2020). Among these rhizosphere microbial communities, bacteria are the most abundant (Kaymak 2010). Recently, much more information has been generated pertaining to diverse rhizospheric communities like in case of *Planctomycetes* (Hol et al. 2010; Jogler et al. 2012). The increased plant vigour by the incorporation of these microorganisms as biological control agents is well documented (Saharan and Nehra 2011; Bhattacharyya and Jha 2012). Microorganisms associated with plant microbiome can benefit the plants either by increasing the nutrient use efficiency, acting as biocontrol agents against various pathogens, or promoting plant growth by production of auxins (Singh et al. 2019).

10.3.1.1 Enhancement of Nutrient Recovery

Different nutrients like N, P, and S are present in most of the soils but are not readily available for plant uptake. Microorganisms convert these organic nutrients into available forms via different processes like production of extracellular enzymes which solubilises and fixes nutrients into available forms.

They also enhance the solubility of certain nutrients such as Fe due to various oxidation and reduction reactions. They are also reported to release organic acids which are responsible for releasing nutrients from minerals and increased rate of weathering (Coyne and Mikkelson 2015).

The most classical example is the symbiotic relationship between arbuscular mycorrhizal and roots of higher plants thereby increasing the supply of various nutrients such as copper, iron, nitrogen, phosphorus, and zinc (Prasad et al. 2017). Arbuscular mycorrhizal fungi are also reported to produce many enzymes responsible for solubilising organic P and, thus, can absorb soluble P from the soil at very low concentrations compared to plant roots alone.

Microorganisms play a vital role in nitrogen cycle as they can convert atmospheric nitrogen into fixed nitrogen. Above 90% nitrogen is fixed by nitrogen-fixing bacteria which fulfils up to 1/5th of the N required by cultivated crops. Nitrogen-fixing bacteria are either free-living/nonsymbiotic (*Azotobacter*, *Cyanobacteria*, *Beijerinckia*, *Clostridium*, and *Anabaena* species) or mutualistic/symbiotic (*Azospirillum*, *Frankia*, and *Rhizobium* species).

Azospirillum is an example of free-living nitrogen-fixing bacteria and is known to enhance rice yields (Tejera et al. 2005). Mutually beneficial relationship between bacteria and plant roots is popularly called as symbiotic. Such mutual relationship starts with the entry of bacteria into root hair followed by multiplication and formation of nodules on the host roots and enlargement of both plant and bacterial cells in association with each other. *Rhizobium*, *Bradyrhizobium*, *Sinorhizobium*, and *Mesorhizobium* comprise of large group of rhizobacteria known to establish symbiotic relationship by formation of nodules on roots of leguminous plants, thus fixing nitrogen to ammonia and solubilising it into available form. *Frankia* has been reported to be associated with non-leguminous host plants (Bhattacharyya and Jha 2012).

Besides N fixation, the ability of rhizosphere microorganisms for P solubilisation is another important trait associated with plant phosphate nutrition. Phosphorus (P) is one of the major macronutrients required by plants. It is applied to soil as chemical phosphatic fertilisers. However, a major part of this phosphatic fertiliser applied to soil gets immobilised, precipitates as orthophosphate and gets adsorbed by Fe and Al oxides via legend exchange and becomes unavailable to plants.

Growth-promoting phosphate solubilising bacterial strains employ following strategies to convert phosphorus into soluble form:

(a) Lowering of soil pH by producing low molecular weight mineral-dissolving compounds such as organic acids (mainly gluconic and keto-gluconic acids), O_2/CO_2 exchange, protons, and hydroxyl ions through which their hydroxyl and carboxyl groups chelate the Fe, Al, and Ca ions bound to phosphate, thereby competing for adsorption sites with phosphate. Inorganic acids are less effective compared to organic acids for phosphate solubilisation at same pH level (Kim et al. 1997). Many times, phosphate starvation may enhance phosphate solubilisation (Gyaneshwar et al. 1999).
(b) Release of extracellular enzymes (biochemical phosphate mineralisation). Key mode of action for mineralisation of organic P in soil is the release of nonspecific acid phosphatases (NSAPs) or alkaline phosphomonoesterases (Jorquera et al. 2008) which is responsible for dephosphorylation of phosphoester or phosphoanhydride bonds of organic matter; phytases causing phytate degradation; phosphonatases; and C–P lyases to split the C–P bond of organophosphonates. These enzymes hydrolyse the organic P into the soil.
(c) Substrate degradation leading to liberation of phosphate and mineralisation (McGill and Cole 1981).

Microorganisms engaged in phosphorus solubilisation are primarily arbuscular mycorrhizae and phosphorus-solubilising bacteria (Fankem et al. 2006). Out of the total microbiota prevalent in the soil, phosphorus-solubilising bacteria (PSB) account for 1–50%, while phosphorus-solubilising fungi (PSF) are meagre (0.1–0.5%) for P solubilisation capability. Among bacteria, ectorhizospheric types (*Pseudomonas striata*, *Bacillus megaterium*, *B. circulans*, *B. subtilis*, *B. polymyxa*, *Bacillus sircalmous*, and *Enterobacter*) and endosymbiotic *Rhizobia* have been reported as most efficient phosphate solubilisers (Igual et al. 2001; Subbarao 1988; Kucey et al. 1989). Besides bacteria, fungi (*Penicillium* and *Aspergillus*), actinomycetes, algae (*Cyanobacteria*), mycorrhiza, and a nematophagous fungus *Arthrobotrys oligospora* are well documented for their P solubilisation activity (Whitelaw 2000; Duponnois et al. 2006). Phosphate-solubilising bacteria belong to genera *Arthrobacter*, *Bacillus*, *Beijerinckia*, *Burkholderia*, *Enterobacter*, *Erwinia*, *Flavobacterium*, *Microbacterium*, *Pseudomonas*, *Rhizobium*, *Rhodococcus*, and *Serratia* (Ahmad et al. 2008). *Kocuria turfanensis* strain 2M4 acts as phosphate solubiliser which is also reported to produce IAA and siderophores (Goswami et al. 2014).

Iron (Fe) holds fourth rank among the most abundant nutrient on the earth; still Fe deficiency is a widespread problem particularly in arid and semi-arid regions. This is because Fe is not readily assimilated due to presence of rarely soluble ferric ion (Fe^{+3}). Hence, the amount of iron available for assimilation is very low (Aloni et al. 2006). Rhizospheric microorganisms play a key role to overcome Fe deficiencies and greatly influence the Fe uptake by production siderophores, which form Fe^{3+} chelates and are transported through plasma membrane (Awad et al. 1994; Sabry et al. 1997; De Felipe and Fijación 2006). They are mostly produced by microorganisms growing under low stress of iron. Most of the siderophores produced by bacterial genera such as *Streptomyces* and *Pseudomonas* are catecholates, whereas siderophores produced by fungi are hydroxamates (Das et al. 2007).

Siderophores have been associated with improved plant vigour, e.g. PGPR (*Aeromonas*, *Azadirachta*, *Azotobacter*, *Bacillus*, *Burkholderia*, *Pseudomonas*, *Rhizobium*, *Serratia*, and *Streptomyces* sp.) increased chlorophyll level as in comparison with uninoculated plants (Berg et al. 1980).

10.3.1.2 Plant Growth Regulators

A huge and diverse variety of rhizosphere microbiome is capable of producing growth hormones such as auxins, cytokinins, gibberellins, and ethylene. Plant growth-promoting rhizobacteria (PGPR) produce organic substances also known as exogenous hormones (phytohormones) in extremely low amounts which can control different morphological and physiochemical processes of the plants. PGPR are well documented for production of phytohormones influencing root system architecture (RSA) with an increasing rate of nutrient and water uptake. The most common effect is a reduction of growth rate of primary root and increased number of lateral roots and root hairs. Further, PGPR can also modify chemical makeup and structure of cell wall of root (Zhang et al. 2007). PGPR are also known to produce growth hormones along with secondary metabolites which interfere with the plant auxin pathway (Prasad et al. 2005). Exogenous production of IAA influences a vast range of plant processes. For example, low concentrations of IAA can encourage elongation of primary root, whereas high concentration of IAA encourages lateral roots formation, decreased root length, and increased number of root hairs (Remans et al. 2008). The PGPR (*Pseudomonas*, *Rhizobium*, *Bradyrhizobium*,

Agrobacterium, *Enterobacter*, and *Klebsiella*) produce IAA in plants majorly through indole-3-pyruvic acid and indole-2-acetic aldehyde pathways (Joo et al. 2005). Root growth promotion by the free-living rhizobacteria like *Enterobacter cloacae*, *Alcaligenes faecalis*, *Acetobacter diazotrophicus*, *Azospirillum* sp., *Pseudomonas* sp., and *Xanthomonas* sp. is reported to be associated with IAA secretion. Auxins like 2,4-diacetylphloroglucinol (DAPG) produced by biocontrol fluorescent pseudomonads induce systemic resistance in plants (Bakker et al. 2007), enhance exudation by roots (Phillips et al. 2004), and stimulate branching of roots (Walker et al. 2011).

PGPR such as *Arthrobacter giacomelloi*, *Azospirillum brasilense*, *Bradyrhizobium japonicum*, *Bacillus licheniformis*, *Pseudomonas fluorescens*, and *Paenibacillus polymyxa* are well documented to produce cytokinins. Cytokinins play a vital role in enhancing cell division, differentiation of root meristem, and root hair proliferation Besides they also prevent formation of lateral roots and elongation of primary root (Riefler et al. 2006). Several PGPR are known to produce both cytokinins and gibberellins, e.g. *Pseudomonas* and *Bacillus* sp. (Han and Lee 2005). Thus, plant organogenesis and root architecture are regulated by ratio of auxins and cytokinins (Aloni et al. 2006).

Another key phytohormone is ethylene which is produced in small quantities but is responsible for inhibiting elongation of roots, promoting senescence and abscission of different plant organs, and ripening of fruits (Perrig et al. 2007). Higher concentration of ethylene leads to inhibited plant growth by inducing premature senescence, thus leading to poor plant growth and yield (Li et al. 2005). Various biotic and abiotic stresses induce plants to synthesise 1-aminocyclopropane-1-carboxylate (ACC). The increased level of ethylene in response to stress conditions causes termination of important cellular mechanisms such as elongation of roots and bacterial nitrogen fixation in leguminous crop plants thereby causing premature senescence (Jackson 1991; Glick 2012; Ahmad et al. 2013).

PGPR are also capable to produce phytohormones like abscisic acid or gibberellic acid (Dodd et al. 2010). ABA is responsible for causing stomatal closure thereby reducing rate of transpiration and plays a key role under drought stress conditions (Bauer et al. 2013). Besides this it is known for its key role in root development particularly the lateral roots (De Smet et al. 2006). Whereas gibberellins encourage both elongation of primary root and extension of lateral roots (Yaxley et al. 2001). A number of PGPR, viz. *Achromobacter xylosoxidans*, *Acinetobacter calcoaceticus*, *Azospirillum* spp., *Azotobacter* spp., *Bacillus* spp., *Herbaspirillum seropedicae*, *Gluconacetobacter diazotrophicus*, and *Rhizobium*, have been well documented for gibberellins production (Bottini et al. 2004; Dodd et al. 2010). Additionally, both of these phytohormones also induce resistance in plants. Thus, PGPR produced phytohormones and can regulate jasmonate and salicylic acid pathways involved in plant resistance.

In spite the fact that production of phytohormones by PGPR has been well studied, the genetic factors responsible for biosynthesis are still not, and the knowledge about role of bacterial mutants in phytohormones production is scanty. Thus, the role of bacterial synthesised towards plant hormonal balance is not well studied.

Within the rhizosphere, both living (like root exudates) and non-living (like pH, oxygen, type of soil, metals, etc.) factors may influence the effect of PGPR's favourable traits that can lead to distinct expression patterns with different effects on host plant (Drogue et al. 2013). Root exudation by plants may influence the expression of PGPR genes particularly those encoding for plant favouring traits. The expression of phlA gene of *Pseudomonas protegens* was enhanced four times near the roots of maize and wheat compared to those of legumes and cucumber. Sugar content in the root exudates largely influenced synthesis of antimicrobial compounds by fluorescent pseudomonas like DAPG, pyoluteorin, and pyrrolnitrin (Duffy and Défago 1999). Root exudates play a key role in tryptophan biosynthesis pathways (Spaepen et al. 2007). In the absence of exogenous tryptophan supply, bacterial IAA biosynthesis is insignificant. In addition, root exudates also contain vitamins and organic acids which can significantly increase IAA biosynthesis in PGPR (Shukla et al. 2011; Zakharova et al. 2000; Somers et al. 2004). Thus, the composition of root exudates can precisely modify genes encoding for plant favouring traits.

In conclusion, rhizosphere microbiota may lessen impact of various living and non-living stresses on plants. But performance of microorganisms is affected by numerous factors such as not consistent efficiency under geographically different climatic conditions and narrow shelf life. More basic knowledge needs to be generated pertaining to beneficial interactions among microorganisms and plant roots both at cellular and molecular level, as well as the effect of beneficial microbes on virulence spectrum of phytopathogens.

10.3.2 The Deleterious Rhizo-Microorganisms

The rhizosphere in addition to beneficial microorganisms also attracts soil-borne pathogens which are deleterious for plant vigour and well-being. It provides a battlefield where the complex rhizosphere microbiome interacts with soil-borne pathogens and influences the phytopathogens and impacts their pathogenicity causing major yield reductions of food, feed, and fibre crops. Predominant soil-borne plant pathogens include nematodes (*Meloidogyne*, *Heterodera*, *Longidorus*, *Paratrichodorus*, etc.), true fungi, bacteria, and viruses. Among these, viruses need vectors (nematodes, insects, and fungi) to invade plant roots (Campbell 1996).

Plants show a variety of symptoms in response to infection with fungal soil-borne pathogens such as pre- and post-emergence damping off of seedlings (*Pythium* sp.), root rots (*Rhizoctonia* sp.), wilts (*Fusarium* sp.) etc. In response to infection by root rot causing phytopathogens, there is increased mortality of root tips causing reduced root growth leading to poor absorption of water and nutrients. This could be due to destruction of root hairs. As a result, plant shows retarded growth and may exhibit nutritional deficiency.

Two important phenomena, viz. fungistasis and rhizodeposition, that occur in the rhizosphere determine the fate of the pathogen and its ability to initiate infection. Fungistasis refers to reduced rate of germination due to exogenously induced dormancy (Dobbs and Hinson 1953; Lockwood 1977). The fungistatic mechanisms facilitated by soil microorganisms include the presence of volatile or soluble inhibitory substances or lack of essential nutrients for spore germination (Ko and Lockwood 1967; Liebman and Epstein 1992).

Soil-borne fungal plant pathogens, viz. *Rhizoctonia*, *Fusarium*, *Sclerotium*, *Pythium*, and *Phytophthora*, perpetuate in the soil as chlamydospores, oopsores, sclerotia, or hyphae or survive on plant debris (Bruehl 1987). The root exudates stimulate these resting propagules to germinate and invade roots of susceptible plants. This stimulus is particularly important for less virulent phytopathogens which remain in dormant phase in the absence of either nutrients or fungistasis. The microorganisms prevalent in rhizosphere are governed by the resistance or susceptibility level of the crop varieties grown.

For the successful establishment of the pathogen in the rhizosphere, host signals play a vital role. The life cycle of fungi like formation of surviving structures (oospores, sclerotia, etc.) is influenced by various parameters such as pH level of soil, soil type, and root exudates. At low concentrations, phenolic compounds present in root exudates like p-hydroxybenzoic, Gallic, coumaric, cinnamic, ferulic, salicylic, cinnamic acids, etc. stimulate germination of spores of pathogenic fungi; however, presence of these compounds at higher concentrations causes inhibition of spore germination (Wu et al. 2008). This has been demonstrated in *Veratrum taliense* (Liliaceae), *Phytophthora capsici*, and *Rhizoctonia cerealis* combination (Zhou et al. 2003). Both soil physical characteristics and microbes diversity influence the alkaloid makeup (retrorsine and retrorsine N-oxide) in roots and shoots of *Jacobaea vulgaris* (Joosten et al. 2009). These compounds prevent hyphal growth of various phytopathogenic fungi, for example, *Fusarium* and *Trichoderma* sp. (Hol and Veen 2002). Saponins among diverse group of glycosides also adversely affect plant pathogenic fungi by interfering with membrane integrity (González-Lamothe et al. 2009; Osbourn et al. 2011). Evidence of role of saponins (avenacin) in plants defence against root rot fungi *Gaeumannomyces graminis* var. *avenae* has been well presented (Bednarek and Osbourn 2009; González-Lamothe et al. 2009; Osbourn et al. 2011). Avenacin is also responsible for eliciting other processes in the plant such as callose deposition (Bednarek and Osbourn 2009) thereby inducing resistance.

Oomycete pathogens reproduce asexually by production of motile biflagellate zoospores which invade plant roots and cause infection. The composition of root exudates influence chemotactic attraction of zoospores of *Phytophthora* sp. van West et al. (2002) demonstrated that the zoospores that get attracted towards roots is also due to electrotaxis, whereby roots produce electric currents in response to ion exchange at growing ends. Thus, electrotaxis play a critical role in colonisation of roots by zoosporic pathogens.

10.3.2.1 Nematodes

Plant pathogenic nematodes are either free living, ectoparasitic, or endoparasitic. Nematodes move towards roots of plants in response to root exudates (Bird 1959; Young et al. 1996). Perry (2005) and Curtis et al. (2009) have designated these attractants as long distance (attract nematodes to the general root area), short distance (attract nematodes to the roots themselves), and local attractants (orient the nematodes to the preferred penetration site). In the complex rhizosphere, both volatile (long-distance chemotaxis) and water-soluble compounds (short-distance chemotaxis) act as important signals for nematode scavenging (Rasmann et al. 2012).

Carbon dioxide act as major signal released by plant roots to attract nematodes (Jogler et al. 2012). Besides carbon dioxide, many other compounds like naturally occurring hydroxamic acid (DIMBOA), glutamic, and ascorbic acid can induce chemotaxis in nematodes. On the basis of chemotactic property of root attractants, trap crops have been exploited to attract the nematodes to non-host plants (Franco et al. 1999). For example, *Asparagus officinalis* and *Tagetes* species attract a vast variety of nematodes by releasing glycosides (Bilgrami 1997).

10.4 Role of Root–Microbe Communication in Shaping Rhizo-Microbiome

Being a dynamically active habitat for microorganisms, microbiome not only influences plant growth but also affects human or animal health. Efforts are being made to redesign rhizosphere microbiome favouring plant growth-promoting microorganisms and inhibiting phytopathogens. Several studies conducted in the last 20 years evidently showed that the plant genetic makeup and soil properties are two major factors that outline microbial communities in rhizosphere (Berg and Smalla 2009; Bakker et al. 2012).

The information on rhizodeposition and its role in activating and attracting soil-borne plant pathogens are limited and patchy. The inadequate understanding of the cross-talk between plants and soil-borne phytopathogens is mainly due to a scanty knowledge of the complex physiochemical conditions in vicinity of rhizosphere (Weston et al. 2012). The chemical composition and spatiotemporal production of root exudates also called as "ecometabolomics" needs to be elucidated to understand these complex interactions. The comparatively young researchable area of ecometabolomics is the utilisation of metabolomics techniques to biology with the target to characterise biochemical communications among organisms across various spatiotemporal patterns.

References

Ahmad F, Ahmad I, Khan MS (2008) Screening of free-living rhizospheric bacteria for their multiple plant growth promoting activities. Microbiol Res 163:173–181

Ahmad M, Zahir ZA, Khalid M (2013) Efficacy of rhizobium and pseudomonas strains to improve physiology, ionic balance and quality of mung bean under salt-affected conditions on farmer's fields. Plant Physiol Biochem 63:170–176

Aloni R, Aloni E, Langhans M (2006) Role of cytokinin and auxin in shaping root architecture: regulating vascular differentiation, lateral root initiation, root apical dominance and root gravitropism. Ann Bot 97:883–893

Awad E, Romheld V, Marschner (1994) Effect of root exudates on mobilization in the rhizosphere and uptake of iron by wheat plants. Plant Soil 165:213–218

Bais HP, Park SW, Weir TL, Callaway RM, Vivanco JM (2004) How plants communicate using the underground information superhighway. Trends Plant Sci 9:26–32

Bais HP, Weir TL, Perry LG, Gilroy S, Vivanco JM (2006) The role of root exudates in rhizosphere interactions with plants and other organisms. Annu Rev Plant Biol 57:233–266

Bakker PA, Pieterse CM, van Loon LC (2007) Induced systemic resistance by fluorescent Pseudomonas spp. Phytopathology 97:239–243. https://doi.org/10.1094/PHYTO-97-2-0239

Bakker MG, Manter DK, Sheflin AM, Weir TL, Vivanco JM (2012) Harnessing the rhizosphere microbiome through plant breeding and agricultural management. Plant Soil 360:1–13

Barriuso J, Solano BR, Lucas JA, Lobo AP, Villaraco AG, Manero FJG (2008) Ecology, genetic diversity and screening strategies of plant growth promoting rhizobacteria (PGPR). In: Ahmad I, Pichtel J, Hayat S (eds) Plant–bacteria interactions: strategies and techniques to promote plant growth. Wiley, Weinheim, pp 1–17

Bauer H, Ache P, Lautner S, Fromm J, Hartung W, Al-Rasheid Khaled AS (2013) The stomatal response to reduced relative humidity requires guard cell-autonomous ABA synthesis. Curr Biol 1:53–57. https://doi.org/10.1016/j.cub.2012.11.022

Bednarek P, Osbourn A (2009) Plant-microbe interactions: chemical diversity in plant defense. Science 324(5928):746–748. https://doi.org/10.1126/science.1171661

Bender SF, Wagg C, van der Heijden MGA (2016) An underground revolution: biodiversity and soil ecological engineering for agricultural sustainability. Trends Ecol Evol 31:440–452. https://doi.org/10.1016/j.tree.2016.02.016

Berendsen RL, Pieterse CM, Bakker PA (2012) The rhizosphere microbiome and plant health. Trends Plant Sci 17(8):478–486. https://doi.org/10.1016/j.tplants.2012.04.001

Berg G, Smalla K (2009) Plant species and soil type cooperatively shape the structure and function of microbial communities in the rhizosphere. FEMS Microbiol Ecol 68:1–13

Berg RH, Tyler ME, Novick NJ (1980) Biology of azospirillum-sugarcane association: enhancement of nitrogenase activity. Appl Environ Microbiol 39:642–649

Berg G, Eberl L, Hartmann A (2005) The rhizosphere as a reservoir for opportunistic human pathogenic bacteria. Environ Microbiol 7(11):1673–1685. https://doi.org/10.1111/j.1462-2920.2005.00891.x

Bhattacharyya PN, Jha DK (2012) Plant growth-promoting rhizobacteria (PGPR): emergence in agriculture. Wood J Microb Biotechnol 28:1327–1350

Bilgrami AL (1997) Nematode biopesticides. Aligarh Muslim University, Aligarh, 262 pp

Bird AF (1959) The attractiveness of roots to the plant-parasitic nematodes *Meloidogyne javanica* and *M. hapla*. Nematologica 4:322–335. https://doi.org/10.1163/187529259X00534

Bonkowski M, Villenave C, Griffiths B (2009) Rhizosphere fauna: the functional and structural diversity of intimate interactions of soil fauna with plant roots. Plant Soil 321(1–2):213–233

Bottini R, Cassán F, Piccoli P (2004) Gibberellin production by bacteria and its involvement in plant growth promotion and yield increase. Appl Microbiol Biotechnol 65:497–503. https://doi.org/10.1007/s00253-004-1696-1

Bruehl GW (1987) Soilborne plant pathogens. Macmillan, New York

Campbell RN (1996) Fungal transmission of plant viruses. Annu Rev Phytopathol 34:87–108

Cook RJ, Thomashow LS, Weller DM, Fujimoto D, Mazzola M, Bangera G, Kim DS (1995) Molecular mechanisms of defense by rhizobacteria against root diseases. Proc Natl Acad Sci USA 4197

Coyne BMS, Mikkelson R (2015) Soil microorganisms contribute to plant nutrition and root health. Better Crops 99(1):18–20

Curtis RHC, Forest R, Perry R (2009) Hatch and host location. In: Perry R, Moens M, Starr J (eds) Root-knot nematodes. CABI, Wallingford, pp 139–162

Das A, Prasad R, Srivastava A, Giang PH, Bhatnagar K, Varma A (2007) Fungal siderophores: structure, functions and regulation. In: Microbial siderophores, vol 12. Springer, Berlin, pp 1–42

De Felipe, Fijación MR (2006) Biológica de dinitrógeno atmosférico en vida libre. In: Bedmar E, Gonzálo J, Lluch C (eds) Fijación de Nitrógeno: Fundamentos y Aplicaciones. Granada: Sociedad Española de Microbiología. Sociedad Española de Fijación de Nitrógeno, Granada, pp 9–16

De Smet I, Zhang H, Inzé D, Beeckman T (2006) A novel role for abscisic acid emerges from underground. Trends Plant Sci 11:434–439. https://doi.org/10.1016/j.tplants.2006.07.003

Dennis PG, Miller AJ, Hirsch PR (2010) Are root exudates more important than other sources of rhizo deposits in structuring rhizosphere bacterial communities? Microbiol Ecol 72(3):313–327. https://doi.org/10.1111/j.1574-6941.2010.00860.x

Deshmukh P, Shinde S (2016) Beneficial role of rhizosphere mycoflora in the field of agriculture: an overview. Int J Sci Res 5(8):529–533

Dobbs CG, Hinson WH (1953) A widespread fungistatis in soils. Nature 172:197–199

Dodd IC, Zinovkina NY, Safronova VI, Belimov AA (2010) Rhizobacterial mediation of plant hormone status. Ann Appl Biol 157:361–379. https://doi.org/10.1111/j.1744-7348.2010.00439.x

Drogue B, Combes-Meynet E, Moënne-Loccoz Y, Wisniewski-Dyé F, Prigent-Combaret C (2013) Control of the cooperation between plant growth-promoting rhizobacteria and crops by rhizosphere signals. In: de Bruijn FJ (ed) Molecular microbial ecology of the rhizosphere, vol 1 and 2. Wiley, Hoboken, pp 281–294. https://doi.org/10.1002/9781118297674.ch27

Duffy BK, Défago G (1999) Environmental factors modulating antibiotic and siderophore biosynthesis by *Pseudomonas fluorescens* biocontrol strains. Appl Environ Microbiol 65:2429–2438

Duponnois R, Kisa M, Plenchette C (2006) Phosphate solubilizing potential of the nematofungus *Arthrobotrys oligospora*. J Plant Nutr Soil Sci 169:280–282

Egamberdieva D, Kamilova F, Validov S, Gafurova L, Kucharova Z, Lugtenberg B (2008) High incidence of plant growth stimulating bacteria associated with the rhizosphere of wheat grown on salinated soil in Uzbekistan. Environ Microbiol 10:1–9

Fankem HD, Nwaga AD, Dieng L, Merbach W, Etoa FX (2006) Occurrence and functioning of phosphate solubilizing microorganisms from oil palm tree (*Elaeis guineensis*) rhizosphere in Cameroon. Afr J Biotech 5:2450–2460

Foster RC (1988) Microenvironments of soil microorganisms. Biol Fertil Soils 6:189–203

Franco J, Main G, Oros R (1999) Trap crops as a component for the integrated management of Globodera spp. (potato cyst nematodes) in Bolivia. Nematropica 29:51–60

Glick BR (1995) The enhancement of plant growth by free-living bacteria. Can J Microbiol 41:109–117

Glick BR (2012) Plant growth promoting bacteria: mechanisms and applications. Scientifica 2012:1–15. https://doi.org/10.6064/2012/963401

González-Lamothe R, Mitchell G, Gattuso M, Diarra MS, Malouin F, Bouarab K (2009) Plant antimicrobial agents and their effects on plant and human pathogens. Int J Mol Sci 10 (8):3400–3419. https://doi.org/10.3390/ijms10083400

Goswami D, Pithwa S, Dhandhukia P, Thakker JN (2014) Delineating *Kocuria turfanensis* 2M4 as a credible PGPR: a novel IAA-producing bacteria isolated from saline desert. J Plant Interact 9 (1):566–576. https://doi.org/10.1080/17429145.2013.871650

Gyaneshwar P, Parekh LJ, Archana GP, Podle S, Collins MD, Hutson RA, Naresh KG (1999) Involvement of a phosphate starvation inducible glucose dehydrogenase in soil phosphate solubilization by *Enterobacter asburiae*. FEMS Microbiol Lett 171:223–229

Han HS, Lee KD (2005) Phosphate and potassium solubilizing bacteria effect on mineral uptake, soil availability and growth of eggplant. Res J Agric Biol Sci 1:176–180

Hartmann A, Rothballer M, Schmid M (2008) Lorenz Hiltner, a pioneer in rhizosphere microbial ecology and soil bacteriology research. Plant Soil 312(1):7–14. https://doi.org/10.1007/s11104-007-9514-z

Hinsinger P, Gobran GR, Gregory PJ, Wenzel WW (2005) Rhizosphere geometry and heterogeneity arising from root-mediated physical and chemical processes. New Phytologist 168 (2):293–303. https://doi.org/10.1111/j.1469-8137.2005.01512.x

Hol WHG, Veen JA (2002) Pyrrolizidine alkaloids from *Senecio jacobaea* affect fungal growth. J Chem Ecol 28:1763–1772. https://doi.org/10.1023/A:1020557000707

Hol WHG, de Boer W, de Hollander M, Kuramae EE, Meisner A, van der Putten WH (2010) Context dependency and saturating effects of loss of rare soil microbes on plant productivity. Front Plant Sci 6:485. https://doi.org/10.3389/fpls.2015.00485

Hooper DU, Hapin FS, Ewel J, Hector A, Inchausti P, Lavorel S, Lawton JH, Lodge M, Loreau M, Aeem S, Schmid B, Setala H, Symstad AJ, Andermeer JV, Wardle DA (2005) Effects of biodiversity on ecosystem functioning: a consensus of current knowledge. Ecol Monogr 75 (1):3–35

Igual JM, Valverde A, Cervantes E, Velázquez E (2001) Phosphate-solubilizing bacteria as inoculants for agriculture: use of updated molecular techniques in their study. Agronomie 21:561–568

Jackson MB (1991) Ethylene in root growth and development. In: Matoo AK, Suttle JC (eds) The plant hormone ethylene. CRC, Boca Raton, pp 159–181

Jogler C, Waldmann J, Huang X, Jogler M, Glockner FO (2012) Identification of proteins likely to be involved in morphogenesis, cell division and signal transduction in Planctomycetes by comparative genomics. J Bacteriol 194(23):6419–6430

Jones DL, Nguyen C, Finlay RD (2009) Carbon flow in the rhizosphere: carbon trading at the soil-root interface. Plant Soil 321:5–33. https://doi.org/10.1007/s11104-009-9925-0

Joo GJ, Kim YM, Kim JT (2005) Gibberellins-producing rhizobacteria increase endogenous gibberellins content and promote growth of red peppers. J Microbiol 43:510–515

Joosten L, Mulder PPJ, Klinkhamer PGL, Veen JA (2009) Soil-borne microorganisms and soil-type affect pyrrolizidine alkaloids in *Jacobaea vulgaris*. Plant Soil 325:133–143. https://doi.org/10.1007/s11104-009-9963-7

Jorquera MA, Hernandez MT, Rengel Z, Marschner P, Mora MD (2008) Isolation of culturable phosphor bacteria with both phytate-mineralization and phosphate-solubilization activity from the rhizosphere of plants grown in a volcanic soil. Biol Fertil Soils 44:1025–1034. https://doi.org/10.1007/s00374-008-0288-0

Kaymak DC (2010) Potential of PGPR in agricultural innovations. In: Maheshwari DK (ed) Plant growth and health promoting bacteria. Springer, Berlin

Kim KY, Jordan D, McDonald GA (1997) Solubilization of hydroxyapatite by *Enterobacter agglomerans* and cloned *Escherichia coli* in culture medium. Biol Fert Soils 24:347–352

Kloepper JW, Ryu CM, Zhang S (2004) Induced systemic resistance and promotion of plant growth by Bacillus spp. Phytopathology 94:1259–1266

Ko WH, Lockwood JL (1967) Soil fungistasis: relation to fungal spore nutrition. Phytopathology 57:894–901

Kogel KH, Franken P, Heuckelhoven R (2006) Endophyte or parasite-what decides? Curr Opin Plant Biol 9:358–363. https://doi.org/10.1016/j.pbi.2006.05.001

Kucey RMN, Janzen HH, Legget ME (1989) Microbial mediated increases in plant available phosphorus. Adv Agron 42:199–228

Kumar M, Kumar V, Prasad R (2020) Phyto-microbiome in stress regulation. Springer, Singapore. ISBN 978-981-15-2576-6. https://www.springer.com/gp/book/9789811525759

Lau JA, Lennon JT (2011) Rapid responses of soil microorganisms improve plant fitness in novel environments. Proc Natl Acad Sci USA 109(35):14058–14062. https://doi.org/10.1073/pnas.1202319109

Li Q, Saleh-Lakha S, Glick BR (2005) The effect of native and ACC deaminase containing *Azospirillum brasilense* Cdl843 on the rooting of carnation cuttings. Can J Microbiol 51:511–514

Liebman JA, Epstein L (1992) Activity of fungistatic compounds from soil. Phytopathology 82:147–153

Lladó S, López-Mondéjar R, Baldrian P (2017) Forest soil bacteria: diversity, involvement in ecosystem processes, and response to global change. Microbiol Mol Biol Rev 81:e00063–e00016. https://doi.org/10.1128/MMBR.00063-16

Lockwood JL (1977) Fungistatis in soils. Biol Rev 2(1):1–43. https://doi.org/10.1111/j.1469-185X.1977.tb01344.x

Lu T, Ke MJ, Peijnenburg WJGM, Zhu YC, Zhang M, Sun LW (2018) Investigation of rhizospheric microbial communities in wheat, barley, and two rice varieties at the seedling stage. J Agric Food Chem 66:2645–2653

Lynch JM, Whipps JM (1990) Substrate flow in the rhizosphere. Plant Soil 129(1):1–10

McGill WB, Cole CV (1981) Comparative aspects of cycling of organic C, N, S and P through soil organic matter. Geoderma 26(4):267–286. https://doi.org/10.1016/0016-7061(81)90024-0

Mendes R, Garbeva P, Raaijmakers JM (2013) The rhizosphere microbiome: significance of plant beneficial, plant pathogenic, and human pathogenic microorganisms. FEMS Microbiol Rev 37:634–663. https://doi.org/10.1111/1574-6976.12028

Narula N, Deubel A, Gans W, Behl RK, Merbach W (2006) Paranodules and colonization of wheat roots by phytohormone producing bacteria in soil. Plant Soil Environ 52:119–129

Nath M, Bhatt D, Bhatt MD, Prasad R, Tuteja N (2018) Microbe-mediated enhancement of nitrogen and phosphorus content for crop improvement. In: Prasad R, Gill SS, Tuteja N (eds) Crop improvement through microbial biotechnology. Elsevier, Amsterdam, pp 291–301

Osbourn A, Goss RJ, Field RA (2011) The saponins: polar isoprenoids with important and diverse biological activities. Nat Prod Rep 28:1261–1268. https://doi.org/10.1039/c1np00015b

Perrig D, Boiero ML, Masciarelli OA, Penna C, Ruiz OA, Cassán FD (2007) Plant-growth-promoting compounds produced by two agronomically important strains of *Azospirillum brasilense*, and implications for inoculant formulation. Appl Microbiol Biotechnol 75:1143–1150. https://doi.org/10.1007/s00253-007-0909-9

Perry RN (2005) An evaluation of types of attractants enabling plant-parasitic nematodes to locate plant roots. Russ J Nematol 13:83–88

Phillips DA, Fox TC, King MD, Bhuvaneswari TV, Teuber LR (2004) Microbial products trigger amino acid exudation from plant roots. Plant Physiol 136:2887–2894. https://doi.org/10.1104/pp.104.044222

Pieterse CMJ, Van Pelt JA, Van Wees SCM, Ton J, Léon-Kloosterziel KM, Keurentjes JJB, Verhagen BWM, Knoester M, Van der Sluis I, Bakker PAHM (2001) Rhizobacteria-mediated induced systemic resistance: triggering, signalling and expression. Eur J Plant Pathol 107:51–61

Prasad R (2008) Studies on interaction between a symbiotic fungus (*Piriformospora indica*), rhizobacteria and selected plants. Ph. D. thesis. CCS University, Meerut

Prasad R, Garg AP, Varma A (2005) Interaction of medicinal plants with plant growth promoting rhizobacteria and symbiotic fungi. In: Podila GK, Varma A (eds) Basic research and applications of Mycorrhizae, vol 1. IK International, Delhi, pp 363–407

Prasad R, Kumar M, Varma A (2015) Role of PGPR in soil fertility and plant health. In: Egamberdieva D, Shrivastava S, Varma A (eds) Plant growth-promoting rhizobacteria (PGPR) and medicinal plants. Springer International, Cham, pp 247–260

Prasad R, Bhola D, Akdi K, Cruz C, Sairam KVSS, Tuteja N, Varma A (2017) Introduction to mycorrhiza: historical development. In: Varma A, Prasad R, Tuteja N (eds) Mycorrhiza. Springer International, Cham, pp 1–7

Prasad R, Chhabra S, Gill SS, Singh PK, Tuteja N (2020) The microbial symbionts: potential for the crop improvement in changing environments. In: Tuteja N, Tuteja R, Passricha N, Saifi SK (eds) Advancement in crop improvement techniques. Elsevier, Amsterdam, Netherlands, pp 233–240

Prashar P, Kapoor N, Sachdeva S (2013) Rhizosphere: its structure, bacterial diversity and significance. Rev Environ Sci Biotechnol 13:63. https://doi.org/10.1007/s11157-013-9317-z

Qiang X, Weiss M, Kogel KH, Schafer P (2012) Piriformospora indica a mutualistic basidiomycete with an exceptionally large plant host range. Mol Plant Pathol 113:508–518

Rasmann S, Ali JG, Helder J, van der Putten WH (2012) Ecology and evolution of soil nematode chemotaxis. J Chem Ecol 38:615–628

Remans R, Beebe S, Blair M, Manrique G, Tovar E, Rao IM (2008) Physiological and genetic analysis of root responsiveness to auxin-producing plant growth-promoting bacteria in common bean (*Phaseolus vulgaris* L.). Plant Soil 302:149–161. https://doi.org/10.1007/s11104-007-9462-7

Riefler M, Novak O, Strnad M, Schmülling T (2006) *Arabidopsis* cytokinin receptor mutants reveal functions in shoot growth, leaf senescence, seed size, germination, root development and cytokinin metabolism. Plant Cell 18:40–54. https://doi.org/10.1105/tpc.105.037796

Sabry SRS, Saleh SA, Batchelor CA (1997) Endophytic establishment of *Azorhizobium caulinodans* in wheat. Proc Biol Sci 264:341–346

Saharan BS, Nehra V (2011) Plant growth promoting rhizobacteria: a critical review. Life Sci Med Res 21:1–30

Schmidt MWI, Torn MS, Abiven S, Dittmar T, Guggenberger G, Janssens IA, Kleber M, Kogel-Knabner I, Lehmann J, Manning DAC, Nannipieri P, Rasse DP, Weiner S, Trumbore SE (2011) Persistence of soil organic matter as an ecosystem property. Nature 478:49–56. https://doi.org/10.1038/nature10386

Schnitzer SA, Klironomos JN, HilleRis LJ, Kinkel LL, Reich PB, Xiao K, Rillig MC, Sikes BA, Callaway RM, Mangan SA, van Nes EH, Scheffer M (2011) Soil microbes drive the classic plant diversity-productivity pattern. Ecology 92:305

Shrivastava S, Prasad R, Varma A (2014) Anatomy of root from eyes of a microbiologist. In: Morte A, Varma A (eds) Root engineering, vol 40. Springer, Berlin, pp 3–22

Shukla KP, Sharma S, Singh NK, Singh V, Tiwari K, Singh S (2011) Nature and role of root exudates: efficacy in bioremediation. Afr J Biotechnol 10:9717–9724

Singh D, Raina TK, Kumar A, Singh J, Prasad R (2019) Plant microbiome: a reservoir of novel genes and metabolites. Plant Gene. https://doi.org/10.1016/j.plgene.2019.100177

Somers E, Vanderleyden J, Srinivasan M (2004) Rhizosphere bacterial signalling: a love parade beneath our feet. Crit Rev Microbiol 30:205–240

Spaepen S, Vanderleyden J, Remans R (2007) Indole-3-acetic acid in microbial and microorganism-plant signaling. FEMS Microbiol Rev 31:425–448. https://doi.org/10.1111/j.1574-6976.2007.00072.x

Subbarao NS (1988) Phosphate solubilizing micro-organism. In: Biofertilizer in agriculture and forestry. Regional Biofert. Dev. Centre, Hisar, pp 133–142

Tejera N, Lluch C, Martínez-Toledo MV (2005) Isolation and characterization of *Azotobacter* and *Azospirillum* strains from the sugarcane rhizosphere. Plant Soil 270:223–232

Van der Heijden MGA, Klironomos J, Ursic M, Moutoglis P, Streitwolf-Engel R, Boller T, Wiemken A, Sanders IR (1998) Mycorrhizal fungal diversity determines plant biodiversity, ecosystem variability and productivity. Nature 396(6706):69. https://doi.org/10.1038/23932

Van der Heijden MGA, Streitwolf-Engel R, Riedl R, Siegrist S, Neudecker A, Ineichen K, Boller T, Wiemken A, Sanders IR (2006) The mycorrhizal contribution to plant productivity, plant nutrition and soil structure in experimental grassland. New Phytol 172:739–752

Van der Heijden MGA, Bardgett RD, Van Straalen NM (2008) The unseen majority: soil microbes as drivers of plant diversity and productivity in terrestrial ecosystems. Ecol Lett 11:296–310

Van West P, Morris BM, Reid B, Appiah AA, Osborne MC, Campbell TA, Gow NAR (2002) Oomycete plant pathogens use electric fields to target roots. Mol Plant-Microbe Interact 15 (8):790–798. https://doi.org/10.1094/MPMI.2002.15.8.790

Wagg C, Jansa J, Schmid B, Van der Hinjen MGA (2011) Belowground biodiversity effects of plant symbionts support aboveground productivity. Ecol Lett 14(10):1001–1009. https://doi.org/10.1111/j.1461-0248.2011.01666.x
Walker V, Bertrand C, Bellvert F, Moënne-Loccoz Y, Bally R, Comte G (2011) Host plant secondary metabolite profiling shows a complex, strain-dependent response of maize to plant growth-promoting rhizobacteria of the genus Azospirillum. New Phytol 189:494–506. https://doi.org/10.1111/j.1469-8137.2010.03484.x
Weston LA, Ryan PR, Watt M (2012) Mechanisms for cellular transport and release of allelochemicals from plant roots into the rhizosphere. J Exp Bot 63:3445–3454. https://doi.org/10.1093/jxb/ers054
Whitelaw MA (2000) Growth promotion of plants inoculated with phosphate solubilizing fungi. Adv Agron 69:99–151
Wu Y, Fang W, Zhu S, Jin K, Ji D (2008) The effects of cotton root exudates on the growth and development of *Verticillium dahliae*. Front Agric China 2(4):435–440
Yaxley JR, Ross JJ, Sherriff LJ, Reid JB (2001) Gibberellin biosynthesis mutations and root development in pea. Plant Physiol 125:627–633. https://doi.org/10.1104/pp.125.2.627
Young IM, Griffiths BS, Robertson WM (1996) Continuous foraging by bacterial-feeding nematodes. Nematologica 42:378382. https://doi.org/10.1163/004425996X00100
Zakharova EA, Iosipenko AD, Ignatov VV (2000) Effect of water-soluble vitamins on the production of indole-3-acetic acid by *Azospirillum brasilense*. Microbiol Res 155:209–214. https://doi.org/10.1016/S0944-5013(00)80034-8
Zhang H, Kim MS, Krishnamachari V, Payton P, Sun Y, Grimson M (2007) Rhizobacterial volatile emissions regulate auxin homeostasis and cell expansion in *Arabidopsis*. Planta 226:839–851. https://doi.org/10.1007/s00425-007-0530-2
Zhou CX, Liu JY, Ye WC, Liu CH, Tan RX (2003) Neoverataline A and B, two antifungal alkaloids with a novel carbon skeleton from *Veratrum taliense*. Tetrahedron 59:5743–5747

Chapter 11
On the Possibility of Accelerating Succession by Manipulating Soil Microorganisms

Virgil Iordache

Abstract Soil microorganisms have effects on brown and green food webs and influence processes up to the ecosystem scale. It is difficult to separate and quantify such effects at site and landscape scale because of negative and positive feedbacks between the various communities and the large numbers or variables. There is not a single research strategy to reduce the dimensionality of the system, which leads to complementary traditions and bodies of knowledge. The objective of this chapter is to provide an image on the structure of the existing relevant knowledge and extract information supporting the use of microorganisms for accelerating succession at ecosystem scale. After describing the succession patterns by groups of organisms (bacteria, fungi, plants, invertebrates, small mammals, and other vertebrates), the literature about the strength of the interaction between community scale processes is reviewed, and key variables or ecological objects influenced by soil microorganisms are identified. Then complementary conceptual tools useful for the practical application of this knowledge are introduced: hotspots, state-and-transition models, network analysis, and coupling models. The last part includes a scheme for a comprehensive multifunctional practical approach putting the use of fungi in the general context of landscape sensitive restoration. When needed soil microbial manipulations could be distributed in space with discretization units having the size controlled by the heterogeneity of abiotic conditions and the spatial structure of vegetation supporting an optimal production of ecosystem services.

V. Iordache (✉)
Research Center for Ecological Services – CESEC, University of Bucharest, Bucharest, Romania
e-mail: virgil.iordache@g.unibuc.ro

A. Varma et al. (eds.), *Plant Microbiome Paradigm*,
https://doi.org/10.1007/978-3-030-50395-6_11

11.1 Introduction

Epistemic strategies for complex socio-ecological problems involve families of scientific sub-discplines dealing with processes of increasing complexity: ecophisiology, population ecology, community ecology, evolutionary ecology, ecosystem ecology and systems ecology. Each one is a strategy for the reduction of complexity, of dimenstionality in terms of variables and relations in between. Restoration ecology is such a complex socio-ecological problem. Thinking restoration or remediation in relation with ecological succession is already a well established approach (Walker et al. 2007). The problem has a large theoretical and methodological heterogeneity because it crosscuts many disciplinary fields, as shown in the first phrase, with different strategies to reduce the dimensionality of the natural complexity.

First of all succession and community assembly are different and complementary theoretical frameworks relevant for this problem (Chang and HilleRisLambers 2016). Community assembly studies investigate the rules and mechanisms relating local diversity patterns and the regional species pool; are characterized by key concepts like species co-occurrence, functional traits, and dispersion; and usually lack a temporal dimension. How the relative importance of niche and neutral processes varies among taxa, along environmental gradients, and across scale is a strategic research direction in community assembly theory (Weiher et al. 2011). Succession studies have key concepts like disturbance, ecosystem development, legacy effects, and threshold effects and place the processes controlling the community structure in a temporal context. According to Chang and HilleRisLambers (2016), common concepts relating the two research traditions could be species pool, priority effects, dispersal filters, abiotic filters, and biotic filters.

In an integrated model, one would expect community assembly theory contributing more to the first phases of community dynamic, controlled mainly by dispersal and abiotic variables, and succession theories more in the later phases, with larger influence of intraspecific and interspecific biotic interactions. The continuum hypothesis states that both deterministic and stochastic processes contribute to the assembly of ecological communities (Powel et al. 2015).

Understanding the relationship of species richness with space and time involves deep conceptual issues (Scheiner et al. 2011) related to the type of entities (ecological objects and processes), their scales, and the research methodologies. For instance, interpreting measurement at one moment as the result of processes in time is only an option: "in the absence of information about historical trajectories, assembly rules are better thought of as patterns of co-occurrence that are statistically different from patterns that are produced by randomly sampling—'assembling'—species from the appropriately delimited species pool" (Falconer et al. 2015). When using the information for management, one has to be aware of such issues.

Holistic ecosystem models based on energy flow and biogeochemical stocks and fluxes, with strongly reduced dimensionality, may be useful for monitoring the success of ecological restoration (Pietrzykowski 2014), but they cannot support

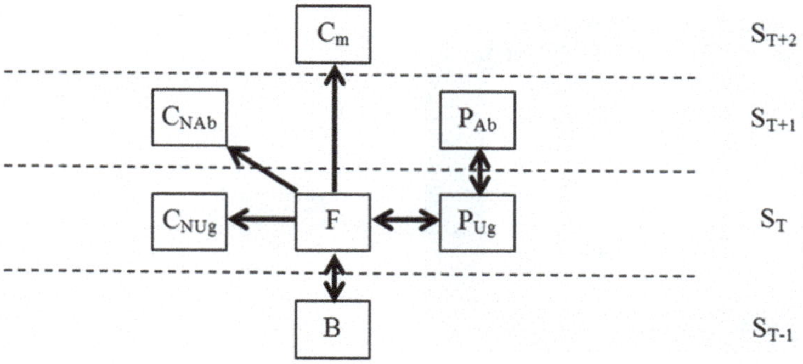

Fig. 11.1 Structural model accounting for the role of fungi in the local network of interactions between organisms involved in the successional processes at community and ecosystem scale. Each square represents a group of statistical populations of different species (trophic dynamic module) including one or more services providing units (smaller groups of populations differentiated in function of their ecological role). *Legend: F* fungi, *B* bacteria, *PUg* underground parts of plants, *PAb* aboveground parts of plants, *C* consumers, C_{NUg} fungivorous underground invertebrates, C_{NAb} fungivorous aboveground invertebrates, C_m fungivorous small mammals, *ST* target scale (characteristic to fungi), $S_{T-1,\ +1,\ +2}$ scales smaller and larger than the target scale

operational measures about species in the active restoration phase. Farther complexity comes from the fact that succession (or community assembly) studies focus on a limited group of organisms or on several groups of organisms and rarely target the whole ecosystem to be restored. Also, besides the classic species composition and diversity, it developed toward the intensive research of intra- and interpopulational functional traits diversity. A general assembly framework based on functional traits and species richness is proposed, for example, by D'Amen et al. (2015), from which one could devise particular adapted schemes for various groups of organisms.

In this context the objective of this chapter is to screen the literature about the succession of soil microorganisms and of groups directly related in the trophic network with soil microorganisms (Fig. 11.1), about the coupling processes between these groups, and to extract relevant information about how and in what context manipulating the soil microorganism might be useful for the acceleration of succession as a management objective.

The structural model from Fig. 11.2 limits to processes at site scale. Coupling between site scale processes can be done by biotic or abiotic processes of larger scale (Iordache et al. 2012). For instance, Lundberg and Moberg (2003) analyze a case of biotic processes in terms of "mobile link organisms," actively moving in the landscape and connecting habitats in space and time, and identify three functional categories: resource, genetic, and process linkers. They conclude that diversity within these functional groups is a central component of ecosystem resilience and that such knowledge needs to be incorporated in the management and policy-making decisions. This may already lead to the idea that manipulation of microorganism can be just an operational measure in a portfolio of coupled scale-specific management activities serving various restoration objectives.

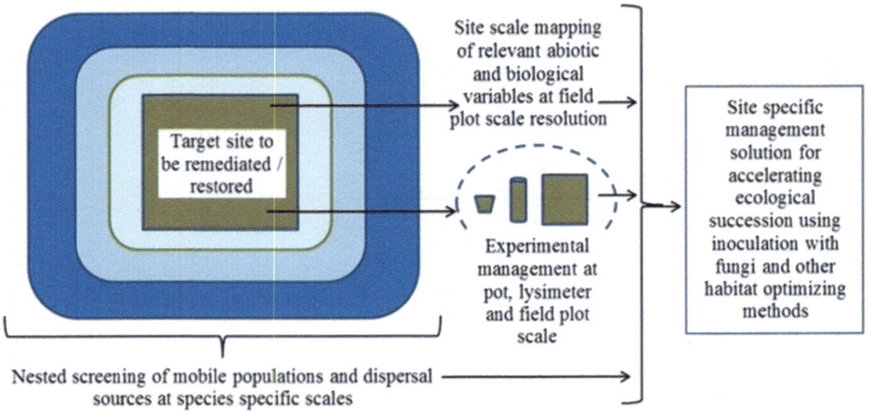

Fig. 11.2 Multiscale field and experimental approach for designing a portfolio of operational measures including soil microbial manipulation for the restoration of a site. It is complementary with the use of non-native species when the construction of a new ecosystem is needed

The structure of the chapter is as follows:

- In the first part, I present patterns of succession by groups of organisms (bacteria, fungi, plants, invertebrates, small mammals, and other vertebrates), and then I compare these patterns.
- The second part describes processes coupling groups of organisms (bacteria and fungi with plants, fungi with underground consumers, underground consumer with aboveground consumers (mediated by fungi and plants), and the case of many groups across many scales) and discusses the strength of these interactions.
- In the third part, I introduce several independent notions which might be useful for the practical application of the knowledge reviewed in the first and second chapters (hotspots, state-and-transition models, network analysis, and coupling models).
- The last part is dedicated to the practical issues.

11.2 Patterns of Succession by Groups of Organisms

11.2.1 Bacteria

In a recent review, Langenheder and Lindstrom (2019) looked for the effects of the ecological context on the relative importance of processes like drift, environmental selection, and dispersal in the assembly of bacterial communities in aquatic and terrestrial ecosystems and found little consistency between studies. They suggested the investigation of how the effects of the relevant factors on the assembly process differ between habitats and organisms. In his review Baldrian (2017) separates 13 microbial microhabitats in a forest ecosystem, atmospheric, foliage, bark surface, wood, ground vegetation, streams, wetlands, roots and rhizosphere, soil, litter, deadwood, rock surface, and invertebrates; discusses their characteristics, the connections in between mediated by fungi during their lifecycles; and compares the timescales of ecological processes affecting each type of microhabitat.

Microbial communities decomposing litter (bacteria and fungi) were stochastically controlled by the species pool in the landscape in the first phase and then more deterministically controlled by the composition of organic matter, water chemistry (when in water), and species interactions in later stages (Fischer et al. 2009). The time after which the maximum taxonomic richness occurred was about 30 days. Copiotrophic microorganisms are more responsive to carbon sources upon availability, while oligotrophic ones are less reactive to abrupt resource availability but are able to exploit nutrient-poor environments (Ho et al. 2017). This two-way of classifying microorganisms for the understanding of their succession evolved toward the three-way continuum between competitors—stress tolerators and ruderals used in the case of plants (Ho et al. 2017).

Stamou and Papatheodorou (2016) looked for the role of 21 variables (selected from 77 and grouped in the categories soil chemical composition, microbial community structure, catabolic activity of microbial community, and enzymatic activity in soil) in depicting the succession course. They found by structural equation modeling that soil chemical background overrides the statistical influence of the other variables. Soil microbial communities in two chronosequences were more influenced by particulate organic carbon than plot age, corresponding with higher level of bioturbation (Bartuska et al. 2015). Due to their short life cycle, bacterial communities can change very fast. A single application of an NP fertilizer caused the soil bacterial community structure of a 3-year-old soil to resemble an 85-year-old soil after 1 year (Knelman et al. 2014). Community adaptability to environmental changes can be decreased by toxic stress. The results of Jacquiod et al. (2018) showed that in Cu-polluted soils the microbial community was less able to adapt to environmental fluctuations, thus losing part of its regulating function. The effect of environmental variables on the community structure differs with the scale. Bacterial diversity at cm scale was very heterogenous, with large changes from sample to sample, but at ecosystem scale (>10 m for these authors), a correlation with fertilization of the plots could be observed (O'Brien et al. 2016).

Soil microbial community structure can remain stable after long-term succession, with only microbial biomass changing across different soil environments corresponding to different chronosequences (Xu et al. 2018). In early stages both biomass and community structure were affected by nutrient status.

Bacterial communities in abandoned tailings contained a genus beneficial to plant growth after 23 years or more of natural attenuation but also genera responsible from the acidification of the tailings and inducing risks for the human health and the environment (Liu et al. 2019). pH, total organic carbon, total nitrogen, As, Pb, and Cu were the main drivers influencing the bacterial community structure, and the authors recommend careful monitoring for detecting the transition between pre-acidification and acidification during natural attenuation to enable timely management.

11.2.2 Fungi

Falconer et al. (2015) provide a detailed analytical framework for understanding the sensitivity of trait groupings/guilds to environmental disturbances at a range of scales and to understand the response of processes controlled by fungi to such disturbances, including steps for a spatially explicit modeling of the fungal community dynamics in the soil. Determining and linking the scales at which to measure the environmental variables, the structure, and functioning of fungal communities are crucial for modeling (Falconer et al. 2015).

While the key determinants of microbial communities are known, the differential response of functional groups is not yet understood, requiring an understanding of major fungal taxa (Falconer et al. 2015). Fungal species have a wide pH optimum, covering 5–9 pH units (Ding et al. 2017). Heavy metals affect the growth of mycelia, initiation of primordia, and development of fruiting bodies of fungi (Dulay et al. 2015). In a 35 years experiment, inorganic fertilizers (NPK) were found to decrease the fungal diversity, a trend counteracted by manure addition (Ding et al. 2017). Soil physical structure and organic matter were the best predictors of changes in fungal diversity along a chronosequence (Dini-Andreote et al. 2016). Complex pollution such as that with garbage leads to a change in the soil abiotic and biotic properties, decreases the mycelial biomass, inhibits the formation of ectomycorrhizal fungal fruiting bodies, and alters their community structure (Sun et al. 2016a).

A general understanding concerning the relationship between the diversity and functioning of fungi is lacking (Falconer et al. 2015). Fungal community composition was found to be significantly related to soil fertility, with *Ascomycetes* dominating in less fertile soils and *Basidiomycetes* increasing under more fertile conditions (Sterkenburg 2016). Fungal traits like spore and sporocarp characteristics are correlated with environmental variables; climate influences the sporocarp phenology and production and has an effect on sporocarp production and species composition (Andrew et al. 2016). Integrating ecophysiological inference into an ecological succession framework of fungi is important (Dini-Andreote et al. 2016).

For fungi in general, the role of competition in structuring the communities is clearly documented. Dong et al. (2016) found a directional replacement model for fungi communities on a 90 years chronosequence, with soil development facilitating arrival of new fungi species, mid-successional diversity maximum that contained both early- and late-successional fungi, and then decrease of overall diversity due to loss of early successional species.

The assembly of mycorrhizal communities in a bare site depends in the first instance on regional stochastic aspects like the spatial distribution and abundance of species around a site and later on deterministic selection in function of plant traits coupled with local stochastic processes, soil conditions, light, and priority effects related to plant species (Chagnon 2015). A similar hierarchical assembly scheme for arbuscular mycorrhizal fungal communities based on traits is proposed by Davison et al. (2016) and Valyi et al. (2016) who underline also the need for a hierarchical spatial structure in the study of arbuscular mycorrhizal fungal communities, accounting for the role and relative importance of different assembly processes. Van der Wal et al. (2012) have a similar approach for the case of fungal succession on logs and the effect of terrestrial decomposition at different scales. Aspects related to saprotrophic fungi succession will be presented in Sect. 2.4 when discussing substrate succession. Below, I have summarized information about mycorrhizal fungi, which are often used in remediation.

Bahram et al. (2015a) in their review did not detect trends in spatiotemporal variation among mycorrhizal types but an important vertical zonation related to soil variables and the distribution of fine roots. Topsoils vary more than lower horizons in the spatial heterogeneity of mycorrhizal fungal communities in correlation with the spatial heterogeneity of abiotic variables. In order to separate stochastic and deterministic (seasonal, successional—environmental filtering and competition) effects of time on the structure of the mycorrhizal community, one needs replicated sampling across 3 or more years (Bahram et al. 2015a) eventually coupled with stratified sampling by depth (confounding effects due to dispersal are less likely than in horizontal stratification). Bahram et al. (2015b) found that for all eukaryotes mass effect and ecological drift are the main drivers of communities at small (50 m) scale in the absence of environmental gradients. When a spatial structure was present for certain groups, it corresponded to the spatial structure of the vegetation.

A review about ectomycorrhizal fungi and upscaling information about them are available in Iordache et al. (2011). Jumpponen and Egerton-Warburton (2005) proposed a theoretical model for the assembly of mycorrhizal fungal communities in successional environments operating mainly on small, local scales and including a host filter (compatibility among host and fungi), an environmental filter, and a biotic filter (facultative and competitive interactions among fungi). Koide et al. (2011) propose similar general principles for the community ecology of ectomycorrhizal fungi in the form of a hierarchical structural model with host filtering, abiotic filtering, and a potential community leading to the realized one after competitive interactions of different strengths (in homogenous habitats they would lead to competitive exclusions, while in heterogenous habitats—eventually resulted from disturbance—they could lead to coexistence of taxa).

Kalucka and Jagodzinski (2016) review the successional traits of ectomycorrhizal fungi in forest reclamation after surface mining pointing out the role of life histories, dispersal, spatial structure, host preferences, and sensitivity to environmental filters. They distinguish a colonization stage (fast increase in species richness, diversity, sporocarp abundance, and biomass production, up to 12–16 years of forest age), a stabilization stage (up to 18–25 years), and a replacement stage (up to 100 years). The study of arbuscular mycorrhizal fungi succession at a single tree scale showed that older trees have more fungal taxa than younger trees, different fungal communities, and that such differences are not observable early in the life cycle (Hart et al. 2014).

The physiological trait supporting the interspecific relation between mycorrhizal fungi and tree seedlings is modulated on environmental gradients from parasitism to mutualism (Ibanez and McCarthy-Neumann 2016). For instance these authors found a positive effect of colonization on plant growth at high light and a negative effect in the case of five plant species at low light. There were light thresholds for shifting from neutral to positive, negative to neutral, and neutral to negative effects.

Hupperts (2016) found no difference in ectomycorrhizal fungal community composition across sites differing in extent of above- and belowground disturbances; the composition was primarily affected by the species of seedling used to assess the soils. Planting a diverse community of trees in reclaimed soils could yield a diverse community of belowground fungi. Seedling biomass was found to be positively correlated with ectomycorrhizal fungi richness up to an optimum and then negatively correlated (Peay and Bruns 2014). Clarifying the role of the interactions between ectomycorrhizal fungal species in relation to the genet size and to the influence of other soil variables is a research priority (Pickles and Anderson 2016).

Current patterns of ectomycorrhizal distribution at very large scale reflect the importance of dispersal barriers, with consequences on how they can be used for remediation, contrary to the pre-molecular view of unlimited dispersal for this group (Peay and Mathney 2017). At landscape scale, the spore dispersal of these fungi is driven by stochastic and deterministic processes, leads to local variation in fungal community structure, and generates variability in plant-fungal interactions (Peay and Bruns 2014). Disturbance effects on variability may go undetected if there is a mismatch between the scale of observation and the scale at which the dominant processes occur (Fraterrigo and Rusak 2008). Increased variability may be a precursor of large abrupt system changes (Fraterrigo and Rusak 2008).

Arbuscular mycorrhizal fungi with ruderal traits (ease of sporulation) can be fast colonizers of early successional habitats from neighboring habitats (De Leon et al. 2016), in less than 1 year. The intra-community interactions between mycorrhizal fungi play a role in the succession of this group in some cases, but in other ones, this may not be the case. Davison et al. (2016) could not find evidence for a role of local competitive interactions between fungi in explaining the structure of the communities. Dumbrell et al. (2010), on the other hand, point out the role of strong interspecific interactions between fungi and of the positive feedback associated to stochastic initial nonspecific colonization of plant roots in the production of communities overdominated by some fungi species.

Although arbuscular mycorrhizal fungi lack saprophytic capability, they preferentially associate with organic substrates, respond by hyphal proliferation, and may enhance the decomposition of organic material by facilitating its physical penetration, local changes in pH, and N release during hyphal turnover (Hodge 2014). These processes modulate also the interaction of arbuscular mycorrhizal fungi with soil invertebrates like protozoa, Collembola, and earthworms (Hodge 2014).

Trace elements interact in mining-affected sites with organic matter in controlling the diversity of arbuscular mycorrhizal fungi (Montiel-Rozas et al. 2016), but depending on the context, the role of trace elements can be secondary, with organic matter content the most significant factor (Montiel-Rozas et al. 2017). The same fungal species were found both in metal-polluted and non-polluted forest soil, but their relative abundance differed (DeBeeck et al. 2015). In this study initially the colonization of tree roots was done by Ascomycota, replaced within 2 years by Basidiomycota.

In similar environments, arbuscular mycorrhizal fungi communities were unpredictable based on environmental conditions suggesting more stochastic processes than those based on environmental filtering (Powel and Bennet 2016). The degree of unpredictability was related to geography and the characteristics of the host plant. Sepp et al. (2019) demonstrated that the interaction network between arbuscular mycorrhizal fungi and plant species can be assembled nonrandomly, with a degree of interaction selectivity larger for forbs than for grasses.

A research priority's relation to community ecology of mycorrhizal fungi is "to clarify the complex and apparently context-dependent responses of arbuscular mycorrhizal fungal communities to anthropogenic activities and disturbances" (Opik and Davison 2016).

11.2.3 Plants

McCook (1994) reviews in detail the causal models and theories with respect to the succession of vegetation, with particular attention to the resource-ratio hypothesis of plant selection (Tilman 1985, 1990). Drivers of succession can be broadly classified in site conditions and history, species availability, and species performance, each of them depending not only on local site conditions but on processes (related to the geographic and evolutionary contexts) occurring at larger scales (Meiners et al. 2015).

The small-scale heterogeneity of soil resources did not influence the vegetation heterogeneity in the early succession but was controlled by interspecific interactions (Collins and Wein 1998). Colonization of bare soil at former mining sites was found to be a nonrandom process with strong positive and negative plant-plant interactions, especially of intraspecific type (Birgit and Wiegand 2008), facilitation being substituted by competition with ongoing succession. In spontaneous succession on post-mining sites, cover of dominant species, total plant cover, and number of species in younger chronosequence sites explained 24% of the variability in species composition in late-successional stages (Mudrak et al. 2016), with consequences on the importance of initial inventory when designing restoration plans.

Grime's theory on the role of traits in the competitive success of plants considers a broad range of traits including reproductive effort, dispersal characteristics, and other attributes affecting the colonization process, while Tilman's theory focuses on resource-use traits (Grace 1991). Plant traits variation on a heavy metal gradient (height, leaf area, specific leaf area, metal concentrations) was correlated with soil concentration at community level, as a result of the process of species turnover, but not at species level (Delhaye et al. 2016). Armesto et al. (1991) suggest a cyclic change in the spatial heterogeneity of plant community during succession as a result of invasions and establishment of species able to monopolize the space (annuals or clonal species) alternating with species exclusion. Shrubs have positive (facilitation) and negative (competition) effects on understory plants, the net interaction effect being modulated by abiotic conditions at the site and their internal heterogeneity in space (Macek et al. 2016).

There are species-specific facilitation interactions between nurse and beneficiary species during plant community succession, controlled also by the beneficiary age (during ontogeny the interaction can shift from positive to neutral to negative interactions, Paterno et al. 2016). June-Wells et al. (2014) found that the dynamic nature of plant population borders in polluted and disturbed habitats varied on a species by species base and that population dynamics of differing life-form guild (forbs/shrub) were not necessarily related to direct competitive interactions (are a result of a mosaic of competitive, neutral, and potentially facilitative interactions). Teste (2016) suggested that the restoration of grassland around remnant patches could be increased by adding local inoculum with arbuscular mycorrhizal fungi at different distances from the patch (5–20 m, facilitating recolonization by plants). The positive effect of inoculum on species diversity in early stages of restoration decreased with the distance from the intact patches, probably related to priority effects (Torrez et al. 2016).

Dispersion limitation may determine the final structure of the community. Vegetation composition in rehabilitated bauxite mines did not become more similar to the unmined forest during 14 years since seeding but reflected the initial species mix (Norman et al. 2008). Li et al. (2016a, b) investigated the convergence of plant communities in a 50-year study of post-agricultural secondary succession and found that dissimilarities decreased in time at larger field scale (hectares), but at plot scale (square meters), there was a divergence both in species composition and abundance. Thus, the importance of deterministic and stochastic processes varied strongly by scale.

On mine pyrite tailings, phosphorus deficiency was found to be the major limiting factor for plant development (Nikolic et al. 2011). Manipulating soil pH was found important for accelerating or directing plant succession on reclaimed coal wastes (Alday et al. 2011). In this study the vegetation approached the composition of native vegetation within 32 years. In favorable conditions (iron-mining sites), the simple revegetation of the sites can lead to an arbuscular mycorrhizal community comparable with that in adjacent pristine sites (Vieira et al. 2018). There are functional consequences of vegetation succession on other groups by the quality of the biomass production and the structure of aboveground and belowground microhabitats. For instance, tree litter in mid and late stages on a primary post-mining chronosequence decomposed faster than the grass litter of the early stage under the control of C and N concentrations in litter (Urbanova et al. 2014).

11.2.4 Invertebrates

Succession of invertebrates takes place at many scales from microhabitat to site in function of the scale of species demographic processes. Bastow (2012) distinguishes space succession (seral succession) and resources (substratum) succession. With respect to **substratum succession** logs, animal dung and carrion are the best studied. Results have pointed out the tight coupling between the successions of fungi of invertebrates. "Fungi increase in abundance relative to bacteria during the decomposition of litter, (reflected by changes in the nematode assemblage)," "among saprotrophic fungi, early successional species (primarily Ascomycota and Zygomycota) utilize more labile or soluble carbon from detritus, including sugars and cellulose, while later successional species (primarily Basidiomycota) utilize more recalcitrant carbon, especially lignin, chitin and tannins" (Bastow 2012). "Saprotrophic fungi are consumed by other parasitic and saprotrophic fungi that occur later during the succession." "There is a delay before microbivores colonize litter, and then a shift from bacteriovorus to fungivorous fauna," fungi feeders including microfauna nematodes, mesofauna Acari (mites), Collembola, Tardigrada, Protura, Enchytraeidae, macrofauna ants (Bastow 2012). Then arrives predatory fauna. The abundance of soil organisms on organic matter is also affected by seasonal change, which interferes with the successional dynamic (Bastow 2012). Future direction in this area includes experimentally distinguishing between detritus as a resource and detritus as a habitat and clarifying the role of early successional species in changing the detritus as a resource or as a habitat (competition vs. facilitation).

The colonization patterns of deadwood differ in function of the cause of death, which controls the moisture and available nutrients (Boulanger and Sirois 2007). One phase of colonization corresponds to standing dead trees, and the second one to fallen dead trees, finally the community converging with the fungivorous and saprophagous one specific to the forest floor (Boulanger and Sirois 2007). Fungal community composition and substratum succession also differed between three wood types during decay in a forest soil (Prewitt et al. 2014). Earthworm space succession in waste sites may occur at a timescale of 10 years with patterns influenced by the litter quality and feces deposited by rabbits from the surrounding areas (Butt and Briones 2017). Ant species richness in an area without trees was not correlated to plant species richness, or a particular plant species, but predicted by moss cover and soil moisture; ant species composition depended on vegetation structure (diversity of microhabitats) and productivity (Dauber and Simmering 2006). In another study Dejean et al. (2008) found that trees have some control over their ant associates, probably by selective attractiveness or the existence of tree filters that screen the ants. In this case the plant individual in itself was a microhabitat for ants. Fungi and ammonifying bacteria are active and better represented in ant nests than in the surrounding soil, which could contribute to the recovery of metal-polluted areas (Grzes 2010).

Factors critical for ant communities' composition on a coal mining spoil were related to successional age (plot age, depth of humus and fermentation layers, cover of bare soil, and soil substrate) and the type of habitat—open or forest (tree cover and presence of shrubs) (Holec and Frouz 2005). During succession of a pasture, ant communities evolved from a fine-grained structure to a broad grain one; spatial heterogeneity increased reflecting community segregation (Zorilla et al. 1986). At successional maturity, the first scale of ant organization is related to geomorphological sectors within the slopes and the second the positions within each sector.

Koehler (1998) did not find a relationship between plant associations and soil mesofaunal succession in a 13 years experiment. The invertebrate succession was controlled by the above- and belowground plant structure providing more diverse microhabitats. Acari subgroups had very different succession trends across forest stands of ages from 16 to 125 years (increasing, decreasing, rise and fall, and U-shaped, Falenczik-Kozirog et al. 2012). In post-mining sites, no group of soil fauna preferred early successional stages, but some of them were tolerant to these stages (Frouz et al. 2008). The density of most guilds increased in intermediate stages, and macrosaprophagous guilds of soil fauna attained the highest density in the oldest sites (Frouz et al. 2008). In a post-mining landscape, collembolan and oribatid species diversity were the most sensitive mesofaunal parameters to evaluate the restoration success at a timescale of 12 years (Andres and Mateos 2006) and are useful as indicators in areas with heavy metal pollution needed for remediation measures (Manu et al. 2017).

Mechanisms explaining the (apparently too) large number of species in soil are predation on the competitively dominant consumer, disturbance, spatial and temporal heterogeneity, favorable abiotic conditions, low resource competition, the large range in body sizes, and specialization of detritivores and decomposers on particular successional stages of their resource (Bastow 2012).

In a unique book dedicated to post-mining succession, Frouz et al. (2014) review the soil macro- and mesofauna succession in post-mining sites and other disturbed areas. Hanel et al. (2014) review in the same book the soil microfauna (body width of less than 0.2 mm). Primary succession begins in their case with random dispersal within a few days and after 1 year reaches a food web structure controlling top-down the microbial populations.

Post-mining areas often include freshwater habitats involved in successional processes and preserving large species diversity of invertebrates useful for the whole landscape (Harabis 2016). Minor disturbances to prevent the excessive overgrowing of vegetation and restart succession promote the habitat heterogeneity and the preservation of high biodiversity (Harabis 2016). The species richness of various invertebrate groups depends on different soil properties separately or in interaction with microclimate and management history of post-mining forests (Hendrychova et al. 2012). A combination of artificial plantation and spontaneous forest development toward a mosaic of habitats with microhabitat diversity (including small not reclaimed areas of mining wastes) supports the maximum invertebrate richness (Hendrychova et al. 2012).

The investigation of epigeic beetle communities in 30 years successional habitats of a post-industrial area founded differences from the processes occurring in natural habitats, namely, the larger importance of abiotic factors compared to intraspecific competition during succession (Hodecek et al. 2015), especially in the context of periodic small-scale disturbance of the vegetation cover. Habitat age did not affect the assemblages of beetles on a reclaimed open-cast mining area, but the species richness was influenced by surface structure (not leveled soil, present of pits) and canopy cover (Topp et al. 2010). Courtney et al (2010) reported data about the colonization of amended and unamended bauxite tailings with macro-artropods They found few immobiles in the unamended sites, but a recovery of the trophic structure for the amended vegetated sites.

Catches of ground beetles on a reclaimed mine spoil were positively correlated with the amount of vegetation cover and declined from amelioration with sewage sludge to compost, to mineral fertilizer, and to untreated control (Kielhorn et al. 1999). Shelter and humid microclimate were the determining factor. For some endangered ground beetles, however, open sites without vegetation are more appropriate. Initial recolonization and dominance of beetles on reclaimed surface coal mines was achieved by species rare in the adjacent native vegetation, their species richness and diversity increased during the first 3 years following revegetation and then declined for the next 3 years (Parmenter and Macmahon 1987).

11.2.5 Small Mammals and Other Vertebrates

The patterns of small mammal community succession in disturbed habitats did not closely match the expectations of the habitat accommodation model, suggesting that the conservation of these communities can be influenced by habitat management at large spatial scales (Holmes and Robinson 2016). For mammals there is not a facilitation mode but an overlap of different species-specific patterns and mechanisms, correlated in some cases with the microhabitats opened by vegetation and litter (Fox 1982). Each mammal species enters in the community at a stage where the vegetation succession expressed as vegetation density best meets its requirements (Fox 1990; Monamy and Fox 2000). Small mammal species do not modify the local physical conditions at the timescale of the community succession but occur in function of the optimal range for the species as perceived by the organisms at their specific scale. By habitat manipulation experiments, the same author demonstrated the role of vegetation density inducing a retrogression in the small-mammal succession (Fox et al. 2003). Two mycophagous mammals of similar size showed a strong preference for specific microhabitats (fine-scale vegetation assemblages), while a third one did not prefer any of the investigate habitat gradients (Vernes 2003).

Swihart and Slade (1990) pointed out the importance of timescale when analyzing the succession of small mammals. Only a monitoring of 15 years (compared with 2–3 years) was appropriate for clarifying the relationships between the patterns of several species as controlled by the heterogeneity and successional stage of habitat and by interspecific relations.

Reptile recolonization in post-mining restoration sites was influenced by canopy height, litter cover, coarse woody debris volume, and not by landscape factors, indicating that the habitat suitability is the main barrier for such species in landscapes with permeable matrices (Triska et al. 2016).

Larger-scale organisms like birds are totally decoupled from the patterns of microbial processes. For instance, the presence of needed habitat types at landscape scale (100 ha) was the most important variable controlling the fungivores, insectivores, and canopy feeder birds, while at site scale (1 ha), the environmental heterogeneity in the surrounding landscape was the most important variable (Burgess and Maron 2016). Species-specific home range, foraging, and nesting requirements control the scale of response to the environment and provide unique dynamics to each local community of such organisms (Burgess and Maron 2016).

In general the effects of habitat heterogeneity of animal distribution vary between species groups and depend on the perception of vegetation structural attributes as heterogeneity or as fragmentation (Tews et al. 2004). The effect of heterogeneity on a species differs in relation to the spatial scale (<100 m^2, 100 m^2–1 ha, 1 ha–1 km^2, >1 km^2), and in some cases, there are "keystone" structures of vegetation (defined as "distinct spatial structure providing resources, shelter or 'goods and services' crucial for other species") determining the presence of animal species diversity (Tews et al. 2004).

11.2.6 Comparison of Successional Patterns

Habitat turnover was the primary driver of bacterial community turnover in soil with its importance decreasing with increasing isolation and disturbance (Powel and Bennet 2016), while fungal communities were independent of disturbance, with highly stochastic assembly process. The contribution of deterministic and stochastic processes varied depending on the ecological context in which the processes were active (Powel and Bennet 2016), impinging on the use of spatially distributed data sets to detect the effects of climatic gradients on the structure of the communities. This may be a general conclusion for phenomena resulting from the coupling of multiscale processes.

A fungal community specific to a tree species formed in the first year after planting and a bacterial one in the second year (Rigg et al. 2017). The successional trajectories for bacteria and fungi may be quite different (Schmidt et al. 2014) one reason being that smaller microbes are less dispersal limited than larger microbes leading to more deterministic community assembly patterns for bacteria. Another reason is related to the different adaptability of bacteria to some environments compared to fungi.

Sun et al. (2016b) found maximal diversity of bacterial communities in mid-age sites on a successional chronosequence and a similar trend for fungal communities. They explained these patterns by less environmental stress and more niches in the middle successional stages. Changes in soil prokaryotic community diversity and taxa abundance paralleled changes in plant community in a forest succession and could be inferred from changes in soil organic matter properties (Shao et al. 2019). Zhang et al. (2018a) reported that plant and bacterial secondary succession took place in parallel and faster than fungal succession in restored grassland after grazing exclusion. Bacterial and fungal succession were controlled in this case by organic carbon, total nitrogen, nitrate, and also moisture in the case of fungi, in correlation with plant diversity. Vegetation cover and soil moisture controlled most of the fungal secondary succession following agricultural abandonment (Zhang et al. 2018b), but soil pH also played a role. In this case the soil bacteria were reported to exhibit a different successional pattern than fungi.

In the restoration of mining area, vegetation played a major role in bacterial and archaeal diversity and assemblies, while soil characteristics, especially nitrogen, were important for fungal diversity and assemblies (Li et al. 2016a, b). These authors suggest that "there are different drivers of bacterial, archaeal and fungal succession during secondary succession in a reclaimed open mine."

Ruess and Lussenhop (2012) compare the changes and fungi and fauna during the decomposition process. Fungi succession starts with pioneer species and continues with polymer-degrading ones, degraders of recalcitrant compounds, and secondary opportunistic invaders; Enchytraeids change from fungivores to detritivores; nematodes from bacterivores to fungivores; springtails from fungivores to detritivores; and mites from predatory to fungivores to detritivores and predatory again. Taxonomically related species do not necessarily use similar fungal food resources (Ruess and Lussenhop 2012).

Stursova et al. (2016) compared the small-scale spatial heterogeneity of bacteria, fungi, and vegetation in a forest soil. They found that soil fungi communities were more heterogenous than bacterial communities. Bacterial communities were controlled by soil chemistry, while litter fungal communities were affected mostly by vegetation and decomposition concentrated in hotspots.

Iordache et al. (2010) compared biological communities of different scales in disturbed sites and nearby reference sites and found that the decrease of species richness in the disturbed site compared to the reference sites was larger as the scale of the organisms is smaller. Even soil invertebrates of similar trophic groups appear to respond very differently to secondary successional changes (Scheu and Schulz 1996). Ant abundance and biomass were largest between 5 and 8 years of restoration of a prairie ecosystem and were correlated with plant diversity and plant richness, while earthworm abundance increased linearly and related only to surface litter (Wodika et al. 2014). In a study investigating the presence of many groups, Frouz et al. (2013) found that in post-mining forest chronosequences, microbial diversity increased with age, *Actinobacteria* were associated with prairie sites, and fungi with forest sites, while saprophagous groups, especially earthworms, were absent in shortgrass prairie sites leading to absence of bioturbation.

At mm to dm scale (soil aggregates, root zone), the vertical stratification of soil climate and resources are the main drivers of organism distribution; at m to 100 m scale, the horizontal heterogeneity of species distribution is controlled by the distribution of soil properties; at 100 to 1 km scale soil type, carbon, topography, and plant communities are the main determinants; at 1 to more than 100 km scale, the key factors are gradients in geomorphology, climate, energy input, and biomes (Berg 2012).

Nichols and Nichols (2003) investigated mammals, birds, reptiles, and ants in a population of rehabilitated mining areas. Different recolonization patterns were observed, different groups reflected different aspects of faunal succession, and no single group was found to be suitable as an overall "indicator."

Five relevant scales have been identified in soil function (Lavelle 2012): microbial biofilms (20 μm), micro-food webs inside and outside soil meso-aggregates (100–500 μm), functional domains of ecosystem engineers (e.g., plant roots, leaf litter system of an individual tree, earthworms, 10^{-1} to 10^{1} m horizontally and 10^{-2} to 10^{0} m in depth), mosaics of functional domains at plot scale, and landscape/watershed scale. The species interactions and existence of coevolution should be tested at each scale. Successional processes occur within discretization units characteristic for each scale.

11.3 Coupling Between Groups of Organisms

11.3.1 Bacteria and Fungi with Plants

Asmelash et al. (2016) review the coupling between mycorrhizal fungi and plant development and how they drive vegetation succession. Brown and Jumpponen (2013) reported that the presence of plants rather than their identity controlled the structure of bacterial communities on a chronosequence and had a minor role in the structuring of fungal communities. Microbial community assembly dynamics did not follow plant-based models of succession, and the patterns for fungi and bacteria are distinct (Brown 2014).

Grasses were not stimulated by higher diversity of microbial community, while herbs (dicots) grew better with more diverse microbial communities (Frouz et al. 2016). Late-successional herbs grew better with the late-successional microbial community, and early successional herbs grew well with both early and late microbial community (Frouz et al. 2016). C3 grasses are usually less mycorrhizal dependent than most perennial herbs (Frouz et al. 2016). In agriculture arbuscular mycorrhizal fungi can directly suppress weak host weeds (10% root length colonization threshold) and indirectly suppress strong ones by competitive effects exerted by strong host crops (Li et al. 2017).

Low level of arbuscular mycorrhizal fungi in late succession phases in forests may be due to high carbon costs for maintaining symbiosis in soils having sufficient nutrient amounts (Zangaro and Rondina 2016). Mycorrhizal symbiosis expands environmental ranges (requirement niche) and influences resource use (impact niche) for both plants and fungi (Peay 2016), with important consequences on the successional processes of both groups (by the competitive advantage of fungi and plants involved in symbiosis). Arbuscular mycorrhizal fungi act as mediators of interactions between plants of different successional stages and may accelerate the succession by facilitating late-successional species (Kikvidze et al. 2010). Positive feedbacks between plants and soil microbes are central in early successional communities, while in late-successional ones, the negative feedbacks contribute to species replacement, coexistence, and diversification (Kikvidze et al. 2010).

An attempt to link the disturbance of fungi with ecosystem functioning (to upscale it) is done by Morris et al. (2016). They state that changes in fungal community structure will have an impact on ecosystem functioning when the fungi affect plant diversity. "All scales of ecosystem disturbance ranging from landslides to fire, to introduced species, to animal burrowing, to hyphal grazing by microarthropods can disrupt critical points in the hyphal network" which later "changes nutrient availability and transfer to plant hosts, which in turn alters ecosystem productivity" (Morris et al. 2016).

The coupling between fungi and plants is hydrogeomorphologically sensitive. Soil fungal diversity, community composition, and the relationship between fungal and woody plant assemblages differed between ridge and valley habitats in a forest ecosystem (Gao et al. 2017). The richness values of saprotrophic, pathogenic, mycoparasitic, and animal parasitic fungi were significantly higher in valley than in ridge habitats, probably due to more resources, but ectomycorrhizal and arbuscular mycorrhizal fungi did not differ significantly (Gao et al. 2017). In terms of patterns of diversity in space, this leads to a lack of resemblance of plant and microbial diversities along altitudinal gradients (Gao et al. 2017).

Fungi may diminish the effects of large abiotic heterogeneity controlling vegetation in early successional stages (by increasing the scale of the interactions between plants and abiotic resources (Baasch et al. 2009).

Late-successional plants in grasslands can grow up to 24 times larger after being inoculated with arbuscular mycorrhizal fungi; they are sensitive to the identity of soil fungi and grow largest and produce more flowers when inoculated with certain fungal species (Koziol 2016), but "despite their large host range, the efficiency of AMF in promoting plant performance differs strongly among fungal species and isolates, and the ability of the plant to respond to the symbiosis depends on the plant genotype" (Lanfranco et al. 2018), so species-specific decisions should be made.

11.3.2 Fungi with Underground Consumers

Ruess and Lussenhop (2012) make a detailed analysis of the trophic interactions between fungi and animals in terms of a fungal food chain starting from plant root and exudated dissolved organic matter (the first trophic level). The second trophic level includes saprophytic, arbuscular mycorrhizal, ectomycorrhizal, and phytopathogenic fungi, the third one fungivores like nematodes, mites, Collembola, and Protura, and then predators like predacious nematodes and predacious mites.

Soil invertebrates are coupled to fungal and bacterial community composition mainly by mixing soil (bottom-up effect, Crowther et al. 2013) and to smaller extent by grazing effects. Such top-down effect may occur, however, with consequences on the relative importance of fungi and bacteria, and on the abundance of functional groups of fungi involved in the decomposition process. Later on the authors develop a detailed trait-based approach to better explain the dynamic of fungal communities linking the abiotic and biotic filtering concepts with the fitness one and the fundamental niches to the realized niches in various contingent environments (Crowther et al. 2014). Detritus-based ("brown") food webs rely on allochthonous inputs from plants and their consumers and cannot be strongly top-down controlled, because microbial grazing cannot affect the production of organic matter (Crowther and Grossart 2015). The top-down effect in this case propagates as changings in the rates of organic matter decomposition (Crowther and Grossart 2015). Direct trophic interaction between fungi and fungivores can change fungal morphology, physiology, and community structure (Ruess and Lussenhop 2012).

Soil mesofauna seem to be more selective when feeding on fungi than soil macrofauna, and in general fungal-feeding decomposer animals tend to be food generalists rather than specialists (Maraun et al. 2003). Vesicular arbuscular mycorrhizae may be grazed extensively by nonspecialized fungivores (Moore et al. 1985). Collembola grazing increased extra-radical mycelium nutrient sequestration of arbuscular mycorrhiza, particularly phosphorus and stimulated plant performance (Ngosong et al. 2014), but the effect was dependent on the stage of the fungi life cycle, with later stage leading to lower palatability.

Grazing of fungi has effects on fungal growth and organic matter decomposition, with the intensity of the effect depending on the size of the consumers (A'Bear et al. 2014). Macroinvertebrates had stronger effects than micro- and meso-invertebrates; the grazing pressure increased the enzyme activities leading to larger organic matter decay. Fungal community structure, especially by differential palatability, determined the diversity of consumers and their abundance (A'Bear et al. 2014). Oribatid mites can have a preference for feeding on certain ectomycorrhizal species (Schneider et al. 2005).

11.3.3 Underground Consumers with Aboveground Consumers

Coupling between underground and aboveground insects may occur through the preferential feeding on fungi. For instance, Gange (2000) found that Collembola prefer to feed on non-mycorrhizal fungi, indirectly benefiting plants through an enhancement of mycorrhizal functioning and the foliar-feeding insect herbivores. Tiunov and Scheu (2005) also found mycorrhizal mycelium to contribute little to Collembola nutrition, which grazed intensively on saprotrophic fungi. The relation between Collembola and saprotrophic fungi may be, however, species-specific, and it might not be always appropriate to group them in uniform functional components (Tordoff et al. 2008).

11.3.4 Coupling Between Many Groups Across Many Scales

From more analytic approaches, we go now to more holistic ones. Ohgushi et al. (2018) edited recently a book on the coupling of aboveground and belowground process including reviews of many aspects of this field of research. Teste and Dickie (2017) made a synthesis of the knowledge about mycorrhizal succession at ecosystem scale and its main drivers (changing soil abiotic conditions, plant community assembly, plant host specificity, fungal community assembly, and interacting drivers). The interacting driver concept couple three explanatory hypotheses: that both fungal and plant communities are controlled by abiotic condition, that plants shape fungal communities, and that fungi shape plant communities (Teste and Dickie 2017). They apply this concept at scales from cm and hours to globe and millennia and identify the scale-specific processes involved in this interaction. One can cut from this continuum of scales the part corresponding to the management scales (in space and in time).

Heinen et al. (2017) found that soil legacy effects on plants and on a herbivore insect were determined by the composition of the previous community, but not by traits of the vegetation like the dimension of the root system. Later on Heinen and his colleagues (2018) reviewed the field studies manipulating bacteria, fungi, nematodes, and soil arthropodes in order to assess the effects on the interactions of plants with associated aboveground insects. Plant growth-promoting bacteria have, for instance, in general negative effects on herbivore performance and abundance, while arbuscular mycorrhizal fungi have positive effects on sap-sucking herbivores and neutral effects on most of the chewers (positive in some cases on specialist and negative in some cases on generalists, Heinen et al. 2018). In turn some herbivores, insects can affect the root bacterial communities (but not the fungal communities) by changes in the exudates composition as demonstrated experimentally by Ourry et al. (2018). Such findings have direct implications for the managerial use of microorganism to suppress aboveground insect pests (Pineda et al. 2017).

Roubickova (2013) investigates the interactions between soil fauna and plants during succession after coal mining; found positive effect of earthworms on the growth of late-successional plants, negative effect on the germination of small seeds compared to large seeds, and a change in plant community composition after their biomass reduction in the field; and concluded that soil macrofauna may affect the whole plant community and the development of the whole ecosystem.

Schrama et al. (2017) found that diversity of brown webs is decoupled from that of green webs during a primary succession, despite the strong functional connection. Green trophic groups were driven mainly by vegetation parameters (amount of bare soil, vegetation biomass production, vegetation height), while brown groups depended mostly on the production and standing stock of dead organic material and soil development (Schrama et al. 2017).

11.3.5 The Strength of Coupled Processes

This is a key aspect in order to evaluate if action on a microbial process can propagate and how far. The direct effect that species have on each other's demography is termed interaction strength (Morales-Castilla et al. 2015). Three kinds of proxies can be used to obtain indirect information about species interaction: traits, phylogenies, and geographical data (Morales-Castilla et al. 2015). Asymmetric networks of interspecific interactions (one group depending strongly on another group, but not vice versa) were found to enhance long-term coexistence and facilitate species diversity maintenance (Bascompte et al. 2006).

Another way to look at interaction strength is in terms of functional consequences. A meta-analysis of Sacket et al. (2010) showed that an increase in the biomass of soil fauna increased plant productivity across ecosystems by 35% However, in the case of ecosystems dominated by legumes (with low nitrogen requirements), the biomass of soil fauna was not positively correlated with the aboveground productivity.

A classic study reported that the effect of plant diversity dampened with increasing trophic level and degree of omnivory both in aboveground and in belowground food webs (Scherber et al. 2010). Herbivores responded more strongly than carnivores and omnivores to changes in plant diversity. The magnitude of a trophic cascade attenuates down the food chain, and consequently an autotroph may be more sensitive to environmental fluctuations affecting limiting resources like nutrients than to food web interactions (Fraterrigo and Rusak 2008).

Predation is a multistage process including detection, attack, and capture and can induce effects on the prey from the first phase, for instance, by changing the foraging behavior (Sitvarin et al. 2016). Such effects complicate very much the patterns of distribution at higher trophic levels and weaken the coupling between plant productivity or diversity and the diversity of consumers.

Schickmann et al. (2012) investigated the relationship of mycophagous small mammals and ectomycorrhizal fungi (measuring the ingestion of fungal spores) and found that "mycophagy can be considered (1) to contribute to nutrition of small forest mammals, (2) to play a pivotal role for forest regeneration and functioning by providing mycorrhizal inoculum to tree seedling and (3) to be vital for reproduction and diversity of the still poorly known hypogeous fungi."

One line of research about the strength of coupled processes associated with succession is in terms of plant-soil feedback. By growing mixed plant communities on early-, mid-, and late-successional soils, Kardol et al. (2006) observed a negative plant-soil feedback for early successional plant species, neutral feedback for mid-successional species, and positive feedback for late-successional species, all of them more related to soil biota (bacteria, fungi) than to abiotic conditions. Negative plant-soil feedbacks increased with plant abundance in the sense that abundant plant species accrue species-specific soil pathogens to a greater extent than rare species, which prevented them to become more abundant (Maron et al. 2016).

Positive feedback is predicted in systems where microbes provide resources or services to the plants (nutrients, pathogen defense) and negative feedback when resources are in large quantities and opportunist pathogens increase relative to mutualist microbes (Revillini et al. 2016). Besides the exchange of goods and services with the plants, fungi and bacteria may exchange also in between, with net positive or negative interactions (Revillini et al. 2016).

In metal-contaminated soils, the role of positive as opposed to negative plant-soil feedbacks may be more important (Kruminis et al. 2015), which could change the patterns observed in the successional processes occurring in uncontaminated soils. "As ecologist, we do not know how theories of community assembly and succession can be applied to human impacted landscapes" (Kruminis et al. 2015).

Effects of microorganisms can occur independently of plant-soil feedbacks, as demonstrated for fungi and relocated seedlings by Rigg et al. (2016a). Variation of soil microbial community in the wild (due to spatial and/or edaphic factors) may affect only the fungal community, but not the bacterial assembly in tree seedlings (Rigg et al. (2016b)).

Plant-soil feedbacks are relevant also from evolutionary timescale perspectives: "genetic variation in plant traits and the evolution of those traits can affect traits and species composition of soil microbial communities. Soil microbial communities can, in turn alter the evolutionary trajectory of plant traits" (terHorst and Zee 2016). Coevolution of fauna feeding on plant roots may exist, but not of fauna feeding on litter, because there is not direct selective pressure on plants (Bastow 2012). The preferences of detritivores are eventually in function of the litter chemical quality, not for particular plant taxa.

In a nutrient-poor environment, Roy-Bolduc et al. (2016) found a strong correlation between aboveground vegetation and soil fungal community, both responding similarly to soil properties. They interpreted this as strong linkage, positive interaction between the two aboveground and belowground communities. Taxonomically and functionally diverse fungal communities with distinct composition existed in each succession stage of this nutrient-poor environment (Roy-Bolduc et al. 2015).

A review of the effects of mycorrhizal fungi on different groups of organisms was recently done by Antunes and Koyama (2017). They point out the difficulties of investigated complex soil food webs and modeling them. The effects of fungi on higher trophic levels contribute to the "multifunctionality" of ecosystems in terms of the production of ecosystem services. The effects on plant performance are only one aspect, although probably the most important in quantitative terms.

The effect of fertilizers can propagate up to three trophic levels, as proved experimentally (Naeem and Colwell 1991), but its detection depends also on the ability of species to habitat select. "Even when the long-term mean abundance of a limiting resource is kept constant, patterns of short-term heterogeneity can affect the distribution and abundance of species" (Naeem and Colwell 1991).

In a study on the interaction between habitat fragmentation, plant, and small mammal succession, Schweiger et al. (2000) found succession of small mammal community to be a function of shifts in the carrying capacity of the entire landscape or a specific patch size. In the first phase of succession, mammals did not react to vegetation structure in the patches, but as the plant succession progressed, some of them preferred large patches of vegetation.

The effects of tree diversity and individual species on fungi, protists, and animals were context dependent as demonstrated by structural equation modeling (Tedersoo et al. 2016). On a local scale, soil resources and tree species were found to have a stronger effect than tree species richness on soil biota diversity.

The positive relationship between plant diversity and the production of ecosystem services at site scale co-determined by above-belowground multitrophic interactions is a research priority receiving increasing attention (Eisenhauer 2018). It is useful to have both analytical and integrated approaches: analytical ones show the role of each group, and integrated ones show the relative importance of this role compared to other groups in the production of ecosystem services.

11.4 Toward an Improved Framework

In this chapter the reader will find several concepts rarely found in the literature screened in the previous chapter which might be useful when attempting to design the manipulation of successional processes.

11.4.1 Hotspots

While in succession and assembly studies one speaks about structural heterogeneity, functional heterogeneity may be relevant as well from a management point of view. Space-time locations with large process rates can be classified as hotspots/hot moments or more generally control points (Bernhardt et al. 2017): permanent control points, activated control points, export control points, and transport control points. When they occur in 5% of the catchment surface or more, they start to have dramatic consequences on the ecological functioning of the system. Many control points occur at the scales of cm, m, or 10s of meters scale, while most management occurs at the scales of 10s of square km or river basins and regions (1000s of square km, Bernhardt et al. 2017). At pedon scale microbial hotspots (defined by process rates, not by pools, or stocks of elements) are responsible for the ecologically relevant processes in soil (Kuziakov and Blagodatskaya 2015) calling for a revision in the rates of many processes in soil. Maintenance of stoichiometric ratios by accelerated microbial growth in hotspots requires additional nutrients (N, P) leading to their mining from soil organic matter (Kuziakov and Blagodatskaya 2015). Most common hotspots (microhabitats) in soil are associated to the rhizosphere, detritusphere, biopores (resulted from burrowing animals and roots), and aggregate surface, and the hot moments are determined by litter fall, root ingrowth, root death, animal activities, heavy rains, snow melting, freezing/thawing, drying/rewetting, and erosion events (Kuziakov and Blagodatskaya 2015). Hotspots and hot moments in soils have scales larger than those of individual bacterial and fungi organisms (up to mm and months) and make a conceptual connection between autoecological processes and ecological ones at site scale (Kuziakov and Blagodatskaya 2015).

11.4.2 State-and-Transition Models

Grant (2006) uses a state-and-transition concept of succession at a timescale of 20 years for the rehabilitation plan of areas following mining, separates desirable and non-desirable states, and proposes manipulation measures to direct the system to the desirable ones. State-and-transition simulation models can be used to evaluate the potential long-term consequences of alternative land management strategies, with scenarios varying according to the total budget allocated for management and the allocation of the budget between different kinds of management actions (Jarnevich et al. 2015).

These models have behind the idea that communities can shift between relatively stable states of equilibria. Testable causes of changes within a state and between states are proposed in the model. In practice plant communities should be first delineated in space. Their "states" have been defined using multivariate analysis of long-term vegetation data sets, without an understanding of the processes behind these changes (Bestelmeyer et al. 2003), so the correspondence with the theory is at the level of the interpretation. It is unclear whether changes in plant composition can be reversed through facilitating or accelerating practices (Bestelmeyer et al. 2003) unless there is a clear mechanistic understanding of the processes underlying the intrastate dynamic and the transitions.

Phases may help describing the real processes leading changes within states and finally to another state. Phases can be defined as steps of succession (Provencher et al. 2016). Transitions in this framework refer either succession between phases or disturbances, or between states, and can be quantified either probabilistically or deterministically. A further improvement of the approach can be done by discretizing the land units after the clarification of the relevant processes and of their scales.

State-and-transition models can be combined with species distribution models for management purposes (Miller et al. 2015), which is relevant also for devising restoration strategies accounting for the stochastic appearance of new species in the restored sites.

Craig et al. (2015) investigated how the state-and-transition models derived from vegetation succession represent the avian succession and found poor congruence, excepting birds with home ranges similar or smaller than the vegetation landscape units. States and models were found to not be conceived in function of long-term monitoring, ecological processes and thresholds, or criteria important for the structuring of the avian community. This has consequences on the extent to which ecological restoration objectives can be formulated based on vegetation succession only. Similar considerations hold more generally for the fauna rehabilitation in degraded lands (Cristescu et al. 2012): "re-establishing flora might be a pre-condition necessary but not sufficient to promote fauna recolonization." From a methodological perspective, state-and-transition simulation models are only able to track discrete state variables, not continuous ones, and are not able to integrate agent-/individual-based models, which are increasingly used to represent drivers of landscape dynamics (Daniel et al. 2016).

Gallagher et al. (2011) demonstrate the existence of distinct assemblage development trajectories of vegetation above and below a critical soil metal threshold and suggest assembly theory as appropriate to describe the development of alternative states, without a role for Gleasonian interspecific interactions. Below a critical threshold, herb/grass, shrub, and early hardwood tree assemblages developed, while above it the shrub stage was skipped (Gallagher et al. 2011). The accumulation of metals may reinforce the pollution in the topsoil precluding the establishment of less tolerant climax hardwood species. The authors recommend flexible endpoints of restoration in function of the abiotic characteristics of the site (Gallagher et al. 2011).

In order for management to utilize state-and-transition, definitions and concepts should be consistent and universally accepted, with clear ecological interpretation (Stringham et al. 2001).

11.4.3 Network Analysis

Trait-based deterministic frameworks attempt to explain the structure of the communities by habitat filtering and interspecific interaction processes (Bartomeus et al. 2016). Species traits like body size and matching traits for resources and consumers are very useful for predicting the structure of ecological networks beyond the simplified interactions between traditional communities of organisms (Eklof et al. 2013). Body size provides a functional link between individual-level processes and higher-level ecological processes such as the strength and outcome of trophic interactions (Kalinkat et al. 2015). Behavior and the intraspecific trait variability (relevant also from evolutionary perspective) can foster interactions that, from traits, would be predicted to be forbidden and can enhance the interaction degree of species and network connectance (Gonzalez-Varo and Traveset 2016).

Stoichiometric traits (individual measures of elemental composition, assimilation, allocation, or excretion usually defined by content or ratios between elements) improve predictions about how organisms interact with their resources (Leal et al. 2017). In theoretical terms the approach corresponds to an extension of the trophic niche with nutritional terms (Machovsky-Capuska et al. 2016). Knowing such traits can be useful for devising appropriate strategies for accelerating ecological succession by providing the most appropriate conditions for growth and reproduction of the organisms during their life cycle. While originally proposed in biogeochemical approaches, they reflect also in the case of animals issues relevant for behavioral ecology with consequences on the food web structure and on the coupling between energy flow and substance fluxes at organism and population scale (Sperfeld et al. 2017).

Network analysis is an increasing popular method for depicting plant-fungal interactions in complex communities (Opik and Davison 2016). It is recommendable to be use network analysis not only to detect structural patterns but for testing specific substantial hypotheses about the structuring of communities at various scales, in relation with the variable determining the formation of the patterns (Poisot et al. 2016).

11.4.4 Coupling Models

Coupling models refer to a methodological class of models linking variables measurable at different scales and intervening in the control of processes at different scales and are related to data-porting tools for model with different discretization needs (Ion et al. 2014). Fungi variables cannot be directly included in models about the succession of mammals responding to vegetation structure, excepting for mycophagous animals, but they potentially control at small scale the quality of keystone vegetation structures (Sect. 11.2.4). What is the relation between the variable "abundance of keystone structures" as part of "habitat suitability" and "aboveground biomass" of a keystone structure as controlled by the fungi and other soil variables? The answer depends on an upscaling from vegetation plot scale to site scale or a downscaling from site scale to vegetation plot scale modeled by so-called coupling modes. Another example is related to roughness at site scale (relevant for erosion models) and vegetation mechanical properties at plot scale as controlled by active management, potentially including inoculation with fungi.

11.5 Practical Issues

Although there are some risks and controversies with respect to the environmental impact of manipulating based on ecological knowledge the underground part of terrestrial ecosystems (Machado et al. 2017; Bender et al. 2017), so-called underground revolution is a reasonable strategy in the current state of agriculture (Bender et al. 2016; De Vries and Wallenstein 2017) and can be a solution also in the particular case of restoration and remediation projects or the construction of new ecosystems in industrial environments with unique characteristics. Optimizing plant-microbe nutritional interactions for more sustainable agricultural systems based on an understanding of the dynamics of microbial communities due to changes in environmental conditions is a research priority (Jacoby et al. 2017).

There are many reviews on the use of fungi for the remediation of industrial pollution, some of them pointing out the interactions with other groups of organisms than plants (e.g., Gadd 2016, with soil invertebrates). In the case of particular management objectives like phytoextraction, there is a clear benefit from the application of microorganisms (Sheoran et al. 2016). Another example is the so-called mycoremediation (Singh et al. 2015; Prasad 2017, 2018), where the fungal strains can eventually be found suitable also for plant growth promotion. Sequestration of heavy metals in mushrooms (mycoremediation) may be an option to be combined with successional effects where appropriate (Rhodes 2014; Dulay et al. 2015). Complementarily, the increasing use of bacteria for decontamination (Malla et al. 2018) could be combined also with the objective of succession manipulation.

Despite the importance of scale, explicitly incorporating a multiscale perspective into research and management remains a challenge (Nash et al. 2014). Aggregation (coupling) of processes at certain ranges of scale may provide perceivable discontinuities. When such discontinuities are not observable, they may be decided for management purposes in function of the scale of relevant key processes.

In mining areas instead of restoration goals targeting a historical reference system, it may be more appropriate the construction of new ecosystems or of hybrid ecosystems. This is because in mining areas the combined abiotic conditions are unique and rarely comparable with natural ones, and there is not a historical system to be targeted (Doley and Audet 2013). In such a framework, manipulating the soil microbiota by active management could be useful, besides vegetation manipulation.

Using local species for restoration has been regarded as a "dogma" for long time. In the cases when "novel ecosystems" are developed due to the large existing modifications, "local is best" is rather a hypothesis to be tested (Jones et al. 2015), because potentially other species might be more appropriate. Nikolic (2013) had reached a similar conclusion: "primary succession [on sulphidic Cu mining wastes] relies on novel types of early vegetation which comprise not only combinations of species, but also the key role of species which are novel to the affected region, and do not survive outside of the polluted area." Even when spontaneous colonization of mining surface is encouraged, the process may lead to novel species assembly, compared to the surroundings, due to propagule limitation (Tischew et al. 2014). The authors suggest active species reintroduction in function of the landscape matrix and the timescale of the management objectives.

Dietterich (2016) concludes that "efforts to restore contaminated areas should focus on remediating soil conditions and fostering desired plant communities, and that soil microbial communities can likely be left alone with little effects on restoration goals." Especially soil amendments applied at the time when the seeds are planted were found by this author to have long lasting effects. On the other hand, Iordache et al. (2016) reported a significant difference in the plant development and other variables when the soil inoculation with arbuscular mycorrhizal fungi changed from 1% to 2%. The use or lack of use of fungi inoculum depends on the substrate chemistry and dispersal opportunities from the landscape.

Potential steps for devising a restoration/remediation/new ecosystem construction plan including use of fungi for accelerating the ecological succession could be:

- Investigate the site and the surroundings at several scales (Fig. 11.2). The scales selected for the proposed approach depend on the realistic discontinuity analysis of the species distribution classified by functional groups (Angeler et al. 2016), and not on classifications of the landscape units serving other management objectives.
- Estimate the relative importance and dispersal and habitat suitability for species recolonization. Describe the relevant succession processes by the community and

the coupled integrated processes. Describe spatial structure of soil variables in the sites to be restored (seldom described, Jianu et al. (2012) or, for instance, the work of Komnitsas et al. 2010 demonstrating different distribution patterns of pH, organic matter, available phosphorus and potassium, and N-NH_4). Use modeling tools to simulate the potential trajectories of coupled successional processes. Information about the potential solutions for fungi manipulations can be obtained from the study of chronosequences relevant for the management situation (with primary or secondary succession) in the same landscape. Walker et al. (2010) provide a critical assessment of this methodology.

- Evaluate if the site can be a control point in the landscape or its potential functional relation with landscape/catchment control points.
- Evaluate to what extent objectives at site scale can be formulated in terms of states and phase with transitions in between.
- Evaluate to what extent small-scale hot spot and keystone vegetation structure can be included in the site.
- Adopt a multifunctional and multiscale approach in formulating the potential objectives in the design phase: which ecosystems services are to be restored and which species to be recovered with their habitat and microhabitat needs.
- Adopt flexible endpoints of restoration in function of the abiotic and biotic characteristics, in relation states, phases, and uncertainties. Associate specific activities in time with specific phases and states. Devise a plan to navigate the trajectory of the site toward this conceptual landscape.
- Perform multiscale experiments in order to detect the effects of microbial inoculation on the relevant ecophysiological and ecosystem processes involved in succession (Neagoe et al. 2010, 2013, 2014; Nicoară et al. 2014).
- Facilitate dispersion or inoculate in a spatially structured approach, geomorphologically sensitive (valley, ridge, slope) and sensitive to discretization units (service providing unit) scale of fungi (Iordache et al. 2011). In appropriate dispersing conditions, planting a diverse community of trees in reclaimed soils could yield a diverse community of belowground fungi with no need for inoculation. Late microbes can be inoculated from the beginning in the construction of new ecosystems. Mycorrhizal fungi are key for coupling succession processes; they expand the niche of the plants and have effects in cascade toward microhabitats for invertebrates and later mammals.
- Provide environmental conditions in a spatially structured approach to facilitate species or to filter some of them out. Design the trophic niche with nutritional terms based on stoichiometric ecological knowledge. Construct microhabitat heterogeneity for invertebrates with allochthonous particulate organic matter (relevant also for controlling the mobility of heavy metals in mining areas, Neagoe et al. 2012). Reclaimed soil from mine lands used for improving conditions suffers many disturbances related to the distribution of soil microflora community, mycorrhizal fungi, and enzymatic activities in soil (Sheoran et al. 2010). The reuse of such soil may benefit from inoculation with microorganism (bacteria and fungi).

- Create spatially structured vegetation cover for microhabitat diversity. Inoculate spatially structured around plots of vegetation facilitating by priority effects in artificial ecosystems, at potential keystone structures to facilitate vertebrates. Kalucka and Jagodzinski (2016) noted that in most cases an artificial mycorrhization of tree seedlings is unnecessary, unprofitable, or economically unjustified (a natural infection in the nursery may be enough). Creating an arbuscular mycorrhizal fungal inoculum that is tailored to an abiotic stress could be a strategy where agriculture is restricted by such a stress or in restoration efforts (Millar and Bennet 2016). The effects of arbuscular mycorrhizal fungi inocula lead to better results in accelerating succession when early seral native fungi are used in consortia instead of single species of exotic late seral fungi (Asmelash et al. 2016). Research areas in this field are the cheap production of inocula and the in situ management methods for effective restoration. Arbuscular mycorrhizal fungi synergistically interact with plant growth-promoting *Rhizobacteria* (an established solution for the management of agricultural soils, Neagoe et al. 2009; Prasad et al. 2005; Majeed et al. 2018) in driving the succession processes (Asmelash et al. 2016).
- Construct keystone structures with specific fungi communities supporting them for small mammals and other vertebrates. Include management by disturbances in later phases to control the system toward desired states. Spatially extensive disturbances may homogenize preexisting differences within patches and reduce variability, while smaller disturbances may enhance within-patch variability by creating heterogeneity (Fraterrigo and Rusak 2008).
- Ensure connectivity (matrix permeability) with the surrounding landscape elements to enhance the viability of the metapopulations of mobile organisms.

11.6 Conclusions

There is a strong theoretical heterogeneity due to different strategies for reducing the complexity of the studied ecological objects. There is a consistent knowledge about community assembly/succession patterns for separate groups of organisms. Patterns are context dependent, the role of stochastic and deterministic processes depending to some extent on local and landscape context. Comparisons of these patterns between groups at the same site or across chronosequences are revealed in general large differences. The way distinct successional processes are coupled toward the whole ecosystem complexity seems to lead to uniqueness of each site conditions.

The strength of the effects of soil microbial processes to plant and aboveground consumers is rather strong, especially due to the mycorrhizal fungi. By keystone vegetation structures, one could expect to detect the effect also on some mobile organisms, besides fungivorous mammal species. The positive and negative feedbacks between green and brown webs do not allow general recommendation about the use of microorganisms in every situation, but it is reasonable to state that at least in industrial sites and in sites with poor nutrient condition, on eroded sites, the use of microbial inocula is useful in terms of plants performance. If it is also economic is a different question, it depends on the existing biotechnologies on a case by case situation.

A comprehensive multifunctional practical approach putting the use of fungi in the general context of landscape sensitive restoration is suggested, with a potential portfolio of objectives. This can lead to a portfolio of distinct coupled projects corresponding to the interests of various stakeholders from different management scales, from local (for industry) to regional (for soil and water) to national ones (for biodiversity).

It is useful to have both analytical and integrated approaches: analytical ones show the role of each group, and integrated ones show the relative importance of this role compared to other groups in the production of ecosystem services. Integrated site management programs allowing a population of different projects and with funding from specific agencies might capitalize on such scientific complementary.

Acknowledgments I warmly thank to Professors Ajit Varma and Ram Prasad for the invitation to prepare this text. This research was supported by the Romanian National Authority for Scientific Research—UEFISCDI [PN-III-P1-1.2-PCCDI-2017-0721 acronym INTER-ASPA].

References

A'Bear AD, Jones TH, Boddy L (2014) Size matters: what have we learnt from microcosm studies of decomposer fungus-invertebrate interactions? Soil Biol Biochem 78:274–283

Alday JG, Marrs RH, Martinez-Ruiz C (2011) Vegetation succession on reclaimed coal wastes in Spain: the influence of soil and environmental factors. Appl Veg Sci 14:84–94

Andres P, Mateos E (2006) Soil mesofaunal responses to post-mining restoration treatments. Appl Soil Ecol 33:67–78

Andrew C, Heegaard E, Halvorsen R, Martinez-Pena F, Egli S, Kirk PM, Bassler C, Buntgen U, Aldea J, Hoiland K, Boddy L, Kauserud H (2016) Climate impacts on fungal community and trait dynamics. Fungal Ecol 22:17–25

Angeler DG, Allen CR, Barichievy C, Eason T, Garmestani AS, Graham NAJ, Granholm D, Gunderson LH, Knutson M, Nash KL, Nelson J, Nystrom M, Spanbauer TL, Stow CA, Sundstrom SM (2016) Management applications of discontinuity theory. J Appl Ecol 53:688–698

Antunes PM, Koyama A (2017) Mycorrhizas as nutrient and energy pumps of soil food webs: multitrophic interactions and feedbacks. In: Johnson CN, Gehring C, Jansa J (eds) Mycorrhizal mediation of soil: fertility, structure and carbon storage. Elsevier, New York, pp 149–173

Armesto JJ, Picklet STA, McDonnell MJ (1991) Spatial heterogeneity during succession: a cyclic model of invasion and exclusion. In: Kolosa J, Picklett STA (eds) Ecological heterogeneity. Springer, New York, pp 256–269

Asmelash F, Bekele T, Birhane E (2016) The potential role of arbuscular mycorrhizal fungi in the restoration of degraded lands. Front Microbiol 7:1095. https://doi.org/10.3389/fmicb.2016.01095

Baasch A, Tischew S, Bruelheide H (2009) Insights into succession processes using temporally repeated habitat models: results from a long-term study in a post-mining landscape. J Veg Sci 20:629–638

Bahram M, Kohout P, Anslan S, Harend H, Abarenkov K, Tedersoo L (2015a) Stochastic distribution of small eukaryotes resulting from high dispersal and drift in a local environment. ISME J:1–12

Bahram M, Peay KG, Tedersoo L (2015b) Local-scale biogeography and spatiotemporal variability in communities of mycorrhizal fungi. New Phytol 205:1454–1463

Baldrian P (2017) Forest microbiome: diversity, complexity and dynamics. FEMS Microbiol Rev 41:109–130

Bartomeus I, Gravel D, Tyliaanakis JM, Aizen MA, Dickie IA, Bernard-Verdier M (2016) A common framework for identifying linkage rules across different types of interactions. Funct Ecol 30:1894–1903

Bartuska M, Pawlet M, Frouz J (2015) Particulate organic carbon at reclaimed and unreclaimed post-mining soils and its microbial community composition. Catena 131:92–98

Bascompte J, Jordano P, Olsen JM (2006) Asymmetric coevolutionary networks facilitate biodiversity maintenance. Science 312:431–433

Bastow J (2012) Succession, resource processing, and diversity in detrital food webs. In: Wall DH et al (eds) Soil ecology and ecosystem services. Oxford University Press, Oxford, pp 117–135

Bender SF, Wagg C, van der Heijden MGA (2016) An underground revolution: biodiversity and soil ecological engineering for agricultural sustainability. Trends Ecol Evol 31:440–452

Bender SF, Wagg C, van der Heijden MGA (2017) Strategies for environmentally sound soil ecological engineering: a reply to Machado et al. Trends Ecol Evol 32:10–12

Berg MP (2012) Patterns of biodiversity at fine and small spatial scales. In: Wall DH et al (eds) Soil ecology and ecosystem services. Oxford University Press, Oxford, pp 136–152

Bernhardt ES, Blaszczack JR, Ficken CD, Fork ML, Kaiser KE, Seybold EC (2017) Control points in ecosystems: moving beyond the hot spot hot moment concept. Ecosystems https://doi.org/10.1007/s10021-016-0103-y, on-line first

Bestelmeyer BT, Brown JR, Havstad KM, Alexander R, Chavez G, Herrick JE (2003) Development and use of state-and-transition models for rangelands. J Range Manag 56:114–126

Birgit F, Wiegand T (2008) Exploring spatiotemporal patterns in early stages of primary succession on former lignite mining sites. J Veg Sci 19:267–276

Boulanger Y, Sirois L (2007) Postfire succession of saproxylic arthropods, with emphasis on coleopter, in the North boreal forest of Quebec. Environ Entomol 36:128–141

Brown SP (2014) Rules and patterns of microbial community assembly. PhD theses, University of Oregon

Brown SP, Jumpponen A (2013) Contrasting primary successional trajectories of fungi and bacteria in retreating glacier soils. Mol Ecol. https://doi.org/10.1111/mec.12487

Burgess EE, Maron M (2016) Does the response of bird assemblages to fire mosaic properties vary among spatial scales and foraging guilds ? Landsc Ecol 31:687–699

Butt KR, Briones MJI (2017) Earthworms and mesofauna from an isolated, alkaline chemical waste site in Northwest England. Eur J Soil Biol 78:43–49

Chagnon P-L (2015) Determinisme et stochasticite dans l'assemblage des communautes mycorhiziennes. PhD theses, Universite de Sherbrooke, Quebec

Chang C, HilleRisLambers J (2016) Integrating succession and community assembly perspectives. F1000Research: 2294 https://doi.org/10.12688/f1000research.8973.1

Collins B, Wein G (1998) Soil resource heterogeneity effects on early succession. Oikos 82:238–245

Courtney R, O'Neill N, Harrington T, Breen J (2010) Macro-arthropod succession in grassland growing on bauxite residue. Ecol Eng 36:1666–1671

Craig MD, Stokes VL, Fontaine JB, Hardy GEJ, Grigg AH, Hobbs R (2015) Do state-and-transition models derived from vegetation succession also represent avian succession in restored mine pits? Ecol Appl 25:1790–1806

Cristescu RH, Frere C, Banks PB (2012) A review of fauna in mine rehabilitation in Australia: current state and future directions. Biol Conserv 149:60–72

Crowther TW, Grossart H-P (2015) The role of bottom-up and top-down interactions in determining microbial and fungal diversity and function. In: Hanley TC, La Pierre KJ (eds) Experimental limnology. Cambridge University Press, Cambridge, pp 260–287

Crowther TW, Stanton DWG, Thomas SM, A'Bear AD, Hiscox J, Jones TH, Voriskova J, Baldrian P, Boddy P (2013) Top-down control of soil fungal community composition by a globally distributed keystone consumer. Ecology 94:2518–2528

Crowther TW, Maynard DS, Crowther TR, Peccia J, Smith JR, Bradford MA (2014) Untangling the fungal niche: the trait-based approach. Front Microbiol 5:579

D'Amen M, Dubuis A, Fernandes RF, Pottier J, Pellissier L, Guisan A (2015) Using species richness and functional predictions to constrain assemblage predictions from stacked species distribution models. J Biogeogr 42:1255–1266

Daniel CJ, Frid L, Sleeter BM, Fortin M-J (2016) State-and-transition simulation models: a framework for forecasting landscape change. Methods Ecol Evol 7:1413–1423

Dauber J, Simmering D (2006) Ant assemblages in successional stages of Scotch Broom stands (Hymenoptera: Formicidae; Spermatophyta) Myrmecologishe Nachrichten 9:55–64

Davison J, Moora M, Jairus T, Vasar M, Opik M, Zobel M (2016) Hierarchical assembly rules in arbuscular mycorrhizal (AM) fungal communities. Soil Biol Biochem 97:63–70

De Vries FT, Wallenstein MD (2017) Below-ground connections underlying above-ground food production: a framework for optimising ecological connections in the rhizosphere. J Ecol 105:913–920

DeBeeck MO, Ruytinx J, Smits MM, Vangronsveld J, Colpaert JV, Rineau F (2015) Belowground fungal communities in pioneer Scots pine stands growing on heavy metal polluted and non-polluted soils. Soil Biol Biochem 86:58–66

Dejean A, Djieto-Lordon C, Cereghino R, Leponce M (2008) Ontogenetic succession and the ant mosaic: an empirical approach using pioneer trees. Basic Appl Ecol 9:316–323

DeLeon DG, Moora M, Opik M, Jairus T, Neunenkamp L, Vasar M, Bueno CG, Gerz M, Davison J, Zobel M (2016) Dispersal of arbuscular mycorrhizal fungi and plants during succession. Acta Oecol 77:128–135

Delhaye G, Violle C, Seleck M, Ilunga EI, Daubie I, Mahy G, Meerts P (2016) Community variation in plant traits along copper and cobalt gradients. J Veg Sci 27:854–864

Dietterich LH (2016) Plant-soil feedbacks in heavy metal soils. PhD Dissertation, University of Pennsylvania

Ding J, Jiang X, Guan D, Zhao B, Ma M, Zhou B, Cao F, Yang X, Li L, Li J (2017) Influence of inorganic fertilizer and organic manure application on fungal communities in a long-term field experiment of Chinese Mollisols. Appl Soil Ecol 111:114–122

Dini-Andreote F, Pylro VS, Baldrian P, Van Elsas JD, Salles JF (2016) Ecological succession reveals potential signatures of marine-terrestrial transition in salt marsh fungal communities. ISME J 10:1984–1997

Doley D, Audet P (2013) Adopting novel ecosystems as suitable rehabilitation alternatives for former mine sites. Ecol Process 2:22

Dong K, Tripathi B, Moroenyane KW, Li N, Chu H, Adams J (2016) Soil fungal community development in high Arctic glacier foreland follows a directional replacement model, with mid-successional diversity maximum. Sci Rep 6:26360

Dulay RMR, Pascual AHL, Constante RD, Tiniola RC, Areglo JL, Arenas MC, Kalaw SP, Reyes RG (2015) Growth response and mycoremediation activity of Coprinus comatus on heavy metal contaminated media. Mycosphere 6:1–7

Dumbrell AJ, Nelson M, Helgason T, Dytham C, Fitter AH (2010) Idiosyncrasy and overdominance in the structure of natural communities of arbuscular mycorrhizal fungi: is there a role for stochastic processes ? J Ecol 98:419–428

Eisenhauer N (2018) Aboveground-belowground interactions drive the relationship between plant diversity and ecosystem function. Res Ideas Outcomes 4:e23688

Eklof A, Jacon U, Jason K (2013) The dimensionality of ecological networks. Ecol Lett 16:577–583

Falconer RE, Otten W, White NA (2015) Toward modeling the resistance and resilience of "below-ground" fungal communities: a mechanistic trait-based approach. Adv Appl Microbiol 93:1–44
Falenczik-Kozirog K, Kaczmarek S, Marquardt T, Marcysiak K (2012) Contribution to succession of mite (Acari) communities in the soil of Tilio-Carpinetum Tracz. 1962 in northern Poland. Acta Zool Cracov 55:47–57
Fischer H, Bergfur J, Goedkoop W, Tranvik L (2009) Microbial leaf degraders in boreal streams: bringing together stochastic and deterministic regulators of community composition. Freshwater Biol 54:2276–2289
Fox BJ (1982) Fire and mammalian secondary succession in an Australian Coastal Heath. Ecology 63:1332–1341
Fox BJ (1990) Changes in the structure of mammal communities over successional time scale. Oikos 59:321–329
Fox BJ, Taylor JE, Thompson PT (2003) Experimental manipulation of habitat structure: a retrogression of the small mammal succession. J Anim Ecol 72:927–940
Fraterrigo JM, Rusak JA (2008) Disturbance-driven changes in the variability of ecological patterns and processes. Ecol Lett 11:756–770
Frouz J, Prach K, Pizl V, Hanel L, Stary J, Tajovsky K, Materna J, Balik V, Kalcik J, Rehounkova K (2008) Interactions between soil development, vegetation and soil fauna during spontaneous succession in post mining sites. Eur J Soil Biol 44:109–121
Frouz J, Jilkova V, Cajthaml T, Pizl V, Tajovsky K, Hanel L, Buresova A, Simackova H, Kolarikova K, Franklin J, Nawrot J, Groninger JW, Stahls PD (2013) Soil biota in post-mining sites along a climatic gradient in the USA: simple communities in shortgrass prairie recover faster than complex communities in tallgrass prairie and forest. Soil Biol Biochem 67:212–225
Frouz J, Pizl V, Tajovsky K, Stary J, Holec M, Materna J (2014) Soil macro- and mesofauna succession in post-mining sites and other disturbed areas. In: Frouz J (ed) Soil biota and ecosystem development in post mining sites. CRC Press, Boca Raton, pp 216–235
Frouz J, Toyota A, Mudrak O, Jilkova V, Filipova A, Cajthaml T (2016) Effects of soil substrate quality, microbial diversity and community composition on the plant community during primary succession. Soil Biol Biochem 99:75–84
Gadd GM (2016) Fungi and industrial pollutants. In: Druzhinina IS, Kubicek CP (eds) Environmental and microbial relationships. Springer, New York, pp 99–125
Gallagher FJ, Pechman I, Holzapfel C, Grabosky J (2011) Altered vegetative assemblage trajectories within an urban brownfield. Environ Pollut 159:1159–1166
Gange A (2000) Arbuscular mycorrhizal fungi, Collembola and plant growth. TREE 15:369–372
Gao C, Shi NN, Chen L, Ji NN, Wu BW, Wang YL, Xu Y, Zheng Y, Mi XC, Ma KP, Guo LD (2017) Relationships between soil fungal and woody plant assemblages differ between ridge and valley habitats in a subtropical mountain forest. New Phytol 213:1874–1885
Gonzalez-Varo JP, Traveset A (2016) The labile limits of forbidden interactions. Trends Ecol Evol 31:700–709
Grace JB (1991) A clarification of the debate between Grime and Tilman. Funct Ecol 5:583–587
Grant CD (2006) State-and-Transition successional model for bauxite mining rehabilitation in the Jarrah forest of Western Australia. Restor Ecol 14:28–37
Grzes IM (2010) Ants and heavy metal pollution – a review. Eur J Soil Biol 46:350–355
Hanel L, Devetter M, Adl SM (2014) Recovery and colonization at post-mining sites by the soil microfauna. In: Frouz J (ed) Soil biota and ecosystem development in post mining sites. CRC Press, Boca Raton, pp 172–215
Harabis F (2016) High diversity of odonates in post-mining areas: meta-analysis uncovers potential pitfalls associated with the formation and management of valuable. Ecol Eng 90:438–446
Hart MM, Gorzelak M, Ragone D, Murch S (2014) Arbuscular mycorrhizal fungal succession in a long-lived perennial. Botany 92:1–8

Heinen R, Van der Sluijs M, Biere A, Harvey JA, Bezemer TM (2017) Plant community composition but not plant traits determine the outcome of soil legacy effects on plants and insects. Journal of Ecology 106:1217–1229

Heinen R, Van der Sluijs M, Biere A, Harvey JA, Bezemer TM (2018) Effects of soil organisms on aboveground plant-insect interactions in the field: patterns, mechanisms and the role of methodology. Front Ecol Evol 6:106

Hendrychova M, Salek M, Tajovsky K, Rehor M (2012) Soil properties and species richness of invertebrates on afforested sites after brown coal mining. Restor Ecol 20:561–567

Ho A, Di Leonardo DP, Bodelier PLE (2017) Revisiting life strategy concepts in environmental microbial ecology. FEMS Microbial Ecol 93:fix006

Hodecek J, Kuras T, Sipos J, Dolny A (2015) Post-industrial area as successional habitats: long-term changes of functional diversity in beetle communities. Basic Appl Ecol 16:629–640

Hodge A (2014) Interactions between arbuscular mycorrhizal fungi and organic material substrates. Adv Appl Microbiol 89:47–99

Holec M, Frouz J (2005) Ant (Hymenoptera: Formicidae) communities in reclaimed and unreclaimed brown coal mining spoil dumps in the Czech Republic. Pedobiologia 49:345–357

Holmes AL, Robinson WD (2016) Small mammal abundance in mountain big sagebrush communities after fire and vegetation recovery. Western North American Naturalist 76:326–338

Hupperts SF (2016) Ectomycorrhizal fungal community response to disturbance and host phenology. MSc Theses, University of Alberta

Ibanez I, McCarthy-Neumann S (2016) Effects of mycorrhizal fungi on tree seedling growth: quantifying the parasitism-mutualism transition along a light gradient. Canadian J Forest Res 46:48–57

Ion S, Marinescu D, Cruceanu SG, Iordache V (2014) A data porting tool for coupling models with different discretization needs. Environ Model Softw 62:240–252

Iordache V, Onete M, Paucă M, Oromolu L, Honciuc V, Purice D, Cobzaru I, Gomoiu I, Neagoe A (2010) Biological communities in mining areas: scale dependent patterns, organisms' potential as bioindicators, and native plants for remediation. Proceedings of the 7th European Conference on Ecological Restoration, Avignon, France

Iordache V, Kothe E, Neagoe A, Gherghel F (2011) A conceptual framework for up-scaling ecological processes and application to ectomycorrhizal fungi. In: Rai M, Varma A (eds) Diversity and biotechnology of ectomycorrhiza. Springer, New York, pp 255–299

Iordache V, Lăcătusu R, Scrădeanu D, Onete M, Ion S, Jianu D, Bodescu F, Neagoe A, Purice D, Cobzaru I (2012) Contributions to the theoretical foundations of integrated modeling in biogeochemistry and their application in contaminated areas. In: Kothe E, Varma A (eds) Bio-geo-interactions in contaminated soils. Springer, Berlin, pp 385–416

Iordache V, Dumitru AL, Francu B, Iosif I, Cătieșanu A, Onete M, Neagoe A (2016) Optimizing a remediation method on a tailing dam, presentation at: Jena, Germany, Conference: 15th symposium on remediation, From "expert knowledge" to basic science to application: 15 years of bio-geo interactions. https://doi.org/10.13140/RG.2.2.10303.92323

Jacoby R, Peukert M, Succurro A, Koprivova A, Kopriva S (2017) The role of soil microorganisms in plant mineral nutrition – current knowledge and future directions. Front Plant Sci 6:1617

Jacquiod S, Nunes I, Brejnrod A, Hansen MA, Holm PE, Johansen A, Brandt KK, Prierne A, Sorensen SJ (2018) Long-term soil metal exposure impaired temporal variation in microbial metatranscriptomes and enriched active phages. Microbiome 6:223

Jarnevich CS, Holcombe TR, Thomas CC, Frid L, Olsson A (2015) Simulating long-term effectiveness and efficiency of management scenarios for an invasive grass. AIMS Environ Sci 2:427–447

Jianu D, Iordache V, Soare B, Petrescu L, Neagoe A, Iacob C, Orza R (2012) The role of mineralogy and geochemistry in hazard potential assessment of mining areas. In: Kothe E, Varma A (eds) Bio-geo-interactions in contaminated soils. Springer, Berlin, pp 35–79

Jones TA, Monaco TA, Rigby CW (2015) The potential of novel native plant materials for the restoration of novel ecosystems. Elementa: Sci Anthropocene 3:000047

Jumpponen A, Egerton-Warburton LM (2005) Mycorrhizal fungi in successional environments: a community assembly model incorporating host plant, environmental, and biotic filters. In: Dighton J, White JF, Oudemans P (eds) The Fungal Community: its organization and role in the ecosystem. CRC Press, Boca Raton, pp 139–168

June-Wells M, Gallagher F, Holzapfel C (2014) Evaluating population border dynamics among *Artemisia vulgaris* and community constituents in an urban successional plant assemblage. J Torrey Botanic Soc 141:14–28

Kalinkat G, Jochum M, Brose U, Dell AI (2015) Body size and the behavioral ecology of insects: linking individuals to ecological communities. Curr Opin Insect Sci 9:24–30

Kalucka IL, Jagodzinski AM (2016) Successional traits in ectomycorrhizal fungi in forest reclamation after surface mining and agricultural disturbances: a review. Dendrobiology 76:91–104

Kardol P, Bezemer TM, Van der Putten WH (2006) Temporal variation in plant-soil feedback controls succession. Ecol Lett 9:1080–1088

Kielhorn K-H, Keplin B, Huttl RF (1999) Ground beetle communities on reclaimed mine spoil: effects of organic matter application and revegetation. Plant Soil 213:117–125

Kikvidze Z, Armas C, Fukuda K, Martinez-Garcia LB, Miyata M, Oda-Tanaka A, Pugnaire FI, Wu B (2010) The role for arbuscular mycorrhizae in primary succession: differences and similarities across habitats. Web Ecol 10:50–57

Knelman JE, Schmidt SK, Lynch RC, Darcy JL, Castle SC, Cleveland CC, Nemergut DR (2014) Nutrient addition dramatically accelerates microbial community succession. PLoS One 9: e102609. https://doi.org/10.1371/journal.pone.0102609

Koehler H (1998) Secondary succession of soil mesofauna: a thirteen year study. Appl Soil Ecol 9:81–86

Koide RT, Fernandez C, Petprakob K (2011) General principles in the community ecology of ectomycorrhizal fungi. Ann For Sci 68:45–55

Komnitsas K, Guo X, Li D (2010) Mapping of soil nutrients in an abandoned Chinese coal mine and waste disposal site. Miner Eng 23:627–635

Koziol E (2016) The role of arbuscular mycorrhizal fungi in prairie grassland succession and restoration. PhD theses, Indiana University

Kruminis JA, Goodey NM, Gallagher F (2015) Plant-soil interactions in metal contaminated soils. Soil Biol Biochem 80:224–231

Kuziakov Y, Blagodatskaya E (2015) Microbial hotspots and hot moments in soil: concept & review. Soil Biol Biochem 83:184–199

Lanfranco L, Fiorilli V, Gutjahr C (2018) Partner communication and role of nutrients in the arbuscular mycorrhizal symbiosis. New Phytol 220:1031–1046

Langenheder S, Lindstrom ES (2019) Factors influencing aquatic and terrestrial bacterial community assembly. Environ Microbiol Rep 11:306–315. https://doi.org/10.1111/1758-2229.12731

Lavelle P (2012) Soil as a habitat. In: Wall DH et al (eds) Soil ecology and ecosystem services. Oxford University Press, Oxford, pp 7–27

Leal MC, Seehausen O, Mattheews B (2017) The ecology and evolution of stoichiometric phenotypes. Trends Ecol Evol 32:108–117

Li S, Caddote MW, Meiners SJ, Pu Z, Fukami SJ, Pu Z, Fukami T, Jiang L (2016a) Convergence and divergence in a long-term old-field succession: the importance of spatial scale and species abundance. Ecol Lett 19:1101–1109

Li J, Liu F, Chen J (2016b) The effects of various land reclamation scenarios on the succession of soil bacteria, archaea, and fungi over the short and long term. Front Ecol Evol 4:32. https://doi.org/10.3389/fevo.2016.00032

Li M, Jordan NR, Koide RT, Yannarell AC (2017) Meta-analysis of crop and weed growth responses to arbuscular mycorrhizal fungi: implications for integrated weed management. Weed Sci 64:642–652

Liu J, Yao J, Wang F, Min N, Gu J, Li Z (2019) Bacterial diversity in typical abandoned multi-contaminated nonferrous metal(loid) tailings during natural attenuation. Environ Pollut 247:98–107

Lundberg J, Moberg F (2003) Mobile link organisms and ecosystem functioning: implications for ecosystem resilience and management. Ecosystems 6:87–98

Macek P, Prieto I, Mackova J, Piston N, Pugnaire FI (2016) Functional plant types drive plant interactions in a Mediterranean mountain range. Front Plant Sci 7:662. https://doi.org/10.3389/fpls.2016.00662

Machado AS, Valyi K, Rillig MC (2017) Potential environmental impacts of "Underground Revolution": a response to Bender et al. Trends Ecol Evol 32:8–10

Machovsky-Capuska GE, Senior AM, Simpson SJ, Raubenheimer D (2016) The multidimensional nutritional niche. Trends Ecol Evol 31:355–365

Majeed A, Muhammad Z, Ahmad H (2018) Plant growth promoting bacteria: role in soil improvement, abiotic and biotic stress management of crops. Plant Cell Rep 37:1599–1609

Malla MA, Dubey A, Yaclav S, Kumar A, Hashem A, Abd_Allah EF (2018) Understanding and designing the strategies for the microbe-mediated remediation of environmental contaminants using omics approaches. Front Microbiol 9:1132

Manu M, Băncilă RI, Iordache V, Bodescu F, Onete M (2017) Impact assessment of heavy metal pollution on soil mite communities (Acari: Mesostigmata) from Zlatna Depression Transylvania. Process Saf Environ Prot 108:121–134

Maraun M, Martens H, Migge S, Theenhaus A, Scheu S (2003) Adding to 'the enigma of soil animal diversity': fungal feeders and saprophagous soil invertebrates prefer similar food substrates. Eur J Soil Biol 39:85–95

Maron JL, Smith AL, Ortega YK, Pearson DE, Callaway RM (2016) Negative plant-soil feedbacks increase with plant abundance, and are unchanged by competition. Ecology 97:2055–2063

McCook LJ (1994) Understanding ecological community succession: causal models and theories, a review. Vegetation 110:115–147

Meiners SJ, Cadotte MW, Fridley JD, Picket STA, Walker LR (2015) Is successional research nearing its climax ? New approaches for understanding dynamic communities. Funct Ecol 29:154–164

Millar NS, Bennet AE (2016) Stressed out symbiotes: hypotheses for the influence of abiotic stress on arbuscular mycorrhizal fungi. Oecologia 182:625–641

Miller BW, Frid L, Chang T, Piekelek N, Hansen AJ, Morisette JT (2015) Combining state-and-transition simulations and species distribution models to anticipate the effects of climate change. AIMS Environ Sci 2:400–426

Monamy V, Fox BJ (2000) Small mammal succession is determined by vegetation density rather than time elapsed since disturbance. Anim Ecol 25:580–587

Montiel-Rozas MM, Lopez-Garcia A, Kjoller R, Madejon E, Rosendahl S (2016) Organic amendments increase phylogenetic diversity of arbuscular mycorrhizal fungi in acid soil contaminated by trace elements. Mycorrhiza 26:575–585

Montiel-Rozas MM, Lopez-Garcia A, Madejon P, Madejon E (2017) Native soil organic matter as a decisive factor to determine the arbuscular mycorrhizal fungal community structure in contaminated soils. Biol Fertil Soil. https://doi.org/10.1007/s00374-017-1181-5

Moore JC, StJohn TV, Coleman DC (1985) Ingestion of vesicular-arbuscular mycorrhizal hyphae and spores by soil microarthropods. Ecology 66:1979–1981

Morales-Castilla I, Matias MG, Gravel D, Araujo MB (2015) Inferring biotic interactions from proxies. TREE 30:347–356

Morris SJ, Friese CF, Allen MF (2016) Disturbance in natural ecosystems: scaling from fungal diversity to ecosystem functioning. In: Druzhinina IS, Kubicek CP (eds) Environmental and microbial relationships. Springer, Berlin, pp 79–98

Mudrak O, Dolezal J, Frouz J (2016) Initial species composition predicts the progress in the spontaneous succession on post-mining sites. Ecol Eng 95:665–670

Naeem S, Colwell RK (1991) Ecological consequences of heterogeneity of consumable resources. In: Kolosa J, Picklett STA (eds) Ecological heterogeneity. Springer, New York, pp 224–255

Nash KL, Allen CR, Angeler DG, Barichievy C, Eason T, Garmestani AS, Graham NAJ, Grandholm D, Knutsen M, Nelson RJ, Nystrom M, Stow CA (2014) Discontinuities, cross-scale patterns, and the organization of ecosystems. Ecology 95:654–667

Neagoe A, Merten D, Iordache V, Buechel G (2009) The effect of bioremediation methods involving different degrees of soil disturbance on the export of metals by leaching and by plant uptake. Chem Erde-Geochem 69:57–73

Neagoe A, Iordache V, Kothe E (2010) Effects of the inoculation with AM fungi on plant development and oxidative stress in areas contaminated with heavy metals. Presentation at COST870 meeting in Jyvaskyla, Finland, 13–15 December, Book of abstracts, pp 22

Neagoe A, Iordache V, Fărcăşanu IC (2012) The role of organic matter in the mobility of metals in contaminated catchments. In: Kothe E, Varma A (eds) Bio-geo-interactions in contaminated soils. Springer, Berlin, pp 297–325

Neagoe A, Iordache V, Kothe E (2013) Upscaling the biogeochemical role of arbuscular mycorrhizal fungi in metal mobility. In: Goltapeh EM, Danesh ER, Varma A (eds) Fungi as bioremediators. Springer, Berlin, pp 285–311

Neagoe A, Stancu P, Nicoară A, Onete M, Bodescu F, Gheorghe R, Iordache V (2014) Effects of arbuscular mycorrhizal fungi on *Agrostis capillaris* grown on amended mine tailing substrate at pot, lysimeter, and field plot scales. Environ Sci Pollut Res 21:6859–6876

Ngosong C, Gabriel E, Ruess L (2014) Collembola grazing on arbuscular mycorrhiza fungi modulates nutrient allocation in plants. Pedobiologia – J Soil Ecol 57:171–179

Nichols OG, Nichols FM (2003) Long-term trends in faunal recolonization after bauxite mining in the Jarrah Forest of Southwestern Australia. Restor Ecol 11:261–272

Nicoară A, Neagoe A, Stancu P, de Giudici G, Langella F, Sprocati AR, Iordache V, Kothe E (2014) Coupled pot and lysimeter experiments assessing pant performance in microbially assisted phytoremediation. Environ Sci Pollut Res 21:6905–6920

Nikolic N (2013) Ecology of alluvial arable land polluted by copper mine tailings: new insights for restoration. PhD Theses, University of Hohenheim

Nikolic N, Kostic L, Djordjevic A, Nikolic M (2011) Phosphorus deficiency is the major limiting factor for wheat on alluvium polluted by copper mine pyrite tailings: a black box approach. Plant Soil 339:485–498

Norman MA, Koch JM, Grant CD, Morald TK, Ward SC (2008) Vegetation succession after bauxite mining in Western Australia, vol 14, pp 278–288

O'Brien SL, Gibbons SM, Owens SM, Hampton-Marcell J, Johnston ER, Jastrow JD, Gilbert JA, Meyer F, Antonopoulos DA (2016) Spatial scale drives patterns in soil bacterial diversity. Environ Microbiol 18:2039–2051

Ohgushi T, Wurst S, Johnson SN (2018) Aboveground-belowground community ecology. Springer, Cham

Opik M, Davison J (2016) Uniting species- and community-oriented approaches to understand arbuscular mycorrhizal fungal diversity. Fungal Ecol 24:106–113

Opik M, Peay KG (2016) Mycorrhizal diversity: diversity of host plants, symbiotic fungi and relationships. Fungal Ecol 24:103–105

Ourry M, Lebreton L, Chaminade V, Guillerm-Erckelboudt A-Y, Herve M, Linglin J, Marnet N, Ourry A, Paty C, Poinsot D, Cortesero A-M, Mougel C (2018) Influence of belowground herbivory on the dynamics of root and rhizosphere microbial communities. Front Ecol Evol 6:91

Parmenter RR, Macmahon JA (1987) Early successional patterns of arthropod recolonization on reclaimed strip mines in Southwestern Wyoming: the ground-dwelling beetle fauna (Coleoptera). Environ Entomol 16:168–177

Paterno GB, Filho JAS, Ganade G (2016) Species-specific facilitation, ontogenetic shifts and consequences for plant community succession. J Veg Sci 27:606–615

Peay KG (2016) The mutualistic niche: mycorrhizal symbiosis and community dynamics. Annu Rev Evol Syst 47:143–164

Peay KG, Bruns TD (2014) Spore dispersal of basidiomycete fungi at the landscape scale is driven by stochastic and deterministic processes and generates variability in plant–fungal interactions. New Phytol 204:180–191

Peay KG, Mathney PB (2017) The biogeography of ectomycorrhizal fungi – a history of life in the subterranean. In: Martin F (ed) Molecular mycorrhizal symbiosis. Wiley, New York, pp 341–361

Pickles BJ, Anderson IC (2016) Spatial ecology of ectomycorrhizal fungi communities. In: Martin F (ed) Molecular mycorrhizal symbiosis. Wiley, New York, pp 363–386

Pietrzykowski M (2014) Reclamation and reconstruction of terrestrial ecosystems on mine sites – ecological effectiveness assessment. Energy 2:120–151

Pineda A, Kaplan I, Bezemer M (2017) Steering soil microbiomes to suppress aboveground insect pests. Trends Plant Sci 9:770–778

Poisot T, Stouffer DB, Kefi S (2016) Describe, understand and predict: why do we need networks in ecology ? Funct Ecol 30:1878–1882

Powel JR, Bennet AE (2016) Unpredictable assembly of arbuscular mycorrhizal fungal communities. Pedobiologia 59:11–15

Powel JR, Karunaratne S, Campbell CD, Yao H, Robinson L, Singh BK (2015) Deterministic processes vary during community assembly for ecologically dissimilar taxa. Nat Commun 6:8444. https://doi.org/10.1038/ncomms9444

Prasad R (2017) Mycoremediation and environmental sustainability. Vol 1. Springer International Publishing, Switzerland. ISBN 978-3-319-68957-9. https://link.springer.com/book/10.1007/978-3-319-68957-9

Prasad R (2018) Mycoremediation and environmental sustainability, Vol 2. Springer International Publishing, Switzerland. ISBN 978-3-319-77386-5. https://www.springer.com/us/book/9783319773858

Prasad R, Garg AP, Varma A (2005) Interaction of medicinal plants with plant growth promoting rhizobacteria and symbiotic fungi. In: Podila GK, Varma A (eds) Basic research and applications of Mycorrhizae, vol 1. IK International, Delhi, pp 363–407

Prewitt L, Kang Y, Kakumanu ML, Williams M (2014) Fungal and bacterial community succession for three wood types during decay in forest soil. Microb Ecol 68:212–221

Provencher L, Frid L, Czembor C, Morisette JT (2016) State-and-transition models: conceptual versus simulation perspectives, usefulness and breadth of use, and land management applications. In: Germino MJ et al (eds) Exotic brome-grasses in arid and semiarid ecosystems of the Western US. Springer, New York, pp 371–407

Revillini D, Gehring CA, Johnson NC (2016) The role of locally adapted mycorrhizas and rhizobacteria in plant-soil feedback systems. Funct Ecol 30:1086–1098

Rhodes CJ (2014) Mycoremediaton (bioremediation with fungi) – growing mushrooms to clean the earth. Chem Speciat Bioavailab 26:196–198

Rigg JL, Offord CA, Singh BK, Anderson I, Clarke S, Powell JR (2016a) Soil microbial communities influence seedling growth of a rare conifer independent of plant–soil feedback. Ecology 97:3346–3358

Rigg JL, Offord CA, Singh BK, Anderson IC, Clarke S, Powell JR (2016b) Variation in soil microbial communities associated with critically endangered Wollemi pine affects fungal, but not bacterial, assembly within seedling roots. Pedobiologia 59:61–71

Rigg JL, Offord CA, Zimmer H, Anderson IC, Singh BK, Powell JR (2017) Conservation by translocation: establishment of Wollemi pine and associated microbial communities in novel environments. Plant Soil 411:209–225

Roubickova A (2013) Interactions of soil fauna and plants during succession on spoil heaps after brown coal mining. PhD theses, Charles University of Prague

Roy-Bolduc A, Bell TH, Boudreau S, Hijri M (2015) Comprehensive sampling of an isolated dune system demonstrates clear patterns in soil fungal communities across a successional gradient. Environ Microbiol Rep 7:839–848

Roy-Bolduc A, Laliberte E, Boudreau S, Hijri M (2016) Strong linkage between plant and soil fungal communities along a coastal dune system. FEMS Microb Ecol 92:fiw156

Ruess L, Lussenhop J (2012) Trophic interactions of fungi and animals. In: Dighton J, White JF, Oudemans P (eds) The fungal community its organization and role in the ecosystems. CRC Press, Boca Raton, pp 581–598

Sacket TE, Classen AT, Sanders NJ (2010) Linking soil food web structure to above- and belowground ecosystem processes: a meta-analysis. Oikos 119:1984–1992

Scheiner SM, Chiarucci A, Fox GA, Helmus MR, McGlinn DJ, Willig MR (2011) The underpinnings of the relationship of species richness with space and time. Ecol Monogr 81:195–213

Scherber C, Eisenhauer N, Weiser WW, Schmid B, Voigt W, Fischer M, Schulze E-D, Roscher C, Weigelt A, Allan E, 30 other coauthors (2010) Bottom-up effects of plant diversity on multitrophic interactions in a biodiversity experiment. Nature 408:553–556

Scheu S, Schulz E (1996) Secondary succession, soil formation and development of a diverse community of oribatids and saprophagous soil macro-invertebrates. Biodivers Conserv 5:235–250

Schickmann S, Urban A, Krautler K, Nopp-Mayr U, Hacklander K (2012) The interrelationship of mycophagous small mammals and ectomycorrhizal fungi in primeval, disturbed and managed Central European mountainous forests. Oecologia 170:395–409

Schmidt SK, Nemergut DR, Darcy JL, Lynch R (2014) Do bacterial and fungal communities assemble differently during primary succession? Mol Ecol 23:254–258

Schneider K, Renker C, Maraun M (2005) Oribatid mite (Acari, Oribatida) feeding on ectomycorrhizal fungi. Mycorrhiza 16:67–72

Schrama M, Van der Plas F, Berg MP, Off H (2017) Decoupled diversity dynamics in green and brown webs during primary succession in a saltmarsh. J Anim Ecol 86:158–169

Schweiger W, Diffendorfer JE, Holt RD, Pierotti R, Gaines MS (2000) The interaction of habitat fragmentation, plant and small mammal succession in an old field. Ecol Monogr 70:383–400

Sepp S-K, Davison J, Jairus T, Vasar M, Moora M, Zobel M, Opik M (2019) Non-random association patterns in a plant-mycorrhizal fungal network reveal host-symbiont specificity. Mol Ecol 28:635–378

Shao P, Liang C, Rubert-Nason K, Li X, Xie H, Bao X (2019) Secondary successional forests undergo tightly-coupled changes in soil microbial community structure and soil organic matter. Soil Biol Biochem 128:56–65

Sheoran V, Sheoran AS, Poonia P (2010) Soil reclamation of abandoned mine land by revegetation: a review. Int J Soil Sedim Water 3:13

Sheoran V, Sheoran AS, Poonia P (2016) Factors affecting phytoextraction. Pedosphere 26:148–166

Singh M, Srivastava PK, Verma PC, Kharwar RN, Singh N, Tripathi RD (2015) Soil fungi for mycoremediation of arsenic pollution in agriculture soils. J Appl Microbiol 119:1278–1290

Sitvarin MI, Rypstra AL, Harwood JD (2016) Linking the green and brown worlds through nonconsumptive predator effects. Oikos 125:1057–1068

Sperfeld E, Wagner ND, Halvorson HM, Malishev M, Raubenheimer D (2017) Bridging ecological stoichiometry and nutritional geometry with homeostasis concepts and integrative models of organism nutrition. Funct Ecol 31:286–296

Stamou GP, Papatheodorou EM (2016) Studying the complexity of the secondary succession process in the soil of restored open mine lignite areas: the role of chemical template. Appl Soil Ecol 103:56–60

Sterkenburg E (2016) Drivers of soil fungal communities in boreal forests. PhD theses, Swedish University of Agricultural Sciences

Stringham TK, Krueger WC, Shaver PL (2001) States, transitions, and threshold: further refinement for rangeland applications, Special Report 1024, Oregon State University

Sturşova M, Barta J, Santruckova H, Baldrian P (2016) Small-scale spatial heterogeneity of ecosystem properties, microbial community composition and microbial activities in a temperate mountain forest soil. FEMS Microbiol Ecol 92:fiw185

Sun Q, Liu Y, Yuan H, Lian B (2016a) The effect of environmental contamination on the community structure and fructification of ectomycorrhizal fungi. Microbiol Open 6:e00396

Sun H, Wu Y, Zhou J, Bing H (2016b) Variations of bacterial and fungal communities along a primary successional chronosequence in the Hailuogou glacier retreat area (Gongga Mountain, SW China). J Mt Sci 13:1621–1631

Swihart RK, Slade NA (1990) Long-term dynamics of an early successional small mammal community. Am Midl Nat 123:372–382

Tedersoo L, Bahram M, Cajthaml T, Polme S, Hiiesalu I, Anslan S, Harend H, Buegger F, Pritsch K, Koricheva J, Abarenkov K (2016) Tree diversity and species identity effects on soil fungi, protists and animals are context dependent. ISME J 10:346–362

terHorst CP, Zee PC (2016) Eco-evolutionary dynamics in plant-soil feedbacks. Funct Ecol 30:1062–1072

Teste FP (2016) Restoring grasslands with arbuscular mycorrhizal fungi around remnant patches. Appl Veg Sci 19:3–4

Teste FP, Dickie IA (2017) Mycorrhizas across successional gradients. In: Johnson NC, Gehring C, Jansa J (eds) Mycorrhizal mediation of soil fertility, structure and carbon storage. Elsevier, Amsterdam, pp 67–89

Tews J, Brose U, Grimm V, Tielborger K, Wichmann MC, Schwager M, Jeltsch F (2004) Animal species diversity driven by habitat heterogeneity/diversity: the importance of keystone structures. J Biogeogr 31:79–92

Tilman D (1985) The resource-ratio hypothesis in plant succession. Am Nat 125:827–852

Tilman D (1990) Constraints and tradeoffs: toward a predictive theory of competition and succession. Oikos 58:3–15

Tischew S, Baasch A, Grunert H, Kirmer A (2014) How to develop native plant communities in heavily altered ecosystems: examples from large-scale surface mining in Germany. Appl Veg Sci 17:288–301

Tiunov AV, Scheu S (2005) Arbuscular mycorrhiza and Collembola interact in affecting community composition of saprotrophic microfungi. Oecologia 142:636–642

Topp W, Thelen K, Kappes H (2010) Soil dumping techniques and afforestation drive ground-dwelling beetle assemblages in a 25-year-old open-cast mining reclamation area. Ecol Eng 36:751–756

Tordoff GM, Boddy L, Jones TH (2008) Species-specific impacts of collembola grazing on fungal foraging ecology. Soil Biol Biochem 40:434–442

Torrez V, Ceulemans T, Mergeay J, de Meester L, Honnay O (2016) Effects of adding an arbuscular mycorrhizal fungi inoculum and of distance to donor sites on plant species recolonization following topsoil removal. Appl Veg Sci 19:7–19

Triska MD, Craig MD, Stokes VL, Pech RP, Hobbs RJ (2016) The relative influence of in situ and neighborhood factors on reptile recolonization in post-mining restoration sites. Restor Ecol 24:517–527

Urbanova M, Snajdr J, Brabcova V, Merhautova V, Dobiasova P, Cajthami T, Vanek D, Frouz J, Santruckova H, Baldrian P (2014) Litter decomposition along a primary post-mining chronosequence. Biol Fertil Soils 50:827–837

Valyi K, Marhiah U, Rillig MC, Hempel S (2016) Community assembly and coexistence in communities of arbuscular mycorrhizal fungi. ISME J 10:2341–2351

Van der Wal A, Geydan TD, Kuyper TW, de Boer W (2012) A thready affair: linking fungal diversity and community dynamics to terrestrial decomposition processes. FEMS Microbiol Rev 37:477–494

Vernes K (2003) Fine-scale habitat preferences and habitat partitioning by three mycophagous mammals in tropical wet sclerophyll forest, north-eastern Australia. Aust Ecol 28:471–479

Vieira CK, Marascalchi MN, Rodrigues AV, Dutra de Armas R, Stürmer SL (2018) Morphological and molecular diversity of arbuscular mycorrhizal fungi in revegetated iron-mining site has the same magnitude of adjacent pristine ecosystems. J Environ Sci 67:330–343

Walker LR, Walker J, Hobbs RJ (2007) Linking restoration and ecological succession. Springer, New York

Walker LR, Wardle DA, Bardgett RD, Clarkson BD (2010) The use of chronosequences in studies of ecological succession and soil development. J Ecol 98:725–736

Weiher E, Freund D, Bunton T, Stefansky A, Lee T, Bentivenga S (2011) Advances, challenges and a developing synthesis of ecological community assembly theory. Philos Trans R Soc Biol 366:2403–2413

Wodika BR, Klopf RP, Baer SG (2014) Colonization and recovery of invertebrate ecosystem engineers during prairie restoration. Restor Ecol 22:456–464

Xu S, Zhang J, Luo S, Zhou X, Shi S, Tian C (2018) Similar microbial community structure across different environment after long-term succession: evidence from volcanoes of different ages. J Basic Microbiol 58:704–711

Zangaro W, Rondina ABL (2016) Arbuscular mycorrhizas in different successional stages in some Brazilian ecosystems. In: Pagano MC (ed) Recent advances on mycorrhizal fungi. Springer, Berlin, pp 47–62

Zhang C, Liu G, Song Z, Wang J, Guo L (2018a) Interactions of soil bacteria and fungi with plants during long-term grazing exclusion in semiarid grasslands. Soil Biol Biochem 124:47–58

Zhang K, Cheng X, Shu X, Liu Y, Zhang Q (2018b) Linking soil bacterial and fungal communities to vegetation succession following agricultural abandonment. Plant Soil 431:19–36

Zorilla JM, Serrano JM, Casado MA, Acosta FJ, Pineda FD (1986) Structural characteristics of an ant community during succession. Oikos 47:346–354

Chapter 12
Composition and Dynamics of Microbial Communities in Fly Ash-Amended Soil

Ayushi Varshney, Sumedha Mohan, and Praveen Dahiya

Abstract Fly ash (FA), solid waste residue, is a by-product of thermal power plant, and its disposal is considered to be the current biggest challenge faced by the entire world. Various studies have revealed that addition of FA to the soil may improve the physical, chemical, and biological properties of soil mixture. FA is also considered to provide readily available soil micro- and macronutrients. Lower doses of FA enhance the nutritional status of agricultural soil, thereby promoting plant growth, whereas higher doses of FA result in heavy metal pollution leading to deleterious effects which hinder the soil microbial communities and related enzymatic activity. Practically, FA is utilized in agriculture due to its usage as "ecofriendly and cost-effective" soil/fertilizer amendment which depends upon various factors like soil types, climatic conditions, plant selected for growth, etc. However, combined biotechnological approaches (organic and inorganic) need to be applied to soil-FA mixtures in order to reduce the amount of toxic elements present in FA and also to maintain the nutrient status as well as soil microbial activity. Overall, the lower doses of FA application could be utilized to revamp the soil structure, soil microflora, N-cycling process, and enzymatic activity which have a significant role in promoting the growth of plants.

12.1 Introduction

Fly ash (FA) is an inorganic noncombustible waste produced from different thermal power plants during coal combustion process. FA has been regarded as a worldwide major solid waste produced due to the global dependence of burning of coal for electricity production. It also poses countless environmental pollution issues. However, productive utilization of FA to maximum range is of vital importance in its management. FA in the form of fine particle residue is carried away by flue gases in

A. Varshney · S. Mohan (✉) · P. Dahiya
Amity Institute of Biotechnology, Amity University Uttar Pradesh (AUUP), Uttar Pradesh, Noida, India
e-mail: smehta1@amity.edu

A. Varma et al. (eds.), *Plant Microbiome Paradigm*,
https://doi.org/10.1007/978-3-030-50395-6_12

electrostatic precipitators and is then collected via different dry or wet scrubbing methods that require huge amount of area, water, as well as energy for its management. Primarily FA contains silica and alumina, oxides of iron, calcium, carbon, magnesium, titanium, and sulfur but contains no free nitrogen (Sett 2017). The nutrient status of FA (elemental or heavy metal) may vary depending upon the type and source of coal being used (Kumar et al. 2017). Due to the presence of various essential elements present in FA, it has been considered as a prospective amendment to agricultural soil in order to enhance the accessibility of nutrients in improving the growth of plants (Pandey and Singh 2010; Singh et al. 2011a, b). Judicious utilization of FA can bring about favorable alteration in the nutrient status of soil, provided all aspects are constantly examined for overall benefit. The exploitation of FA as fertilizer for crop production is limited in most of the countries as FA contains some non-essential elements (Pb, Cd, Se, As) which adversely affect crop productivity, and on the other hand, it is deficient in both nitrogen N (as nitrogen is oxidized at the time of combustion of coal) and phosphorus P (soluble P is converted to insoluble P due to excessive levels of Fe and Al, which is not easily available for the plants) (Sahu et al. 2017). Many reports have suggested that low application rates of FA in soil result in increased plant growth and also yield because of the availability of essential elements; however, higher dosage of FA application adversely impacts the growth of plants due to toxicity by heavy metals (Selvakumari et al. 2000; Patil et al. 2005; Pandey et al. 2009).

Several investigators reported the utilization of FA in agricultural soil, in enhancing the physical and chemical characteristics of soil as well as soil biological health leading to increased plant productivity (Pandey et al. 2009; Nashine 2014; Panda et al. 2015; Raj et al. 2015; Honghua et al. 2017; Basha et al. 2018). FA application in soil at low concentration has proved to affect the microbial activities in soil (Pandey et al. 2011), N-transformations (Singh et al. 2014), and enzymatic activity of soil mixture (Kohli and Goyal 2010). Similarly, FA along with selected different organic matter also enhances the physico-chemical and biological properties associated with the N-dynamics in case of paddy soils (Pandey and Singh 2010).

12.1.1 Soil Ecosystem

In comparison to soil fertility, soil productivity is regarded as a significant factor which can determine the best suitable agricultural production (Brady and Weil 2012). Soil fertility includes the nutrient status of the soil and its capability to provide the nutrient in the soil, whereas soil productivity is responsible for the higher yield production in soil (Onduru et al. 2006; Behera and Prasad 2020). Soil productivity relies on different factors, out of which the role of soil microbes is eminent.

Soil is regarded as the habitat of diverse life forms. Soil provides shelter for various invertebrates like insects and worms including various microorganisms. Microorganisms are diverse which includes all the bacteria, fungi, archaea, and all the protozoans. These diverse life forms are interrelated with one another and also with their soil environment in order to survive in continuously changing scenarios which may lead to alterations in fertility of soil and its productivity. Globally soil microbes are considered as the key driving force for biogeochemical cycles including carbon, phosphorus, nitrogen, and sulfur and other elements present in nature (Basu et al. 2020). Their activity in soil deals with the soil physiological system functions, including organic matter and nutrient cycling processes, and in the soil structure development, they have important role in all biochemical reactions taking place in soil (Brussaard 2012). Soil microbes are considered as a significant factor in determining the soil environmental quality. The plant root system is capable of absorbing the soil microbes and therefore can provide the nutrients to the plants for their growth (Prosser 2015). In the soil ecosystem, soil microbes possess two major factors. Firstly, the microbes contain specific elements in themselves such as C, N, and P which can be used to adjust the soil nutrients. Secondly, inorganic elements can flow via microbial transformation and promotion of system's metabolic process (Chu 2018). Disintegration of carbon, other nutrients, and the associated cyclic processes are controlled due to the microbial activities taking place in soil ecosystems.

Microbes possess vital function in transforming inert nitrogen present in the atmosphere to ionic nitrogen and also decompose insoluble minerals to soluble minerals that can be absorbed directly to the plants (Subhani et al. 2001). Microbes are not only associated with the degradation of organic and inorganic pollutants which reduces the increased toxicity in plants but are also involved in providing healthy ecological environment for better growth of plant. They can also synthesize few important secondary metabolites which may trigger the growth and development of plant (Chu 2018). Soil microbes in the rhizosphere prepare a physical barrier all around the plant roots, reducing the invasion of pathogens and thus protecting the roots of the plant in these micro-ecological surroundings (Wu and Lin 2003). Plant-promoting bacteria in the rhizospheric soil have the ability to regulate crop vulnerability via atmospheric nitrogen fixing, synthesizing plant growth hormones, stimulating specific enzyme activities, and involving certain carriers and chelating agents which can inhibit pathogenic microorganisms (Khan 2005; Prasad et al. 2015, 2020). Thus, soil microbial activity is essential for healthy soil, and therefore its role in the functioning and evolution of soil ecosystem is predictable, and hence the variations obtained in soil microbial activity serve as an indicator for changes in the soil health (Subhani et al. 2001; Behera and Prasad 2020). Certain factors like soil moisture, organic matter, nutrient availability, temperature, soil pH, and heavy metals present will significantly alter the microbial population and its activity.

12.2 Microbial Communities in Soil Amended with FA

The fundamental component of our earth's ecosystem is soil and its biota which directly affects its sustainability (Wilkinson et al. 2009). Enough studies are available reporting variations in physico-chemical characteristics of soil when FA is added to soil at different doses. FA is reported to show a great impact on soil biological system including soil health, microbial respiration, number, size, soil microbes, soil enzymatic activity, soil fertility, soil nitrogen cycling, etc. (Pandey and Singh 2010). There exists a paucity of information with reference to the influence of FA amendment to agriculture soil leading to change in the soil biological properties (Furlani et al. 2011). Soil microflora is vital factor influencing the soil environment as it leads to the degradation of organic matter present and makes available the nutrients to the other soil organisms. It also favors the formation of soil aggregates and immobilizes the heavy metals and stimulates soil enzymes activity.

FA is devoid of any soil microbes but could improve upon the physical and chemical characteristics of soil, thus improving the soil microbiological activities. There is an urgent need to critically evaluate the soil bioindicators such as soil enzymes, respiration, microbial biomass, earthworms, and many others to ascertain the functional value of FA for soil amendment. Microorganisms due to their co-enzymatic activities are considered as primary decomposers among the decomposer organisms in soil. They possess important role in mineralization and demineralization which will facilitate cycling of minerals (Marumoto et al. 1982). Microbial population can influence carbon or mineral cycles and have the ability to colonize harsh environments. Lignite fly ash application was found to reduce the growth of various soilborne pathogens, and the number of *Rhizobium* and P-solubilizing bacteria was found to increase when amended with farmyard manure, fly ash, or a combination of both (Sen 1997; Panda and Biswal 2018). Similarly, an increase was observed in the population of mycorrhizal fungi and Gram-negative bacteria when bituminous FA was amended at a rate of 505 Mg ha^{-1}. FA (at 100 t/ha)-amended tropical red laterite soil was reported to be safe for soil microbial communities (Roy and Joy 2011). Selective dosage of NPK fertilizers along with FA was found to significantly increase the population of bacteria when compared with the control set which could be due to the complementary impact of FA with NPK fertilizer (Yeledhalli et al. 2007). Vallini et al. (1999) observed an increasing trend in the count of bacteria and actinomycetes when FA was amended to the soil which results in increasing level of soil dehydrogenase activity. The influence of FA with organic or inorganic amendments on microbial communities is presented in Table 12.1 in detail.

Acidic or alkaline FA in the amended soils may have several detrimental effects on the microflora due to extreme pH levels of the soil (Wong and Wong 1987). Alkaline FA when added to the sandy soil decreases microbial respiration and nitrification (Wong and Wong 1986; Cerevelli et al. 1986). Nayak et al. (2015) observed decrease in the population of both fungi and actinomycetes when alkaline FA at pH −7.7 was added to the soil when compared to the control. Acidic FA (at 100 t/ha) amended to soil was not found to possess any significant effect to the heterotrophic microbial activity, whereas high doses (at 400–7000 t/ha) severely impacted the microflora present in soil (Arthur et al. 1984). In an acid soil, FA amended with other liming materials like lime or dolomite can favor the increasing population of soil bacteria and actinomycetes (Nivetha and Sheeba 2017). The application of 50% dolomite +50% FA + RDF (refuse-derived-fuel) results in higher levels of bacterial (67×10^6 cfu gm^{-1} soil) and actinomycete (59×10^3 cfu gm^{-1}

Table 12.1 Influence of fly ash (FA) doses (alone or in combination) on soil microbial communities

S. no.	Fly ash doses (in combination or alone)	Microbial communities present	References
1	1:1 soil-fly ash ratio (50% fly ash)	Increase in the population of *Azotobacter chroococcum*, *Azospirillum brasilense*, and *Bacillus circulans*	Gaind and Gaur (2004)
2	Fly ash at 5%	Presence of metal-tolerant plant growth-promoting bacteria *Enterobacter* sp. NBRI K28	Kumar et al. (2008)
3	Low doses of fly ash	Population of *Sphingomonas* sp. 23 L was found to increase at lower doses of FA	Hrynkiewiez et al. (2009)
4	Coal fly ash + clinch sediments	Increase in the population of *Proteobacteria*	Schwartz et al. (2016)
5	Fly ash + soil + farm yard manure	Combination of FYM leads to increasing levels of *Rhizobium* sp. and P-solubilizing bacteria	Sen (1997)
6	Fly ash at 40 t/ha	Increase in the population of *Pseudomonas striata*	Gaind and Gaur (2002)
7	Low doses of fly ash	Population of both *Bacillus curcas* and *Bacillus subtilis* was found to increase	Banerjee et al. (2015)
8	Alkaline fly ash + lime	*Salmonella* sp. and total coliforms levels were found to be higher	Wong et al. (2001)
9	Fly ash dumping site	Significant levels of *Bacillus* spp. and *Paenibacillus* spp. observed	Rau et al. (2009)
10	FA at lower doses	Increase in the population of arbuscular mycorrhizal fungi	Karpagavalli and Ramabadran (1997)
11	Fly ash at 20%	Decrease observed in the population of bacteria, actinomycetes, and fungi by 57, 80, and 86%	Pichtel and Hayes (1990)
12	Fly ash at 40%	Population of fungi and actinomycetes both decreased at 40% FA dosage, whereas no significant change was observed in aerobic heterotrophic bacterial population	Nayak et al. (2015)
13	Fly ash-amended soil at 505 Mg/ha^{-1}	Gram-positive bacteria *Arthrobacter illicis* population increased from 47 to 71% after FA amendment. A higher number of arbuscular mycorrhizal fungi and Gram-negative bacteria were obtained	Schutter and Fuhrmann (2001)

soil) population when tested in post-harvest soil, whereas the fungal population in post-harvest soil implied that all the treatments imposed recorded lower fungal count compared to that of the control (15×10^4 cfu gm^{-1} soil). Significant decrease was also observed in the number of all the microorganisms when FA was used at higher doses which may be due to the availability of limited C substrate as source of energy for heterotrophic microorganisms as well as lack of an adequate N supply (Klubek et al. 1992). This may also be attributed to higher concentration of toxic heavy metals present (Jala and Goyal 2006) that may hamper the microbial processes taking place under normal conditions.

12.3 FA Management and Soil Microbial Dynamics

Very little information on the impact of FA amendments on soil biological characteristics is available (Schutter and Fuhrmann 2001). For effective incorporation of FA in soil, proper understanding on how the soil microbiota reacts to soil amendment with FA is needed. In agriculture, survival rate and effective functioning of microbes are recognized as the hallmark of soil health and productivity.

The observations from various experimental results discovered that the incorporation of unweathered FA to sandy soil greatly reduced the microbial respiration, number of microbes, size, enzymatic activity, as well as soil nitrogen cycling processes like nitrification and N mineralization (Polat et al. 2002; Sahu et al. 2017). Unweathered FA showed these harmful effects due to the presence of excessive soluble salts and trace elements. Deleterious effects of FA were reduced during leaching in due of course of time due to reduction in concentration of soluble salt and the trace elements (Singh and Siddiqui 2003). Furthermore, the alkaline FA (pH 10–12) could also be the cause for these deleterious effects on soil health and fertility. There are studies where FA at low doses has been used as a potential resource in improving the soil system by increasing microbial diversity and soil enzymatic activity and by improving the soil structure (Gaind and Gaur 2002; Machulla et al. 2004; Kumar et al. 2008).

12.3.1 Physico-Chemical Response and Soil Health

FA incorporation in soil improves the physico-chemical properties of different types of soils (Mupambwa et al. 2015). The impact of FA on soil biological properties is mainly dependent upon the type of coal and soil being used. FA can be acidic or alkaline in nature (depending upon the source and type of coal and coal combustion process) and hence can be used to neutralize the pH of the soil (Kumar et al. 2017). Its application in sandy soil could alter the soil texture, improve porosity, and increase the water holding capacity (Gagnon et al. 2004). The Ca in FA substitutes Na at clay exchange sites which thereby improves the flocculation of clayey soil (Panda and Biswal 2018). The electrical conductivity and metal content increases with increase in FA application. Lime in FA generally counters with the acidic components in the soil and releases certain nutrients such as S, B, and Mo in the form favorable for the plant growth. FA can be used to eliminate the deficiencies of sulfur and boron in the acidic soil (Rautaray et al. 2003). pH ranging between 6.5 and 7.0 is considered favorable for majority of crops as the availability of plant nutrients is maximized. Hence, soil fertility is generally influenced at optimum pH. FA at low doses enhances the soil pH due to the release of Ca, Na, Al, and OH- ions (Pandey and Singh 2010). Many researchers revealed that FA incorporation in soil alters the soil environment, though changes in moisture retention, release and transmission behavior, pH, EC, and organic carbon (Murugan and Vijayarangam 2013).

Amending soil with FA causes addition of many more elements (K, Ca, Mg, Cu, Zn, Mn, etc.) which causes the alteration in chemical properties of soil to which they are added (Yeledhalli et al. 2007). FA contains certain essential nutrients required for the plant growth and development. C and N are usually present in low amount in FA though it is medium in available K and high in available P. These nutrients in FA may prove to be good for agricultural soil as they enrich the soil during the course of time by effectively changing the composition and dynamics of soil microflora (Mahalingam 1973; Plank and Martens 1974; Carry et al. 1986).

In the soil biological system, the enzymes play an essential role in disintegration of organic matter and nutrient cycling processes and hence play a vital role in agriculture (Rao et al. 2017). These enzymes are mainly responsible for catalyzing different chemical reactions vital for life processes of soil microbes and for stabilization of soil structure. Soil enzymes react promptly to changes in environmental conditions and soil management practices. Their activity in soil is largely dependent upon the soil physico-chemical and biological properties. Therefore, soil enzymes are used as indicators for soil microbial status, in determining the physico-chemical conditions of soil, and for the influence of climatic conditions on soil fertility. Good understanding of possible roles of different soil enzymes in maintaining soil condition and its productivity is necessary particularly in agricultural ecosystems.

FA addition to soil can alter the soil biological responses mainly due to the altered physico-chemical changes in the soil. One of the important reasons limiting the microbial activity is alteration in pH, major soluble elements, and electrical conductivity of FA amended with soil (Elliott et al. 1982). In a pot culture experiment, an upsurge in the CO_2 evolution and soil enzyme activity (primarily dehydrogenase and protease) in FA-soil mixture was observed (Elliott et al. 1982). Increase in CO_2 assimilation and enzyme activity in soil was found to be favorable for soil microbial activity. Sarangi et al. (2001) reported an increase in certain enzymes like invertase, amylase, dehydrogenase, and protease with increase in FA application up to 10 t/ha but decrease at high rate of FA application. Similarly, Pichtel and Hayes (1990) reported the decline in soil phosphatase, sulfatase, dehydrogenase, and invertase at high FA application rates.

Pati and Sahu (2004) conducted a pot experiment of FA-soil mixture in different ratios (0, 2.5, 5, 10, 15, 25, and 50% w/w) to study the toxicity test on earthworms (*Drawida willsi*) and CO_2 evolution and soil enzyme activities (dehydrogenase, protease, and amylase). An increase in the soil microbial respiration and enzyme activities was reported up to 2.5% of FA application in soil. But with higher fly ash application, all these activities were significantly decreased. On the other hand, when soils contained earthworms, significant increase up to 5% FA was observed in soil respiration and microbial activities. This might be due to the significant enhancement in the substrate-induced microbial activity formed by the earthworms. Low FA application with earthworms can be used to increase the soil biological responses and thereby improve the nutrient status in deficit soils. There are several reports that show the combination of sludge-FA-soil mixtures mostly improves enzyme activity (Kucharski et al. 2006; Singh et al. 2016a, b). Several researchers have reported the enhancement in activity of soil enzymes such as urease, cellulose, and dehydrogenase on addition of 16% FA (w/w) in soil (Lal et al. 1996). Moreover, soil microbial population also increased with addition of 10% (w/w) FA in soil which in turn positively influenced soil enzyme activity and soil biomass (Kohli and Goyal 2010). Similar results were found with the field experiment carried out by Ramteke et al. (2017) which reported that incorporation of appropriate amount of FA along with manures (FYM) and fertilizers (NPK) has enhanced soil microbial population as well as enzymatic activity in rice-wheat cropping system alfisols and versitols. Another study reported maximum grain yield and soil dehydrogenase activity on combined application of lime+ FA+ FYM in relation to control treatment. Combination of lime, FA, and gypsum was found to be effective for higher dehydrogenase activity, but urease activity was found to be higher in lime and FA (Chandrakar et al. 2015). The detailed explanation of alterations in soil enzymatic activities and its ecological function in fly ash amendments (organic or inorganic) is presented in Table 12.2.

12.3.2 Microbial-Mediated Nutrient Cycling

Microbes are considered as vital components of the soil atmosphere that are accountable for the degradation of organic matter and for cycling of various nutrients (Basu et al. 2020). Due to FA incorporation in the soil system, certain aspects such as pH, salinity, some trace elements, and deprived physical conditions can reduce the growth of microorganisms and plants (Pandey and Singh 2010). Higher FA application rate causes reduced microbial activity due to the presence of insufficient amount of C and N as an energy source of various heterotrophic microorganisms (Klubek et al. 1992). Arthur et al. (1984) revealed that lower FA application can enhance the microbial activity but inhibition of microbial activity at higher application rate was due to the deficiency of C and N source. Schutter and Fuhrmann (2001) concluded that FA amended with degraded subsoil caused an increased concentration of microbial communities in soil. Impact of fly ash amendment on microbial communities, soil enzyme activity, and soil microbial community structure affecting nutrient cycling is shown in Fig. 12.1.

FA contains little or no nitrogen but contains several essential nutrients required for the plant growth and development (Mandre 2006; Patterson et al. 2004; Uckert 2004). If FA is employed in agriculture in a considerable amount, then its probable effects on the microbial facilitated natural processes such as organic manure decomposition and N and P nutrient cycling need to be evaluated further for maintaining the potency and efficiency of soil. Negative impact of FA at high application rate may be due to the change in the chemical properties of the soil mixture (Singh and Yunus 2000). FA poses harmful impact on soil microbes which are mainly involved in N fixation and no nodule formation by *Rhizobium* bacteria due to its high pH and availability of excessive soluble elements at high application rates (Cheung et al. 2000). Furthermore, it was found that FA-resistant *Rhizobium* strain when infected with leguminous plants causes an increase in the nitrogen content of the infertile FA

Table 12.2 Influence of fly ash doses on soil enzymatic activity mentioning its ecological function

S. no.	Fly ash doses	Plant studied	Effect on enzymatic activity	Ecological function	References
1	75% GRD + 40 t FA/ha + 5 t FYM/ha	Rice var. MTU-1010	Significant increase in dehydrogenase activity was found	Reflects oxidative activities of soil microorganisms and takes part in oxidation of soil organic matter	Singh et al. (2016b)
2	100% NPK/ha + 5 t FYM + 20 t FA/ha	Rice-wheat cropping system	Microbial biomass and soil dehydrogenase activity increases		Ramteke et al. (2017)
3	10 t/ha FA	Ground nut (*Arachis hypogea*)	Increase in dehydrogenase activity was found higher at lower doses		Sireesha and Rani. (2014)
4	At 10% FA	*Populus deltoides*	Microbial biomass and soil dehydrogenase activity decreases at 10% fly ash application in soil		Kohli and Goyal (2010)
5	20 t/ha FA	*Vigna radiata* L.	Alkaline phosphatase and beta-galactosidase increases with increase in concentration of fly ash up to 20 t/ha	Phosphatase releases plant PO_4 from organic matter and beta-galactosidase releases glucose for microbial activity	Singh et al. (2016a)
6	Low doses of FA	Borowiak oat and maize crop	Higher activity of dehydrogenase, alkaline, and acid phosphatase were observed in soil contaminated with fly ash in oats than in maize		Kucharski et al. (2006)
7	FA amendment at 20 t/ha with phosphogypsum	*Vigna radiata* L.	Microbial biomass, dehydrogenase activity, alkaline phosphatase, and beta-galactosidase were found significantly higher in 20 t/ha		Singh et al. (2016a)
8	Lime+ FA + gypsum	Maize	Higher dehydrogenase and urease activity were found	Urease belongs to group of enzymes acting on C–N bonds of urea, which acts as a fertilizer source	Chandrakar et al. (2015)

(continued)

Table 12.2 (continued)

S. no.	Fly ash doses	Plant studied	Effect on enzymatic activity	Ecological function	References
9	Low doses of FA	Maize	FA incorporation at lower doses in soil significantly increases the acid phosphatase and dehydrogenase activity and decreasing the alkaline phosphatase activity	Phosphatase releases plant PO_4 from organic matter, and dehydrogenase enzyme reflects oxidative activities of soil microorganism and takes part in oxidation of organic matter	Kalembasa and Symanowicz (2012)
10	FA at 10 kg/m^2 with supplementation of cyanobacteria	Rice	Peroxidase activity and catalase activity increased significantly at 10 kg/m^2 of fly ash in rice crops	These enzymes help in release of oxygen molecule from hydrogen peroxide	Padhy et al. (2016)
11	50% FA application in soil	Rice (*Oryza sativa* L.)	Significant reduction in dehydrogenase, acid phosphatase, beta-galactosidase, and urease was found	Beta-galactosidase releases glucose for microbial activity, phosphatase releases plant PO_4 from organic matter, and urease belongs to group of enzymes acting on C–N bonds of urea, which acts as a fertilizer source	Singh et al. (2015)
12	15 t/ha FA application	Rice (*Oryza sativa* L.)	Fly ash application up to 15 t/ha causes an increase in amylase, invertase, dehydrogenase, and protease activity with respect to control	Amylase converts starch into maltose; invertase hydrolyzes sucrose to glucose and fructose, providing energy for microbial activity; proteases hydrolyzes proteins, releasing amino compounds/ important in N cycling and N mineralization	Sarangi et al. (2001)

landfill (Vajpayee et al. 2000). The elevated silt content of FA has the higher affinity to strengthen and solidify the soil mixture. A study was conducted to evaluate the performance 65 rhizobacterial species of a wild grass (*S. ravennae*) in fly ash

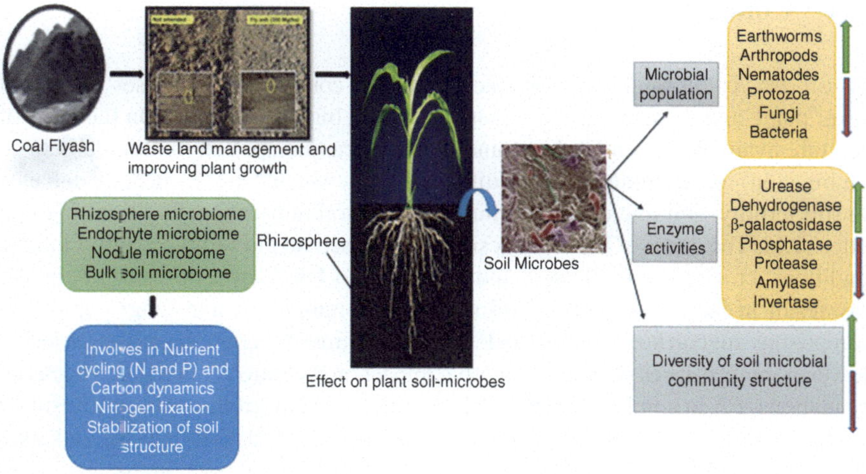

Fig. 12.1 Figure showing impact of fly ash ameliorant on microbial population, soil enzyme activity, and soil microbial community structure

dumping site and observed the growth of all bacterial species in nitrogen-deficient medium. Rau et al. (2009) studied on 65 rhizobacteria of a wild grass (*S. ravennae*) colonizing two FA dumping site of Delhi region and found that almost all bacterial species could grow on nitrogen-deficient medium. FA when used in combination with sewage sludge at 10% causes an increase in the N and P nutrients (Lai et al. 1999).

Rai et al. (2000) have reported some of the application of nitrogen-fixing bacteria (cyanobacteria) inoculants for increasing the N and P concentration and decreasing the heavy metal toxicity of FA. For this study, they have grown seven nitrogen-fixing bacteria (especially blue green algae) on FA and observed an enhanced growth of only alga (*Anabaena doliolum*) on the substrate containing higher level of N, P, and organic matter which thereby improved the plant performance, though *Anabaena doliolum* when grown in FA accumulated enormous amount of toxic metals (Fe, Mn, Ni, Cr, and Ca) leading to its deleterious effect. In addition, the possible utilization of FA with organic matter has been reported so far in improving the nutrient status of the mixture by employing vermicomposting technologies (Bhattacharya and Chattapadhyaya 2004). In the study, FA and cow dung were used alone or in combination in different ratios (1:1, 1:3, and 3:1) and were allowed to mix with epigenic earthworms (*Eisenia foetida*) for 50 days. It was observed that there was a significant rise in the N-fixing bacteria and an improved microbial activity in all the vermicomposted samples. The highest nitrogen content was reported in the combination of vermicomposted FA and cow dung (1:1) mixture.

12.4 Conclusion

Fly ash (FA) is the solid waste obtained from coal combustion processes in thermal power plant, and its disposal is the world's current biggest problem. On the basis of literature available, it can be determined that FA can be applied (individually or in combination) to agricultural soils at different doses which can significantly enhance the soil fertility, soil microbial communities, and soil nutrient cycling resulting in a better yield. The doses of FA to be designed are based on the soil type, agro-climatic conditions, crops to be cultivated, and properties of FA. Lower doses of FA in soil may result in increased existing soil microbial communities like fungi population (comprising mycorrhizal fungi) and bacteria (primarily gram-negative bacteria). However, higher doses showed a negative impact on the ratio of fungal and bacterial populations. FA in combination with FYM, lime, gypsum, earthworms, etc. resulted in increased level of microbial populations, enhanced dehydrogenase, alkaline phosphatase and beta-galactosidase activity, and soil microbial biomass leading to proliferation of microbial communities and thus soil fertility.

References

Arthur MF, Zwick TC, Tolle DA, Vanvoris P (1984) Effect of fly ash on microbial CO_2 evolution from the agricultural soil. Water Air Soil Pollut 22:209–216

Banerjee S, Gothalwal R, Sahu PK, Sao S (2015) Microbial observation in bioaccumulation of heavy metals from fly dyke of thermal power plants of thermal power plants Chhattisgarh, India. Adv Biosci Biotechnol 6:131–138

Basha NAI, James A, Bharose R, Rao PS (2018) Impact of fly ash on soil physical properties under sunflower-spinach-sunflower crop rotation system in Central India. Int J Curr Microbiol App Sci 7(12):1815–1828

Basu S, Kumar G, Chhabra S, Prasad R (2020) Role of soil microbes in biogeochemical cycle for enhancing soil fertility. In: Verma JP, Macdonald C, Gupta VK, Podile AR (eds) New and future developments in microbial biotechnology and bioengineering: phytomicrobiome for sustainable agriculture. Elsevier, Amsterdam, pp 149–157

Behera BK, Prasad R (2020) Strategies for soil management. In: Behera BK, Prasad R (eds) Environmental technology and sustainability. Elsevier, Amsterdam, pp 143–167

Bhattacharya SS, Chattapadhyaya GN (2004) Transformation of nitrogen during vermicomposting of fly ash. Waste Manage Res 22:488–491

Brady NC, Weil RC (2012) The nature and properties of soils, 14th edn. Dorling Kindersley, Noida

Brussaard L (2012) Ecosystem services provided by the soil biota. In: Wall DH (ed) Soil ecology and ecosystem services. Oxford University Press, Oxford, pp 45–58

Carry EE, Gilbert M, Bache CA, Quenmann WH, Lisk DJ (1986) Elemental composition of potted vegetables and millet grown on hard coal bottom ash amended soil. Bull Environ Contom Toxicol 31:418–423

Cerevelli S, Petruzzelli G, Perna A, Menicagli R (1986) Soil nitrogen and fly ash utilization: a laboratory investigation. Agrochemica 30:27–35

Chandrakar T, Jena D, Dash AK, Jena SN, Panda N, Monika M (2015) Soil microbial activity as influenced by application of fly ash and soil amendments to maize crop in acidic alfisols. Int Res J Agric Sci Soil Sci 5(4):120–128

Cheung KC, Wong JPK, Zhang ZQ, Wong JWC, Wong MH (2000) Revegetation of lagoon ash using the legume species *Acacia auriculiformis* and *Leucaena leucocephala*. Environ Pollut 109:75–82

Chu D (2018) Effect of heavy metals on soil microbial communities. IOP Conf Ser: Earth Environ Sci 113:1–5

Elliott LF, Tittemore D, Papendick RL, Cochran VL, Bezidicek DF (1982) The effect of Mount St. Helen's ash on soil microbial respiration and numbers. J Environ Qual 1:164–166

Furlani E, Tonello G, Maschio S, Aneggi E, Minichelli D, Bruckner S, Lucchini E (2011) Sintering and characterization of ceramics containing paper sludge, glass cullet and different types of clayey materials. J Ceramics Int 37:1293–1299

Gagnon B, Nolin MC, Cambouris AN (2004) Combined de-inking paper sludge and poultry manure application on corn yield and soil nutrients. Can J Soil Sci 84:503–512

Gaind S, Gaur AC (2002) Impact of fly ash and phosphate solubilizing bacteria on soybean productivity. Bioresour Technol 85:313–315

Gaind S, Gaur AC (2004) Evaluation of fly ash as a carrier for diazotrophs and phosphobacteria. Bioresour Technol 95:187–190

Honghua H, Dong Z, Pieng Q, Wang X, Fan C, Zhang X (2017) Impact of coal fly ash on plant growth and accumulation of essential nutrients and trace elements by alfalfa (*Medicago sativa*) grown in a loessial soil. J Environ Manage 197:428–439

Hrynkiewiez K, Baum C, Niedojadlo J, Dahm H (2009) Promotion of mycorrhiza formation and growth of willows by the bacterial strain *Sphingomonas* sp. 23L on fly ash. Biol Fertil Soil 45:385–394

Jala S, Goyal D (2006) Fly ash as a soil ameliorant for improving crop production – a review. Bioresour Technol 97:1136–1147

Kalembasa S, Symanowicz B (2012) Enzymatic activity of soil after applying various waste organic materials, ash, and mineral fertilizers. Pol J Environ Stud 21(6):1635–1641

Karpagavalli S, Ramabadran R (1997) Effect of lignite fly ash on the growth and dry matter production (DMP) of soil borne pathogens. In: Souv and abstracts: National Sem on bio-utilization of fly ash, p 11

Khan AG (2005) Role of soil microbes in the rhizosphere of plants growing on trace metal contaminated soils in phytoremediation. J Trace Elem Med Biol 18(4):355–364

Klubek B, Carison CL, Oliver J, Adriano DC (1992) Characterization of microbial abundance and activity from 3 coal ash basins. Soil Biol Biochem 24:1119–1125

Kohli SJ, Goyal D (2010) Effect of fly ash application on some soil physical properties and microbial activities. Acta Agrophysica 16(2):327–335

Kucharski J, Jastrzębska E, Wyszkowska J (2006) Contamination of soil with hard coal ash as modifier of physicochemical and biological properties of soil. EJPAU 9(1):35

Kumar KV, Singh N, Behl HM, Srivastava S (2008) Influence of plant growth promoting bacteria and its mutant on heavy metal toxicity in *Brassica juncea* grown in fly ash amended soil. Chemosphere 72:678–683

Kumar T, Tedia K, Samadhiya V, Kumar R (2017) Review on effect of fly ash on heavy metal status of soil and plants. Int J Chem Stud 5(4):11–18

Lai KM, Ye DY, Wong JWC (1999) Enzyme activities in a sandy soil amended with sewage sludge and coal fly ash. Water Air Soil Pollut 113:261–272

Lal JK, Mishra B, Sarkar AK (1996) Effect of fly ash on soil microbial and enzymatic activity. J Ind Soc Soil Sci 44:77–80

Machulla G, Zikeli S, Kaster M, Jahn R (2004) Microbial biomass and respiration in soils derived from lignite ashes: a profile study. J Plant Nutr Soil Sci 167:449–456

Mahalingam PK (1973) Ameliorative properties of lignatic fly ash in reclaiming saline and alkali soils. Madras Agric J 8:1055–1057

Mandre M (2006) Influence of wood ash on soil chemical composition and biochemical parameters of young scots pine. Proc Estonian Acad Sci Biol Ecol 55(2):91–107

Marumoto T, Anderson JPE, Domsch KH (1982) Mineralization of nutrients from soil microbial biomass. Soil Biol Biochem 14:469–475

Mupambwa HA, Dube E, Mnkeni PNS (2015) Fly ash composting to improve fertilizer value – a review. S Afr J Sci 111:1–6

Murugan S, Vijayarangam M (2013) Effect of fly ash in agricultural field on soil properties and crop productivity – a review. Int J Eng Res Technol 2(12):54–60

Nashine R (2014) Studies on the effect of and plant growth hormones on chlorophyll a, b and total chlorophyll contents in green gram leaves. Am J Phytomed Clin Therapeut 2(11):1333–1337

Nayak AK, Raja R, Rao KS, Shukla A, Mohanty S, Shahid M, Tripathi R, Panda B, Bhattacharyya P, Kumar A, Lal B, Sethi S, Puri C, Nayak D, Swain C (2015) Effect of fly ash application on soil microbial response and heavy metal accumulation in soil and rice plant. Ecotoxicol Environ Saf 114:257

Nivetha E, Sheeba S (2017) Effect of fly ash application on microbial population in acid soil. Int J Curr Microbiol App Sci 6(12):888–894

Onduru DD, Jager AD, Wouters B, Muchena FN, Gachimbi L et al (2006) Improving soil fertility and farm productivity under intensive crop-dairy smallholdings: experiences from farmer field schools in the highlands of Kiambu District, Central Kenya. Middle-East J Sci Res 1(1):31–49

Padhy RN, Nayak N, Dash-Mohini RR, Rath S, Sahu RK (2016) Growth, metabolism and yield of rice cultivated in soils amended with fly ash and cyanobacteria and metal loads in plant parts. Rice Sci 23(1):22–32

Panda RB, Biswal T (2018) Impact of fly ash on soil properties and productivity. Int J Agric Environ Biotechnol 11(2):275–283

Panda SS, Mishra SP, Muduli SD, Nayak BD, Dhal NK (2015) The effect of fly ash on vegetative growth and photosynthetic pigment concentrations of rice and maize. Biologija 61(2):94–100

Pandey VC, Singh N (2010) Impact of fly ash incorporation in soil systems. Agric Ecosyst Environ 136:16–27

Pandey VC, Abhilash PC, Upadhyay RN, Tewari DD (2009) Application of fly ash on the growth performance, translocation of toxic heavy metals within *Cajanus cajan* L.: implication for safe utilization of fly ash for agricultural production. J Hazard Mater 166:255–259

Pandey VC, Singh JS, Singh RP, Singh N, Yunus M (2011) Arsenic hazards in coal fly ash and its fate in Indian scenario. Resour Conserv Recycl 55:819–835

Pati SS, Sahu SK (2004) CO_2 evolution and enzyme activities (dehydrogenase, protease and amylase) of fly ash amended soil in the presence and absence of earthworms (Drawida willsi Michaelsen) under laboratory conditions. Geoderma 118:289–301

Patil PV, Chalwade PB, Solanke AS, Kulkarni VK (2005) Effect of fly ash and FYM on nutrient uptake and yield of onion. J Soils Crops 15(1):187–192

Patterson SJ, Acharya SN, Thomas JE et al (2004) Integrated soil and crop management: barley biomass and grain yield and canola seed yield response to land application of wood ash. Agron J 96(4):971–977

Pichtel JR, Hayes JM (1990) Influence of fly ash on soil microbial activity and populations. J Environ Qual 19:593–597

Plank CO, Martens DC (1974) Boron availability as influenced by application of fly ash to soil. Proc Soil Sci Am 38:974–977

Polat M, Lederman E, Pelly I et al (2002) Chemical neutralization of acidic wastes using fly ash in Israel. J Chem Technol Biotechnol 77(3):377–381

Prasad R, Kumar M, Varma A (2015) Role of PGPR in soil fertility and plant health. In: Egamberdieva D, Shrivastava S, Varma A (eds) Plant growth-promoting rhizobacteria (PGPR) and medicinal plants. Springer International Publishing, Switzerland, pp 247–260

Prasad R, Chhabra S, Gill SS, Singh PK, Tuteja N (2020) The microbial symbionts: potential for the crop improvement in changing environments. In: Tuteja N, Tuteja R, Passricha N, Saifi SK (eds) Advancement in crop improvement techniques. Elsevier, Amsterdam, Netherlands, pp 233–240

Prosser JI (2015) Dispersing misconceptions and identifying opportunities for the use of 'omics' in soil microbial ecology. Nat Rev Microbiol 13:439–446

Rai UN, Tripathi RD, Singh N, Kumar A, Ali MB, Pal A, Singh SN (2000) Amelioration of fly-ash by selected nitrogen fixing blue green algae. Bull Environ Contam Toxicol 64:294–301
Raj S, Dahiya P, Mohan S (2015) Physico-chemical analysis and in-vitro antibacterial activity of *Jatropha Curcas* grown on fly ash amended soil. Int J Appl Environ Sci 10(4):1375–1383
Ramteke LK, Sengar SS, Porte SS (2017) Effect of fly ash, organic manure and fertilizers on soil microbial activity in rice-wheat cropping system in alfisols and versitols. Int J Curr Microbiol App Sci 6(7):1948–1952
Rao CC, Grover M, Kundu S, Desai S (2017) Soil enzymes. In: Encyclopedia of soil sciences. Taylor & Francis, Oxfordshire, pp 2100–2107
Rau N, Mishra V, Sharma M, Das MK, Ahaluwalia K, Sharma RS (2009) Evaluation of functional diversity in rhizobacterial taxa of a wild grass (*Saccharum ravennae*) colonizing abandoned fly ash dumps in Delhi urban ecosystem. Soil Biol Biochem 41:813–821
Rautaray SK, Ghosh BC, Mittra BN (2003) Effect of fly ash, organic wastes and chemical fertilizers on yield, nutrient uptake, heavy metal content and residual fertility in a rice–mustard cropping sequence under acid lateritic soils. Bioresour Technol 90:275–283
Roy G, Joy VC (2011) Dose-related effect of fly ash on edaphic properties in laterite crop land soil. Ecotoxicol Environ Saf 74:769–775
Sahu G, Bag AG, Chatterjee N, Mukherjee AK (2017) Potential use of fly ash in agriculture: a way to improve soil health. J Pharmacogn Phytochem 6(6):873–880
Sarangi PK, Mahakur D, Mishra PC (2001) Soil biochemical activity and growth response of rice *Oryza sativa* in fly ash amended soil. Bioresour Technol 76:199–205
Schutter ME, Fuhrmann JJ (2001) Soil microbial community responses to fly ash amendment as revealed by analyses of whole soils and bacterial isolates. Soil Biol Biochem 33:1947–1958
Schwartz GE, Lauren K, Redfern, Ikuma K, Claudia K, Gunsch LS, Ruhl VA, Heileen HK (2016) Impacts of coal ash on methyl mercury production and the methylating microbial community in anaerobic sediment slurries. Environ Sci 18:1427–1439
Selvakumari G, Baskar M, Jayanthi D, Mathan KK (2000) Effect of integration of fly ash with fertilizers and organic manures on nutrient availability, yield and nutrient uptake of rice in alfisols. J Indian Soc Soil Sci 48(2):268–278
Sen A (1997) Microbial population dynamics in fly ash amended acid lateritic soil. B. Tech. Thesis Indian Institute of Technology, Kharagpur
Sett R (2017) Fly ash characteristics, problems and possible utilization. Adv Appl Sci Res 8 (3):32–50
Singh LP, Siddiqui ZA (2003) Effects of fly ash and *Helminthosporium oryzae* on growth and yield of three cultivars of rice. Bioresour Technol 86:73–78
Singh N, Yunus M (2000) Environmental impacts of FA. In: Iqbal M, Srivastava PS, Siddiqui TO (eds) Environmental hazards – plants and people. C.B.S., New Delhi, pp 60–79
Singh JS, Pandey VC, Singh DP (2011a) Coal fly ash and farmyard manure amendments in dry-land paddy agriculture: effect on N-dynamics and Paddy productivity. Appl Soil Ecol 47:133–140
Singh RP, Singh P, Ibrahim MH, Hashim R (2011b) Land application of sewage sludge: physico-chemical and microbial response. Rev Environ Contam Toxicol 214:41–61
Singh RP, Sharma B, Sarkar A, Sengupta C, Singh P, Ibrahim MH (2014) Biological responses of agricultural soils to fly ash amendments. Rev Environ Contam Technol 232:45–60
Singh PK, Tripathi P, Dwivedi S, Awasthi S, Shri M, Chakrabarty D, Tripathi RD (2015) Fly-ash augmented soil enhances heavy metal accumulation and phytotoxicity in rice (*Oryza sativa* L.); a concern for fly-ash amendments in agriculture sector. Plant Growth Regul 78(1):21–30
Singh K, Pandey VC, Singh B, Dharani PD, Singh RP (2016a) Effect of fly ash on crop yield and physico-chemical, microbial and enzymatic activities of sodic soils. Environ Eng Manag J 15 (11):2433–2440
Singh MP, Tedia K, Samadhiya VK, Kumar T, Sharma P (2016b) Effect of graded dose of fly ash applied with or without FYM on microbial and enzymatic activity of soil. Bioscan 11 (4):2537–2539

Sireesha A, Rani PP (2014) Effect of fly ash and fertilizers on yield, nutrient uptake and soil nutrient ststus of groundnut (*Arachis hypogea*). Asian J Soil Sci 9(2):226–233

Subhani A, Changyong H, Zhengmiao X, Min L, El-ghamry AM (2001) Impact of soil environment and agronomic practices on microbial/dehydrogenase enzyme activity in soil. A review. Pak J Biol Sci 4:333–338

Uckert GB (2004) Versuche zur landbaulichen Verwertung von Holzaschen unter besonderer Berücksichtigung der Knickholzpotenziale Schleswig-Holsteins. Dissertation, University of Kiel

Vajpayee P, Rai UN, Choudhary SK, Tripathi RD, Singh SN (2000) Management of flyash landfills with *Cassia surattensis Burm*: a case study. Bull Environ Contam Toxicol 65:675–682

Vallini G, Vaccari F, Pera A, Agnolucci M, Scatena S, Varallo G (1999) Evaluation of cocomposted coal fly ash on dynamics of microbial populations and heavy metal uptake. Compost Sci Util 7:81–90

Wilkinson MT, Richards PJ, Humphreys GS (2009) Breaking ground: pedological, geological, and ecological implications of soil bioturbation. Earth Sci Rev 97:257–272

Wong MH, Wong JWC (1986) Effects of fly ash on soil microbial activity. Environ Pollut Ser A 40:127–144

Wong JWC, Wong MH (1987) Co-recycling of fly ash and poultry manure in nutrient-deficient sandy soil. Resour Conserv 13:291–304

Wong JWC, Fang M, Jiang R (2001) Persistency of bacterial indicators in biosolids stabilization with coal fly ash and lime. Water Environ Res 73(5):607–610

Wu JF, Lin XG (2003) Effects of soil microbes on plant growth. Soil 35(1):18–21

Yeledhalli NA, Prakash SH, Gurumurthy SB, Ravi MV (2007) Coal fly ash as modifier of physicochemical and biological properties of soil. J Agric Sci 20(3):531–534

Chapter 13
Molecular Insight into Plant-Fungal Pathogen Interaction: Emerging Trends and Implication in Designing Climate-Smart Field Crops

Richa Kamboj, Manoj Nath, Bhavna Thakur, Tapan Kumar Mondal, Deepesh Bhatt, and Deepak Singh Bisht

Abstract Among the varieties of pathogens infecting plants, fungal pathogens are infamous for causing extensive damage to the plants. The establishment of fungal pathogen on plant tissues is largely determined by the cross-reactivity of the pathogen-secreted biomolecules with host immune response. Plants alleviate the stressful environment by employing different stress-adaptive responses. Particularly, to fend off pathogen and to keep them healthy, plants have evolved a highly complex defence response. The robustness of the plant immunity against invading fungus depends on how the components of the shared signalling network are manoeuvred by the plant for its own defence. However, due to the rapid rate of evolution of fungus pathogenesis genes coupled with global climate changes, the conditions are becoming more favourable for growth and progression of the fungal pathogens. Consequently, the incidences of fungus attack on otherwise resistant host species have now escalated. Thus, the important challenge facing global agriculture is to minimise the crop losses incurred due to plant diseases. In this context, a deeper insight into the mechanism of plant-fungi interaction is necessary to combat the invading pathogens. In this chapter, we have touched upon the mechanistic aspect of plant-fungi interaction and how this information can be utilised strategically for designing climate smart crops.

R. Kamboj · T. K. Mondal · D. S. Bisht (✉)
ICAR-National Institute for Plant Biotechnology, New Delhi, India
e-mail: deepak.bisht@icar.gov.in

M. Nath
ICARI, Solan, Himachal Pradesh, India

B. Thakur
G.B. Pant University of Agriculture and Technology, Pantnagar, India

D. Bhatt
Veer Narmad South Gujarat University, Surat, India

A. Varma et al. (eds.), *Plant Microbiome Paradigm*,
https://doi.org/10.1007/978-3-030-50395-6_13

13.1 Plant-Fungal Interaction: An Introduction

Interaction between plants and fungi can be either beneficial or harmful. In beneficial interactions, both the partners are benefited from the relationship. For instance, in case of mycorrhizal fungus, the fungi improve plant performance by increasing bioavailability of mineral nutrients to plant and in turn plant provide sugar and photosynthates to fungus (Buscot et al. 2000; Varma et al. 2017a, b, c), whereas the harmful interactions may result into the destabilization of the host, as the fungi uses host cell machinery for its sustenance, thereby breaking the balance of mutual benefit. The interaction behaviour of fungi with host is largely determined by its genetic makeup, e.g. genome of pathogenic fungi contains virulence effector genes that are responsible for pathogenicity of the fungi (Prasad et al. 2017). Interestingly some of these virulence genes are present across different classes of fungus (Schafer 1994). The process of attack of fungus on the plants to its establishment can be categorised into four different stages, viz. (1) early infection stage—this involved contact of fungus with the plant tissue and germination of spores; (2) penetration stage—penetration using either indirect (wounds and stomatal openings) or direct specialized penetration structures (appressoria); (3) infection or invasion stage—spreading of the fungal hyphae inside the host cells either intercellular or intracellular; and (4) reproduction stage—colonization of fungus inside the host cells and production of large number of fungal spores (Zeilinger et al. 2015). Each stage is accompanied by several physiological and biochemical alterations in both the partners.

The strategy employed by pathogen for establishment inside the host cell depends upon its lifestyle (Fig. 13.1). Depending on their mode of interaction, the pathogenic fungi are classified as biotrophs, hemibiotrophs and necrotrophs (Rodriguez-Moreno et al. 2018). Necrotrophic fungus, like *Bacillus cinerea*, produces a combination of

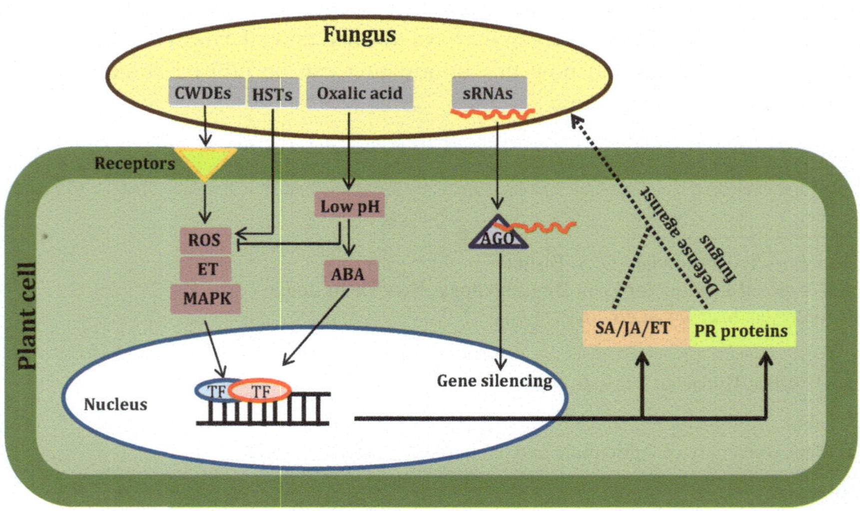

Fig. 13.1 Diagrammatic representation of plant fungal interaction

cell wall degrading enzymes, reactive oxygen species (ROS) and many other host-specific toxin leading to disruption of host cell membrane favouring pathogen penetration and colonization (Wang et al. 2014; Wolpert et al. 2002). On the contrary, biotrophic pathogens in spite of producing toxin secretes effector molecules hijacking the host cell machinery to favour its growth and survival (Perfect and Green 2001). An archetypal example of biotrophic fungus is powdery mildew pathogen, *Erysiphe cichoracearum*, which form an intricate intracellular feeding structure for extracting nutrient from plant for completing their life cycle (Ridout 2009).

13.2 Strategies Adopted by Fungus to Invade Host Cells

During infection process, pathogen uses specialised structures called appressoria and imposes optimum turgor pressure to breach the host cell wall. Additionally, pathogen also secretes diverse array of biomolecules including cell wall degrading enzymes (CWDEs), effector proteins, secondary metabolites and sometimes small RNAs for its progression and establishment on the host tissue. Each one of them is discussed in detail in this section.

13.2.1 Specialized Penetration Structures

The process of host penetration is variable across different classes of fungi. Though exact mechanism of adhesions of the spores on the host tissue surface is not known, presumably some specific interaction of the spores with the host surface via lectins, ionic or hydrophobic interactions favours the adhesion of the spores on host surface. Germination of spores of different fungal pathogens depends on the nature of spore and environmental conditions. After getting appropriate stimulations (host surface hardiness, its topography, hydrophobicity and plant signals), spore mobilizes their food reserves (like lipids, polysaccharides and CHO) to the cell membrane and cell wall for the formation of germ tube. Germ tube further differentiates into the appressorium, infection hyphae, haustoria, infection cushions and finally into rhizomorphs (Mendgen and Deising 1993). The appressorium accumulates turgor pressure that produces mechanical force enabling the penetration of pathogen across the host cell wall. This is followed by enzymatic degradation of the cuticle and the cell wall. Some fungi that lack specialized penetration structures use pre-existing openings in the host like stomatal pores and wounds or other topographical cues for gaining access into the host. They form swollen appressorium to enter through the stomatal aperture and further form a fine penetration hypha to enter the airspace inside the leaf, from where these infection hyphae infects the surrounding cells (Hoch et al. 1987; Brand and Gow 2009).

13.2.2 Cell Wall Degrading Enzymes

To overcome the barrier of plant cell wall, phytopathogenic fungi secrete cell wall degrading enzymes (CWDEs). They belong to carbohydrate-active enzyme family and are involved in the degradation of cellulose, pectin, etc. However, specificity of CWDEs may vary for pathogenic and non-pathogenic fungi. These enzymes are particularly important for those fungal pathogens that do not have specialized structures for the host cell penetration (Kubicek et al. 2014). Generally, all fungal pathogens require these enzymes during late stages of invasion as they utilize the degraded mono- or oligosaccharide residues as a source of energy (Gibson et al. 2011). For instance, the necrotrophic pathogens rely on these CWDEs to initiate colonization by degrading components of cell wall, whereas biotrophic pathogens need these enzymes only at the penetration site of haustoria and make very subtle changes in the cell wall (Bolton et al. 2006). In response, plants defend themselves by secreting inhibitory proteins like cellulose inhibitor, pectinase inhibitor, xylanase inhibitor, etc. that are discussed later in this chapter.

13.2.3 Secondary Metabolites

Secondary metabolites (SMs) are the small bioactive compounds secreted by fungal pathogens during plant-fungal interactions. On the basis of their structure, fungal secondary metabolites have been divided into four major classes: polyketides, terpenoids, shikimic acid-derived compounds and non-ribosomal peptides (Table 13.1). Fusions of these SMs may result in complex forms of SMs like meroterpenoids. The number of genes coding for secondary metabolites varies across different classes of pathogens (Pusztahelyi et al. 2015).

For example, ascomycetes contain more number of SM genes than basidiomycetes and chytridiomycetes. However, it is not necessary that SM genes must be present in all fungal pathogens as neither class of SMs is present in hemiascomycetes and zygomycetes (Collemare et al. 2008). These compounds are not necessary for growth, development and reproduction. However, these compounds make an important class of virulent factors like aflatoxins and antimicrobial compounds like penicillin. Fungal SMs involved in virulence are mainly divided into two major

Table 13.1 Secondary metabolites produced by the fungi upon interaction with the host

Groups	Examples
Polyketides	Aflatoxin, sterigmatocystin and T-toxin
Terpenoids	Mycotoxin, trichothecene mycotoxins, culmorin and aristolochene
Shikimic acid-derived compounds	Chorismic acid, phenylpyruvic acid, lignans, phenylpropenes and coumarins
Non-ribosomal peptides	δ-(L-α-aminoadipyl)-L-cysteinyl-D-valine synthetase, AM toxin, HC toxin and pyrrolopyrazine

classes, i.e. HSTs (host-specific toxins) and non-HSTs (non-host-specific toxins), depending upon their structural specificity and mode of action. HSTs secreted by fungi require specific target in the host plant for pathogenesis (Wolpert et al. 2002). For example, *T-urf13* gene encodes URF-13 responsible for host susceptibility, a mitochondrial membrane protein in maize harbouring Texas cytoplasm for male sterility that displayed extreme sensitivity towards T-toxin and PM-toxin of *Cochliobolus heterostrophus* race T and *Mycosphaerella zeae-maydis*, respectively (Levings III et al. 1995; Wolpert et al. 2002; Tsuge et al. 2013). These interactions cause conformational change in *URF13* and result in pore formation in the mitochondrial membrane. Another HST toxin is victorin, secreted by *Cochliobolus victoriae* that causes victoria blight in oats (Wolpert et al. 1985, 1986; Meehan and Murphy 1947). *Locus Orchestrating Victorin Effects1* (LOV1) provides victorin susceptibility in *Arabidopsis* plants (Gilbert and Wolpert 2013). In contrast to HST, non-HSTs do not require a specific target in the host plant and therefore affect a broad range of organisms. For example, perylenequinone toxin family is the well-studied family of SMs produced by the group *Ascomycota*. Perylenequinone toxins known for the photo-movement in protozoans play a diverse role as defence compounds and pathogenic determinants in fungi. Most prominent member of the family is cercosporin produced by *Cercospora* spp. which gets activated by light, reacts with oxygen to form ROS, causes DNA damage and lipid peroxidation and finally leads to host cell death. It has a very broad toxicity range including plants, animals, bacteria and mostly fungi (Daub and Ehrenshaft 2000; Blokhina et al. 2003; Birben et al. 2012).

13.2.4 Small RNAs

Small RNAs (sRNAs) are the non-coding RNAs that bind with the argonaute proteins and direct the RNA-induced silencing complex to genes with complementary sequences (Castel and Martienssen 2013). While the role of sRNAs in various, cellular processes have been comprehensively studied; the mechanism of trans-kingdom sRNAs that regulated plant immunity was recently discovered in *Botrytis cinerea* (Weiberg et al. 2013). *B. cinerea* is a necrotrophic pathogen that infects more than 200 plant species. Weiberg et al. (2013) showed that *B. cinerea* can transfer its small RNA (Bc-sRNA) into host and cause silencing host immunity genes. Similarly, in wheat, *Puccinia striiformis* f. sp. tritici (Pst) interaction novel microRNA-like RNA (milRNA) from Pst termed microRNA-like RNA 1 (*Pst*-milR1) supresses wheat defence by binding to wheat pathogenesis-related 2 (PR2) gene (Wang et al. 2017). Understanding the role of ‘RNA effectors’ in plant-fungi interaction sets stage for exploring the new avenues of designing control strategies against the devastating fungal pathogens.

13.3 Plant's Defence Response Against Fungal Invasion

In response to fungal invasion, plants have several layers of defence responses to defend themselves against infection. These defence barriers include structural compounds, complex immune network, phytohormone signalling, production of secondary metabolites, ROS production, etc. These are discussed sequentially in the following text.

13.3.1 Mechanical Barriers

Plant protects itself from fungal invasion by depositing lignin, suberin, callose and hydroxyproline-rich glycoproteins at the site of infection. This deposition makes the cell wall more resistant to CWDEs. Callose deposition prevents the formation and penetration of haustoria at the penetration sites. Suberin deposition prevents the colonization of the pathogen at the vascular tissues. The strategies adopted by fungus to overcome host mechanical barrier is discussed in Sect. 13.4.1 (Rodriguez-Moreno et al. 2018).

13.3.2 Immune Responses

The initial pathogen attack is sensed by surface-exposed receptors eliciting the first level of plant defence called as pathogen-triggered immunity (PTI). Some successful pathogens overcome PTI to shuttle specific effectors into the plant cell and hijack the complete cellular machinery for their own sustenance. To counteract, plants have evolved a second layer of defence, called effector-triggered immunity (ETI). Both PTI and ETI activate an array of signal cascade, providing an active defence against the invading pathogen (Abramovitch et al. 2006; Navarro et al. 2004; Tao et al. 2003; Zipfel et al. 2006; Katagiri and Tsuda 2010).

13.3.3 Hypersensitive Response and Phytohormone Signalling

During pathogen attack, plants defend themselves by eliciting hypersensitive response (HR) that includes accumulation of phytoalexins and other PR proteins like chitinases, glucanases, peroxidases, etc. (Lebeda et al. 2001). HR response is characterised by rapid synthesis of nitric oxide and hyperaccumulation of reactive oxygen species (ROS) triggering the cell death. It not only restricts the fungal progression but also primes the plant for further invasion by a mechanism called systemic acquired resistance (SAR). The overexpression of PR genes in transgenic plants have been shown to have enhanced resistance against pathogen attack. For example, pectin methylesterase inhibitor gene overexpressed in transgenic wheat lines showed a good level of resistance against hemibiotrophic pathogen *Fusarium graminearum* and *Bipolaris sorokiniana* (Volpi et al. 2011). Similarly, transgenic wheat lines overexpressing xylanase inhibitor and polygalacturonase inhibitor gene had shown resistance against *Fusarium* head blight (Tundo et al. 2016).

Plant hormones are generally known for providing protection against a broad spectrum of pathogens. For example, salicylic acid (SA) and ethylene (ET) provide protection against biotrophic pathogens while jasmonic acid (JA) against necrotrophic pathogens. Basically, SA or JA and ET induction takes place primarily during biotic stress depending upon the pathogen's lifestyle and provides local resistance which further induces SAR (systemic acquired resistance) or ISR (induced systemic resistance) against a broad spectrum of pathogens. Induction of SAR and ISR not only depends upon the SA/JA/ET but also on small secondary metabolites. Pathogenesis-related protein like chitinases, glucanases, etc. plays a major role in the induction of SAR and ISR. Some other defence-related enzymes also get activated in parallel to the ISR induction by the effect of methyl JA and SA.

13.3.4 Plant Secondary Metabolites

Plants may also respond through the secretion of secondary metabolites (SMs) against the fungal attack. SMs are used in signalling and regulation of primary metabolic pathways, thereby acting as regulatory molecules in some of the developmental processes as well as defence response against pathogen attack (Pusztahelyi et al. 2015). Generally, plant SMs are classified into three major groups on the basis of their biosynthetic origin: (a) flavonoids; (b) terpenoids; and (c) nitrogen-containing alkaloids and sulphur-containing compounds (Table 13.2).

Flavonoids are the water-soluble phenolic compounds found in the vacuoles of the plant cells. Along with the other physiological processes like flower colouration, UV filtration and symbiotic nitrogen fixation, they also play very important role in plant resistance against pathogenic bacteria and fungi. Antifungal activity of flavonoid compounds against different fungi *Verticillium alba-atrum*, *Fusarium oxysporum*, etc. has been very well documented (Picman et al. 1995; Galeotti et al. 2008). Vitamin P, citrin, lignin and tannin are the examples of flavonoids and other phenolic compounds having broad spectrum antifungal properties.

Terpenoids are the derivatives of polymeric isoprene and get synthesized from acetate via the mevalonic acid pathway. They form the biggest class of secondary metabolites acting as defence molecules such as toxins. Further, terpenes are subdivided into monoterpenes, sesquiterpenes, diterpenes, triterpenes, tetraterpene, polyterpenes, etc. Pyrethroid, abietic acid, phorbol, ABA, coumarins, etc. are examples of terpenes which are known to play an important role in defence against fungus and other pathogens. Some of terpenoids are used as fungicides to protect crop plants

Table 13.2 Secondary metabolites produced by the plants

Groups	Examples
Phenols	*With one ring*: Phenol, hydroquinone, pyrogallol acid, gallic acid, salicylic acid *With two rings*: Mangostin, resveratrol, chlorophorin, quercetin, glyceollin, sakuranetin
Quinones	Alizarin, emodin
Terpenoids	Isoprene (hemiterpene), pyrethroid (monoterpene), abietic acid (diterpene), friedelin (triterpene), ABA (sesquiterpene), carotenoids (tetraterpene) and ubiquinones (polyterpene)
Nitrogen-containing alkaloids	Allosecurinine, tomatine, solanine, nicotine
Glucosinolates	Sinigrin, glucobrassicin
Non protein amino acids	L-canavanine
Amines	Phenylethylamine, tyramine, morphine
Cyanogenic glycosides	Amygdalin, sambunigrin, linamarin

(Kabera et al. 2014). Huffaker et al. (2011) reported that optimum amount of zealexins (terpene) showed antifungal activity against *F. graminearum*, *A. flavus* and *R. microspores*.

SMs that contain basic nitrogen atom are referred as alkaloids. Oxygen, sulphur and rarely other elements such as chlorine, bromine and phosphorus may also be present in alkaloids along with basic atoms carbon, hydrogen and nitrogen. These compounds are known to have diverse pharmacological effects and are generally toxic to other organisms. Allosecurinine from *Phyllanthus amarus* has shown strong antifungal activity against *Curvularia lunata*, *Curvularia* sp., *Colletotrichum* sp., *C. musae* and *Heterosporium* sp. (Singh et al. 2007). Phytoalexins, GSL, GSH, thionins, defensin and alliins are the sulphur-rich compounds known to get activated after pathogen attack or under other stressful environment and play important role in plant defence against the pathogens. For example, thionin gets accumulated in the cell wall of spikelets of resistant wheat cultivar after infection with *Fusarium culmorum* (Kang and Buchenauer 2003).

13.3.5 Small RNAs

The natural transfer of sRNA from plant host to pathogen was not known until recently when Zhang et al. (2016) showed that cotton plant produces sRNAs, miR166 and miR159 upon infection with *Verticillium dahliae* and exports them into the fungal hyphae of *V. dahliae* causing silencing of pathogenicity genes. This is the only report present till date on natural sRNA transfer from host to fungal pathogen. Exact mechanism of sRNA from host to fungal pathogen is not clearly understood. However, few studies suggest that the transfer may occur through extracellular vesicles (Cai et al. 2018). Moreover, in spite of the knowledge of exact mechanism of transfer of sRNA into pathogen, the HIGS (host-induced gene silencing) technology has been successfully utilised in protecting plants against the fungal pathogens. Overexpression of hpRNAs in *Arabidopsis* and tomato plants resulted in the silencing of *Bc-Dcl 1* and *Bc-Dcl-2* genes, thereby reducing the fungal infection (Wang et al. 2016). Silencing of chitin synthase gene (*chs*) by the overexpression hpRNA resulted in the enhanced resistance against *Fusarium* in wheat and *S. sclerotiorum* in tobacco (Cheng et al. 2015; Andrade et al. 2016). HIGS of *PsCPK1* gene has resulted in reduced pathogenicity of *Puccinia striiformis* f. sp. tritici (Qi et al. 2017). Two pathogenicity genes (*RPMK-1* and *RPMK-2*) of *R. solani* were targeted by transforming rice plants with HD-RNAi construct, and partial resistance has been obtained against sheath blight disease (Tiwari et al. 2017).

13.4 Strategies Adopted by Fungus for Successful Pathogenesis

To colonize plants and cause disease, pathogenic fungi use diverse strategies. The mechanism adopted by fungal pathogen to subvert the host defence response is discussed in the following section.

13.4.1 Prevention from Host Recognition

Recognition of pathogen by host is one of the first steps in activating the host defence response (Felix et al. 1993; Shibuya et al. 1993; Cosio et al. 1996; Cote et al. 2000). Enzymes like chitinases and glucanases (gets activated after pathogen recognition) act on fungal cell wall and release oligomers which are further recognised by the extracellular receptors of the host (Sanchez-Vallet et al. 2015). To bypass the host recognition, pathogen keeps on modifying the composition of cell wall and the secreted effector molecules. For example, *Magnaporthe oryzae* accumulates α-1,3-glucans in response to the epidermal wax component 1,16-hexadecanediol at the surface of the cell wall to prevent the chitin degradation by plant chitinases (Fujikawa et al. 2012). In addition to the modification of the cell wall content, fungal pathogens secrete some carbohydrate-binding effector proteins that suppress chitin-triggered host defence responses. For example, *Cladosporium fulvum* secretes the LysM-containing effector Ecp6 that binds chitin with high specificity (Sanchez-Vallet et al. 2013). Ecp6 conceals the fungal chitin which would otherwise be recognised by the host immune receptors. Another strategy is secretion of the proteases like chitinase-modifying proteins (CMPs) by fungal pathogens to prevent the degradation of fungal chitin by extracellular host chitinases (Jashni et al. 2015; Rodriguez-Moreno et al. 2018).

13.4.2 Depreciating ROS Damage

During the pathogen invasion, plant cell exhibits a rapid synthesis of ROS (reactive oxygen species) by the activation of membrane-bound NADPH oxidases and other cell wall and membrane-associated oxidases (Bolwell et al. 2002; Sasaki et al. 2004; Bindschedler et al. 2006). This oxidative burst is an important signal mediating defence gene activation. To counteract this cellular response, fungal pathogen secretes ROS scavenging molecules. For instance, DES-1 (defence suppressor 1) is a scavenging enzyme secreted by *M. oryzae* extracellulary to counter the basal defence achieved by ROS production in the host cell (Chi et al. 2009). Similarly, ROS-scavenging enzymes Yap-1 (yes-associated protein) and Pep-1 (apoplastic effector) secreted by *U. maydis* prevent the accumulation of hydrogen peroxide and inhibit the heme peroxidase reaction, thereby subverting the oxidative damage caused by ROS production (Molina and Kahmann 2007; Doehlemann et al. 2009; Hemetsberger et al. 2012). Transcription factors associated with stress-activated protein kinase and mitogen-activated protein kinase pathway also play an important role in the induction of antioxidants and secretion of secondary metabolites required by the fungus to counteract the defence response in the host plant (Strehmel et al. 2017).

13.4.3 pH Manipulation

The ability to manipulate the pH of the host cells surrounding the infection site is one of the key mechanisms generally used by fungal pathogens to counter host defence response for their survival inside the host tissue. Some fungi cause acidification, while other causes alkalinisation of the host tissue. Fungi like *Sclerotinia sclerotiorum*, *Penicillium* sp. and *Aspergillus* sp. secretes oxalic acid, gluconic acid and citric acid, respectively, leading to acidification of the infected cells that eventually leads into death of the host tissue (Ruijter and Visser 1999; Manteau et al. 2003; Prusky and Yakoby 2003; Bolton et al. 2006). In contrast, pathogens like *F. oxysporum* cause alkalinisation of the extracellular pH by the secretion of peptides homologous to the RALFs (rapid alkalinisation factors) to colonize inside the host cells (Murphy and De Smet 2014; Masachis et al. 2016). Alkaline pH is suitable for disease progression as it favours germination, hyphal growth and formation of fruiting bodies (Vylkova 2017). Presence of RALFs in most of the fungal pathogens suggests that alkalinisation at the infection site is a common mechanism utilised by the fungal pathogen to supress the host immune response (Masachis et al. 2016; Thynne et al. 2017).

13.4.4 Destroying Host Proteases

Fungal pathogens secrete various proteins which act upon host proteases and protect fungus against the immune response generated by the host cells. Avr2 effector of pathogen *C. fulvum* inhibits Rcr3 protease present in the tomato apoplast and other proteases in other hosts by changing its conformation which finally inhibits further downstream host immune responses (Kruger et al. 2002; Rooney et al. 2005; van Esse et al. 2008; Doehlemann et al. 2011; Mueller et al. 2013). Similarly, Pit2 effector of *Ustilago maydis* inhibits host proteases and thereby protects fungus from the immune response generated by host cells (Doehlemann et al. 2011; Mueller et al. 2013).

13.4.5 Depreciating Phytohormone Signalling

Plant hormones like SA, JA and ET plays a pivotal role in regulating the immune responses against the invading pathogen. Targeting the host immune signalling via altering the biosynthesis of signalling molecules is a straightforward approach utilised by few classes of fungus for their sustenance inside the host tissue. For example, *U. maydis* the causal agent of corn smut produces an enzyme called chorismite mutase that affects the biosynthesis of the plant immune signal salicylic acid by channelling chorismite into the phenylpropanoid pathway (Djamei et al. 2011). Similar alteration in the SA biosynthesis is also caused by effector (isochorismatase activity) secreted by *V. dahliae* to hydrolyse isochorismate (Liu et al. 2014). Likewise, perturbation of JA signalling by the fungal effectors MiSSP7 (mycorrhiza-induced small secreted protein 7) secreted by *Laccaria bicolor* has been reported to be important for its infection on *Populus trichocarpa* (Plett et al. 2014).

13.5 Impact of Climate Change in Plant Pathogen Interaction

Climate change refers to the changes in temperature, wind pattern, degree of rainfall, increased pollutants, increasing atmosphere concentrations of ozone, SO_2, CO_2, etc. (Harvell et al. 2002; Ramanan et al. 2020). Changes in climatic patterns have become a major factor limiting the global agricultural productivity (Chakraborty et al. 2000; Anderson et al. 2004; Garrett et al. 2006; Altizer et al. 2013). The increase in global temperature, in particular, provides a conducive environment for survival of pathogen consequently thereby have detrimental effect on plant health. For instance, increase in global temperature has resulted in frequent disease outbreaks in crops like rice and wheat (Olsen et al. 2011). Similarly, in the United Sates, the geographical distribution and establishment of the tree pathogen *Phytopthora ramorum* is reported to be affected by shift in the climatic patterns (Rizzo et al. 2011; Venette 2009). Some other factors like changes in the gaseous (CO_2, SO_2 and O_3) composition in the air also have direct or indirect impact on disease epidemics and are interrelated with the warmer temperatures. The effect of elevated CO_2 on pathogen establishment and disease progression is considerably variable and largely depends on the behaviour and life cycle of the pathogens (Das et al. 2016). For instance in case of rust pathogens, the elevated CO_2 concentration is reported to increase the infection of *P. graminis* and *P. recondite* in wheat and rye crops respectively. However, for aspen tree rust pathogen, *Melampsora meduase the* elevated levels of CO_2 had no effect on disease infection and its severity (Gassner and Straib 1930; Karnosky et al. 2002; Manning and von Tiedemann 1995; Percy et al. 2002). Like CO_2, the increase in atmospheric concentration of SO_2 also reported to alter the host parasite relationship.

A long-term (170 years) experiment was conducted by Rothamsted Research, UK, to study the relation of emitted SO_2 and disease progression by *P. nodorum* and *M. graminicola* on wheat. This study showed a strong correlation between the changes in the ratio of the pathogens with changes in atmospheric SO_2 concentration over a period of 160 years (Bearchell et al. 2005). It was found that growth of both the pathogens was favoured by the presence of SO_2, but concentration of SO_2 was responsible for the balance between the populations of the two pathogens as these two pathogens share the same niche and require leaf wetness for which they compete to survive when infecting the same tissue (Bearchell et al. 2005; Shaw et al. 2008; Chandramohan 2010). Extreme changes in the climate like from drought to floods, heat waves to winds, rainfall, storms or other natural calamities drastically affect the plant health and plant pathosystems (Boland et al. 2004; Hegerl et al. 2011; Peng et al. 2011). For example, floods can spread water-borne diseases, storms can spread airborne diseases, heat waves can help the disease progression faster, etc. Increase in temperature in colder regions may favour the growth of some of the pathogens like *Fusarium circinatum* (the causal agent of pine pitch canker) in Europe (Watt et al. 2011). In contrast, the pathogens that need colder temperatures to infect the host will no longer be able to cause disease with sudden rise in average minimum temperature e.g. *Seiridium cardinal*e on Cyprus species (Garbelotto 2008). Many evidences suggests that such climatic shifts have already affected the phenology, abundance and diversity of species (Körner and Basler 2010; Matesanz et al. 2010).

13.6 Effective Management Strategies

For maintaining a continuous global food supply, it is imperative to develop crops resilient to the climatic changes. Data generated from long-term experiments designed to study effect of climate on plant-fungi interaction can be used for simulating prediction models, to predict behaviour of pathogen in response to the climatic fluctuations (Roos et al. 2010; La Porta et al. 2008; Watt et al. 2010; Chakraborty and Newton 2011; Luck et al. 2011; Seidl et al. 2011). For instance, a computer-simulated model has predicted that disease incidences on oil seed rape will increase in Germany due to the presence of favourable conditions for pathogens like *Alternaria brassicae*, *Sclerotinia sclerotiorum* and *Verticillium longisporum* (Siebold and von Tiedemann 2012).

To minimize the negative effects of climate change on crops, climate-smart field crops can be developed by adopting climate-smart agricultural (CSA) practices that include technologies that reduce greenhouse gas emissions, enhance resilience to climatic stress, increase productivity and provide opportunities to overcome the negative impact of climate change on plant pathosystems (FAO 2010; Behera and Prasad 2020). For the reduction of greenhouse gas emissions, effective crop rotation should be done which can be achieved by rotating crops high in carbon with the crops high in nitrogen (Corsi et al. 2012). This allows the carbon to accumulate in the soil and enables the nitrogen in the decaying surface residues to be released slowly for being uptaken by the next crop. If the amount of nitrogen in the crop residues is too low, microorganisms use the mineral nitrogen existing in the soil (nitrogen immobilization), which reduces the amount of nitrogen available to the growing crop until (weeks) the carbon in the crop residues starts to deplete (Gál et al. 2007). Crop rotation can be integrated with the legume crops to support carbon sequestration. Nitric oxide emissions can be reduced by integrated nutrient management and with the applications of the precise amount of mineral fertiliser. In addition to this, crops having large amount of root biomass can be used to avoid anaerobic conditions in the soil and improve drainage (Kuzyakov et al. 2000; Fontaine et al. 2004; Sisti et al. 2004; Fontaine 2007).

Researchers are more focussed towards cereals, particularly maize, wheat and rice, and legumes, such as groundnut and soybean. However, for a healthy cropping system and climate-smart approach, diversity among crops and other living organisms is an important criterion to enhance resilience and provide economic stability and profitability (Glover et al. 2010). Although it seems to be a labour-intensive technique and very costly, it is one of best and eco-friendly approaches to manage pests and diseases. Multipurpose crop varieties can be integrated whose biomass can be further used for food, biofuel, feed and fibre. Crops should not compete for basic resources such as light, nutrient, water, etc. (FAO 2019).

13.7 Conclusion

Pathogen attack and climatic drifts are the two major concerns related to food security. Extreme climate change affect the plant-fungal interaction as discussed in the previous sections. Developing countries are more vulnerable towards risk of climate change and food security compared to developed countries as they have less means of agricultural tools, policies and institutions to address the challenges of high productivity and efficiency gap. However, they have more potential for mitigation and adaptation. To address these problems, climate-smart agriculture (CSA) is one of the most economic and eco-friendly approaches. CSA refers to natural auto-controlled systems for growing healthy crops which can be achieved by having knowledge of the effect of different climatic conditions on a particular plant-pathogen interaction. A multidisciplinary approach has to be targeted to develop CSA practices for efficient and integrated management of crops, soil, water and nutrients. Researchers need to do experiments for the study of interaction of a particular plant-pathogen system under different climatic conditions and should record the data for generating model system of a particular plant-pathogen system under each climatic condition used in the experiment. By using this approach, it would be possible to combat the menace of pathogen invasion and can support the growing demands of increasing population.

References

Abramovitch RB, Anderson JC, Martin GB (2006) Bacterial elicitation and evasion of plant innate immunity. Nat Rev Mol Cell Biol 7:601–611

Altizer S, Ostfeld RS, Johnson PTJ, Kutz S, Harvell CD (2013) Climate change and infectious diseases: from evidence to a predictive framework. Science 341(6145):514–519. https://doi.org/10.1126/science.1239401

Anderson PK, Cunningham AA, Patel NG, Morales FJ, Epstein PR, Daszak P (2004) Emerging infectious diseases of plants: pathogen pollution, climate change and agrotechnology drivers. Trends Ecol Evol 19:535–544. https://doi.org/10.1016/j.tree.2004.07.021

Andrade CM, Tinoco MLP, Rieth AF, Maia FCO, Aragão FJL (2016) Host-induced gene silencing in the necrotrophic fungal pathogen *Sclerotinia sclerotiorum*. Plant Pathol 65:626–632

Bearchell SJ, Fraaije BA, Shaw MW, Fitt BDL (2005) Wheat archive links long-term fungal pathogen population dynamics to air pollution. Proc Natl Acad Sci USA 102:5438–5442. https://doi.org/10.1073/pnas.0501596102

Behera BK, Prasad R (2020) Environmental technology and sustainability. Elsevier. ISBN 9780128191033. https://www.elsevier.com/books/environmental-technology-and-sustainability/behera/978-0-12-819103-3

Bindschedler LV, Dewdney J, Blee KA, Stone JM, Asai T, Plotnikov J et al (2006) Peroxidase-dependent apoplastic oxidative burst in Arabidopsis required for pathogen resistance. Plant J 47:851–863

Birben E, Sahiner UM, Sackesen C, Erzurum S, Kalayci O (2012) Oxidative stress and antioxidant defense. World Allergy Organ J 5:9–19

Blokhina O, Virolainen E, Fagerstedt KV (2003) Antioxidants, oxidative damage and oxygen deprivation stress: a review. Ann Bot 91:179–194

Boland GJ, Melzer MS, Hopkin A (2004) Climate change and plant diseases in Ontario. Can J Plant Pathol 26:335–350. https://doi.org/10.1080/07060660409507151

Bolton MD, Thomma BP, Nelson BD (2006) *Sclerotinia sclerotiorum* (lib.) de Bary: biology and molecular traits of a cosmopolitan pathogen. Mol Plant Pathol 7:1–16

Bolwell GP, Bindschedler LV, Blee KA, Butt VS, Davies DR, Gardner SL et al (2002) The apoplastic oxidative burst in response to biotic stress in plants: a three-component system. J Exp Bot 53:1367–1376

Brand A, Gow NA (2009) Mechanisms of hypha orientation of fungi. Curr Opin Microbiol 12:350–357

Buscot F, Munch JC, Charcosset JY, Gardes M, Nehls U, Hampp R (2000) Recent advances in exploring physiology and biodiversity of ectomycorrhizas highlight the functioning of these symbioses in ecosystems. New Phytol 24:601–614

Cai Q, Qiao L, Wang M, He B, Lin FM, Palmquist J et al (2018) Plants send small RNAs in extracellular vesicles to fungal pathogen to silence virulence genes. Science 360 (6393):1126–1129

Castel SE, Martienssen RA (2013) RNA interference in the nucleus: roles for small RNAs in transcription, epigenetics and beyond. Nat Rev Genet 14:100–112

Chakraborty S, Newton AC (2011) Climate change, plant diseases and food security: an overview. Plant Pathol 60:2–14

Chakraborty S, Tiedemann AV, Teng PS (2000) Climate change: potential impact on plant diseases. Environ Pollut 108:317–326. https://doi.org/10.1016/S0269-7491(99)00210-9

Chandramohan P (2010) Causal relationships between sulphur emissions and pathogen abundance. University of Reading, PhD thesis, Reading

Cheng W, Song XS, Li HP, Cao LH, Sun K, Qiu XL et al (2015) Host-induced gene silencing of an essential chitin synthase gene confers durable resistance to Fusarium head blight and seedling blight in wheat. Plant Biotechnol J 13:1335–1345

Chi MH, Park SY, Kim S, Lee YH (2009) A novel pathogenicity gene is required in the rice blast fungus to suppress the basal defences of the host. PLoS Pathog 5:e1000401

Collemare J, Billard A, Böhnert HU, Lebrun MH (2008) Biosynthesis of secondary metabolites in the rice blast fungus *Magnaporthe grisea*: the role of hybrid PKS-NRPS in pathogenicity. Mycol Res 112:207–215. https://doi.org/10.1016/j.mycres.2007.08.003

Corsi S, Friedrich T, Kassam A, Pisante M, de Moraes Sà J (2012) Soil organic carbon accumulation and carbon budget in conservation agriculture: a review of evidence. FAO Integrated CROP Management, 16

Cosio EG, Feger M, Miller CJ, Antelo L, Ebel J (1996) High-affinity binding of fungal b-glucan elicitors to cell membranes of species of the plant family Fabaceae. Planta 200:92–99

Cote F, Roberts KA, Hahn MG (2000) Identification of high-affinity binding sites for the hepta-b-glucoside elicitor in membranes of the model legumes *Medicago truncatula* and *Lotus japonicus*. Planta 211:596–605

Das T, Hajong M, Majumdar D, Devi RKT, Rajesh T (2016) Climate change impacts on plant diseases. SAARC J Agri 14(2):200–209

Daub ME, Ehrenshaft M (2000) The photoactivated Cercospora toxin cercosporin: contributions to plant disease and fundamental biology. Annu Rev Phytopathol 38:461–490

Djamei A, Schipper K, Rabe F, Ghosh A, Vincon V, Kahnt J et al (2011) Metabolic priming by a secreted fungal effector. Nature 478:395–398

Doehlemann G, van der Linde K, Assmann D et al (2009) Pep1, a secreted effector protein of *Ustilago maydis*, is required for successful invasion of plant cells. PLoS Pathog 5:e1000290

Doehlemann G, Reissmann S, Assmann D, Fleckenstein M, Kahmann R (2011) Two linked genes encoding a secreted effector and a membrane protein are essential for *Ustilago maydis*-induced tumour formation. Mol Microbiol 81:751–766

FAO (2010) Climate smart agriculture: policies, practices and financing for food security, adaptation and mitigation. Food and Agriculture Organization of the United Nations (FAO), Rome

FAO (2019) Climate-smart agriculture: Sourcebook; Food and Agriculture Organization of the United Nations. http://www.fao.org/climate-smart-agriculture-sourcebook/production-resources/en/

Felix G, Regenass M, Boller T (1993) Specific perception of subnanomolar concentrations of chitin fragments by tomato cells: induction of extracellular alkalinization, changes in protein phosphorylation, and establishment of a refractory state. Plant J 4:307–316

Fontaine S (2007) Stability of organic carbon in deep soil layers controlled by fresh carbon supply. Nature 450:277–280

Fontaine S, Bardoux G, Abbadie L, André M (2004) Carbon input to soil may decrease soil carbon content. Ecol Lett 7:314–320

Fujikawa T, Sakaguchi A, Nishizawa Y, Kouzai Y, Minami E, Yano S et al (2012) Surface a-1,3-glucan facilitates fungal stealth infection by interfering with innate immunity in plants. PLoS Pathog 8:e1002882

Gál A, Vyn TJ, Michéli E, Kladivko EJ, McFee WW (2007) Soil carbon and nitrogen accumulation with long-term no-till versus moldboard plowing overestimated with tilled-zone sampling depths. Soil Till Res 96:42–51
Galeotti F, Barile E, Curir P, Dolci M, Lanzotti V (2008) Flavonoids from carnation (*Dianthus caryophyllus*) and their antifungal activity. Phytochem Lett 1:44–48. https://doi.org/10.1016/j.phytol.2007.10.001
Garbelotto M (2008) Molecular analysis to study invasions by forest pathogens: examples from Mediterranean ecosystems. Phytopathol Mediterr 47:183–203
Garrett KA, Dendy SP, Frank EE, Rouse MN, Travers SE (2006) Climate change effects on plant disease: genomes to ecosystems. Annu Rev Phytopathol 44:489–509. https://doi.org/10.1146/annurev.phyto.44.070505.143420
Gassner G, Straib W (1930) Untersuchungen uber die Abbangigkeit des Infecktionsverhaltens der getreiderostpilze vom kohlensouregehalt der luft. J Phytopathol 1:1–30
Gibson DM, King BC, Hayes ML, Bergstrom GC (2011) Plant pathogens as a source of diverse enzymes for lingo cellulose digestion. Curr Opin Microbiol 14:264–270
Gilbert BM, Wolpert TJ (2013) Characterization of the LOV1-mediated, victorin-induced, cell-death response with virus-induced gene silencing. Molec Plant-Microbe Interact 26:903–917
Glover JD, Reganold JP, Bell LW, Borevitz J, Brummer EC, Buckler ES et al (2010) Increased food and ecosystem security via perennial grains. Science 328:1638–1639
Harvell CD, Mitchell CE, Ward JR, Altizer S, Dobson AP, Ostfeld RS et al (2002) Climate warming and disease risks for terrestrial and marine biota. Science 296:2158–2162
Hegerl GC, Hanlon H, Beierkuhnlein C (2011) Climate science: elusive extremes. Nat Geosci 4:142–143. https://doi.org/10.1038/ngeo1090
Hemetsberger C, Herrberger C, Zechmann B et al (2012) The *Ustilago maydis* effector Pep1 suppresses plant immunity by inhibition of host peroxidase activity. PLoS Pathog 8:e1002684
Hoch HC, Staples RC, Whitehead B, Comeau J, Wolf ED (1987) Signaling for growth orientation and cell differentiation by surface topography in *uromyces*. Science 235:1659–1662
Huffaker A, Kaplan F, Vaughan MM, Dafoe NJ, Ni X, Rocca JR et al (2011) Novel acidic sesquiterpenoids constitute a dominant class of pathogen-induced phytoalexins in maize. Plant Physiol 156:2082–2097. https://doi.org/10.1104/pp.111.179457
Jashni MK, Dols IH, Iida Y, Boeren S, Beenen HG, Mehrabi R et al (2015) Synergistic action of a metalloprotease and a serine protease from *Fusarium oxysporum f. sp. lycopersici* cleaves chitin-binding tomato chitinases, reduces their antifungal activity, and enhances fungal virulence. Mol. Plant Microbe Interact 28:996–1008
Kabera JN, Semana E, Mussa AR, He X (2014) Plant secondary metabolites: biosynthesis, classification, function and pharmacological properties. J Pharm Pharmacol 2(7):377–392
Kang Z, Buchenauer H (2003) Immunocytochemical localizations of cell wall bound thionins and hydroxyproline-rich glycoproteins in Fusarium culmorum-infected wheat spikes. J Phytopathol 151:120–129
Karnosky DF, Percy KE, Xiang B, Brenda C, Noormets A, Mankovska B et al (2002) Interacting elevated CO_2 and tropospheric O_3 predisposes aspen (*Populus tremuloides* Michx.) to infection by rust (*Melampsora medusae* f. sp. *tremuloidae*). Glob Change Biol 8:329–338
Katagiri F, Tsuda K (2010) Understanding the plant immune system. Mol Plant-Microbe Interact 23 (12):1531–1536. https://doi.org/10.1094/MPMI-04-10-0099
Körner C, Basler D (2010) Phenology under global warming. Science 327:1461–1462. https://doi.org/10.1126/science.1186473
Kruger J, Thomas CM, Golstein C, Dixon MS, Smoker M, Tang S et al (2002) A tomato cysteine protease required for Cf-2-dependent disease resistance and suppression of autonecrosis. Science 296:744–747
Kubicek CP, Starr TL, Glass NL (2014) Plant cell wall–degrading enzymes and their secretion in plant-pathogenic fungi. Annu Rev Phytopathol 52:427–451. https://doi.org/10.1146/annurev-phyto-102313-045831
Kuzyakov Y, Friedel JK, Stahr K (2000) Review of mechanisms and quantification of priming effects. Soil Biol Biochem 32:1485–1498

La Porta N, Capretti P, Thomsen IM, Kasanen R, Hietala AM, Von Weissenberg K (2008) Forest pathogens with higher damage potential due to climate change in Europe. Can J Plant Pathol 30:177–195

Lebeda A, Luhova L, Sedlarova M, Jancova D (2001) The role of enzymes in plant–fungal pathogens interactions. J Plant Dis Protect 108:89–111

Levings CS III, Rhoads DM, Siedow JN (1995) Molecular interactions of *Bipolaris maydis* T-toxin and maize. Can J Bot 73:483–489

Liu T, Song T, Zhang X, Yuan H, Su L, Li W et al (2014) Unconventionally secreted effectors of two filamentous pathogens target plant salicylate biosynthesis. Nat Commun 5:4686

Luck J, Spackman M, Freeman A, Trebicki P, Griffiths W, Finlay K et al (2011) Climate change and diseases of food crops. Plant Pathol 60:113–121. https://doi.org/10.1111/j.1365-3059.2010.02414.x

Manning WJ, von Tiedemann A (1995) Climate change: potential effects of increased atmospheric carbon dioxide (CO2), ozone (O3), and ultraviolet-B (UV-B) radiation on plant diseases. Environ Pollut 88:219–245. https://doi.org/10.1016/0269-7491(95)91446-R

Manteau S, Abouna S, Lambert B, Legendre L (2003) Differential regulation by ambient pH of putative virulence factor secretion by the phytopathogenic fungus *Botrytis cinerea*. FEMS Microbiol Ecol 43(3):359–366. https://doi.org/10.1111/j.1574-6941.2003.tb01076.x

Masachis S, Segorbe D, Turra D, Leon-Ruiz M, Fürst U, Ghalid ME et al (2016) A fungal pathogen secretes plant alkalinizing peptides to increase infection. Nat Microbiol 1:16043

Matesanz S, Gianoli E, Valladares F (2010) Global change and the evolution of phenotypic plasticity in plants. Ann NY Acad Sci 1206:35–55. https://doi.org/10.1111/j.1749-6632.2010.05704.x

Meehan F, Murphy H (1947) Differential phytotoxicity of metabolic by products of Helminthosporium victoriae. Science 106:270–271

Mendgen K, Deising H (1993) Infection structures of fungal plant pathogens – a cytological and physiological evaluation. New Phytol 124:193–213

Molina L, Kahmann R (2007) An *Ustilago maydis* gene involved in H_2O_2 detoxification is required for virulence. Plant Cell 19:2293–2309

Mueller AN, Ziemann S, Treitschke S, Abmann D, Doehlemann G (2013) Compatibility in the *Ustilago maydis*-maize interaction requires inhibition of host cysteine proteases by the fungal effector Pit2. PLoS Pathog 9:e1003177

Murphy E, De Smet I (2014) Understanding the RALF family: a tale of many species. Trends Plant Sci 19:664–671

Navarro L, Zipfel C, Rowland O, Keller I, Robatzek S, Boller T et al (2004) The transcriptional innate immune response to flg22. Interplay and overlap with *Avr* gene-dependent defense responses and bacterial pathogenesis. Plant Physiol 135:1113–1128

Olsen L, Choffnes ER, Relman DA, Leslie Pray R (2011) Fungal diseases: an emerging threat to human, animal, and plant health: workshop summary. Forum on microbial threats. National Academic, Washington, DC

Peng C, Ma Z, Lei X, Qiuan Z, Huai C, Weifeng W et al (2011) A drought-induced pervasive increase in tree mortality across Canada's boreal forests. Nat Clim Chang 1:467–471. https://doi.org/10.1038/nclimate1293

Percy KE, Awmack CS, Lindroth RL, Kubiske ME, Kopper BJ, Isebrands GJ et al (2002) Altered performance of forest pests under atmospheres enriched by CO2 and O3. Nature 420:403–407. https://doi.org/10.1038/nature01229

Perfect SE, Green JR (2001) Infection structures of biotrophic and hemibiotrophic fungal plant pathogens. Mol Plant Pathol 2:101–108

Picman AK, Schneider EF, Pieman J (1995) Effect of flavonoids on mycelial growth of *Verticillium albo-atrum*. Biochem Syst Ecol 23:683–693

Plett JM, Daguerre Y, Wittulsky S, Vayssières A, Deveau A, Melton SJ et al (2014) Effector MiSSP7 of the mutualistic fungus *Laccaria* bicolor stabilizes the *Populus* JAZ6 protein and represses jasmonic acid (JA) responsive genes. Proc Natl Acad Sci USA 111:8299–8304

Prasad R, Gupta N, Satti U, Wang S, Ahmed ISA, Varma A (2017) Management of fungal pathogens by mycorrhiza. In: Varma A, Prasad R, Tuteja N (eds) Mycorrhiza. Springer International Publishing AG, Switzerland, pp 179–194

Prusky D, Yakoby N (2003) Pathogenic fungi: leading or led by ambient pH? Mol Plant Pathol 4 (6):509–516. https://doi.org/10.1046/j.1364-3703.2003.00196.x

Pusztahelyi T, Holb IJ, Pócsi I (2015) Secondary metabolites in fungus-plant interactions. Front Plant Sci 6:573. https://doi.org/10.3389/fpls.2015.0057

Qi T, Zhu X, Tan C, Liu P, Guo J, Kang Z et al (2017) Host-induced gene silencing of an important pathogenicity factor PsCPK1 in *Puccinia striiformis* f. sp. tritici enhances resistance of wheat to stripe rust. Plant Biotechnol J 16(3):797–807. https://doi.org/10.1111/pbi.12829

Ramanan VV, Shah S, Prasad R (2020) Global climate change and environmental policy: agriculture perspectives. Springer Singapore. ISBN 978–981–13-9570-3. https://www.springer.com/gp/book/9789811395697

Ridout CJ (2009) Profiles in pathogenesis and mutualism: powdery mildews. In: Deising HB (ed) The mycota V, part 1, plant relationships. Springer, Berlin, pp 51–68

Rizzo DM, Meentemeyer RK, Garbelotto M (2011) The emergence of *Phytophthora ramorum* in North America and Europe. In: Olsen LA, Relman DA, Choffnes ER, Pray L (eds) Fungal diseases: an emerging challenge to human, animal and plant health. Institute of Medicine of the National Academies, Washington, DC, pp 312–324

Rodriguez-Moreno L, Ebert MK, Bolton MD, Thomma BPHJ (2018) Tools of the crook- infection strategies of fungal plant pathogens. Plant J 93:664–674. https://doi.org/10.1111/tpj.13810

Rooney HC, Van't Klooster JW, van der Hoorn RA, Joosten MH, Jones JD, de Wit PJ (2005) Cladosporium Avr2 inhibits tomato Rcr3 protease required for Cf-2-dependent disease resistance. Science 308:1783–1786

Roos J, Hopkins R, Kvarnheden A, Dixelius C (2010) The impact of global warming on plant diseases and insect vectors in Sweden. Eur J Plant Pathol 129:9–19. https://doi.org/10.1007/s10658-010-9692-z

Ruijter GJ, Visser J (1999) Characterization of *Aspergillus niger* phosphoglucose isomerase. Use for quantitative determination of erythrose 4-phosphate. Biochimie 81(3):267–272

Sanchez-Vallet A, Saleem-Batcha R, Kombrink A, Hansen G, Valkenburg DJ, Thomma BP et al (2013) Fungal effector Ecp6 out competes host immune receptor for chitin binding through intrachain LysM dimerization. elife 2:e00790

Sanchez-Vallet A, Mesters JR, Thomma BP (2015) The battle for chitin recognition in plant-microbe interactions. FEMS Microbiol Rev 39:171–183

Sasaki K, Iwai T, Hiraga S, Kuroda K, Seo S, Mitsuhara I et al (2004) Ten rice peroxidases redundantly respond to multiple stresses including infection with rice blast fungus. Plant Cell Physiol 45:1442–1452

Schafer W (1994) Molecular mechanisms of fungal pathogenicity to plants. Annu Rev Phytopathol 32:461–477

Seidl R, Fernandes PM, Fonseca TF, Gillet F, Jönssong AM, Merganičová K et al (2011) Modelling natural disturbances in forest ecosystems: a review. Ecol Model 222:903–924. https://doi.org/10.1016/j.ecolmodel.2010.09.040

Shaw MW, Bearchell SJ, Fitt BDL, Fraaije BA (2008) Long-term relationships between environment and abundance in wheat of *Phaeosphaeria nodorum* and *Mycosphaerella graminicola*. New Phytol 177:229–238

Shibuya N, Kaku H, Kuchitsu K, Maliarik MJ (1993) Identification of a novel high-affinity binding site for N-acetylchitooligosaccharide elicitor in the membrane fraction from suspension-cultured rice cells. FEBS Lett 329:75–78

Siebold M, von Tiedemann A (2012) Potential effects of global warming on oilseed rape pathogens in Northern Germany. Fungal Ecol 5:62–72. https://doi.org/10.1016/j.funeco.2011.04.003

Singh AK, Pandey MB, Singh UP (2007) Antifungal activity of an alkaloid allosecurinine against some fungi. Mycobiology 35(2):62–64. https://doi.org/10.4489/MYCO.2007.35.2.062

Sisti CPJ, dos Santos HP, Kohhann R, Alves BJR, Urquiaga S, Boddey RM (2004) Change in carbon and nitrogen stocks in soil under 13 years of conventional or zero tillage in southern Brazil. Soil Till Res 76:39–58

Strehmel N, Hoehenwarter W, Mönchgesang S, Majovsky P, Krüger S, Scheel D et al (2017) Stress-related mitogen-activated protein kinases stimulate the accumulation of small molecules and proteins in *Arabidopsis thaliana* root exudates. Front Plant Sci 8:1292. https://doi.org/10.3389/fpls.2017.01292

Tao Y, Xie Z, Chen W, Glazebrook J, Chang HS, Han B et al (2003) Quantitative nature of *Arabidopsis* responses during compatible and incompatible interactions with the bacterial pathogen *Pseudomonas syringae*. Plant Cell 15:317–330

Thynne E, Saur IML, Simbaqueba J, Ogilvie HA, Gonzalez-Cendales Y, Mead O et al (2017) Fungal phytopathogens encode functional homologues of plant rapid alkalinisation factor (RALF) peptides. Mol Plant Pathol 18:811–824

Tiwari IM, Jesuraj A, Kamboj R, Devanna BN, Botella JR, Sharma TR (2017) Host delivered RNAi, an efficient approach to increase rice resistance to sheath blight pathogen (*Rhizoctonia solani*). Sci Rep 7:7521. https://doi.org/10.1038/s41598-017-07749-w

Tsuge T, Harimoto Y, Akimitsu K, Ohtani K, Kodama M, Akagi Y et al (2013) Host-selective toxins produced by the plant pathogenic fungus *Alternaria alternata*. FEMS Microbiol Rev 37:44–66

Tundo S, Kalunke R, Janni M, Volpi C, Lionetti V, Bellincampi D et al (2016) Pyramiding PvPGIP2 and TAXI-III but not PvPGIP2 and PMEI enhances resistance against *Fusarium graminearum*. Mol Plant-Microbe Interact 29:629–639

van Esse HP, Van't Klooster JW, Bolton MD, Yadeta KA, van Baarlen P, Boeren S et al (2008) The *Cladosporium fulvum* virulence protein Avr2 inhibits host proteases required for basal defense. Plant Cell 20:1948–1963

Varma A, Prasad R, Tuteja N (2017a) Mycorrhiza: function, diversity and state-of-art. Springer International Publishing, Switzerland. ISBN 978-3-319-53064-2. http://www.springer.com/us/book/9783319530635

Varma A, Prasad R, Tuteja N (2017b) Mycorrhiza: eco-physiology, secondary metabolites, nanomaterials. Springer International Publishing, Switzerland. ISBN 978-3-319-57849-1. http://www.springer.com/us/book/9783319578484

Varma A, Prasad R, Tuteja N (2017c) Mycorrhiza: nutrient uptake, biocontrol, ecorestoration. Springer International Publishing, Switzerland. ISBN 978-3-319-68867-1. http://www.springer.com/us/book/9783319688664

Venette RC (2009) Implication of global climate change on the distribution and activity of *Phytophthora ramorum*. In: McManus K, Gottschalk KW (eds) Proceedings 20th U.S. Department of Agriculture interagency research forum on invasive species 2009. USDA FS, GTR NRS-P-51, pp 58–59

Volpi C, Janni M, Lionetti V, Bellincampi D, Favaron F, D'Ovidio R (2011) The ectopic expression of a pectin methyl esterase inhibitor increases pectin methyl esterification and limits fungal diseases in wheat. Mol Plant-Microbe Interact 24:1012–1019

Vylkova S (2017) Environmental pH modulation by pathogenic fungi as a strategy to conquer the host. PLoS Pathog 13(2):e1006149. https://doi.org/10.1371/journal.ppat.1006149

Wang X, Jiang N, Liu J, Liu W, Wang GL (2014) The role of effectors and host immunity in plant–necrotrophic fungal interactions. Virulence 5:722–732

Wang M, Weiberg A, Lin FM, Thomma BP, Huang HD, Jin H (2016) Bidirectional cross-kingdom RNAi and fungal uptake of external RNAs confer plant protection. Nat Plants 2:16151

Wang B, Sun Y, Song N, Zhao M, Liu R, Feng H et al (2017) *Puccinia striiformis* f. sp. *tritici* microRNA like RNA 1 (*Pst*-milR1), an important pathogenicity factor of *Pst*, impairs wheat resistance to *Pst* by suppressing the wheat pathogenesis-related 2 gene. New Phytol 215 (1):338–350

Watt MS, Stone JK, Hood IA, Palmer DJ (2010) Predicting the severity of Swiss needle cast on Douglas-fir under current and future climate in New Zealand. Forest Ecol Manage 260:2232–2240. https://doi.org/10.1016/j.foreco.2010.09.034

Watt MS, Ganley RJ, Kriticos DJ, Manning LK (2011) Dothistroma needle blight and pitch canker: the current and future potential distribution of two important diseases of Pinus species. Can J For Res 41:412–424. https://doi.org/10.1139/X10-204

Weiberg A, Wang M, Lin FM (2013) Fungal small RNAs suppress plant immunity by hijacking host RNA interference pathways. Science 342:118–123

Wolpert TJ, Macko V, Acklin W, Jaun B, Seibl J, Meili J et al (1985) Structure of Victorin-C, the major host-selective toxin from *Cochliobolus victoriae*. Experientia 41:1524–1529

Wolpert TJ, Macko V, Acklin W, Jaun B, Arigoni D (1986) Structure of minor host-selective toxins from *Cochliobolus victoriae*. Experientia 42:1296–1299

Wolpert TJ, Dunkle LD, Ciuffetti LM (2002) Host-selective toxins and avirulence determinants: what's in a name? Annu Rev Phytopathol 40:251–285

Zeilinger S, Gupta VK, Dahms TES, Silva RN, Singh HB, Upadhyay RS et al (2015) Friends or foes? Emerging insights from fungal interactions with plants. FEMS Microbiol Rev 40:182–207

Zhang T, Zhao YL, Zhao JH, Wang S, Jin Y, Chen ZQ et al (2016) Cotton plants export microRNAs to inhibit virulence gene expression in a fungal pathogen. Nat Plants 2 (10):16153. https://doi.org/10.1038/nplants.2016.153

Zipfel C, Kunze G, Chinchilla D, Caniard A, Jones JD, Boller T et al (2006) Perception of the bacterial PAMP EF-Tu by the receptor EFR restricts agrobacterium-mediated transformation. Cell 125:749–760

Chapter 14
Biochemical Dynamics of Plant-Microbe Interactions

Priyanka Lonakadi, Renitta Jobby, Nitin Desai, and Pamela Jha

Abstract Plants and microbes coexist and compete for survival in their myriad interactions, which plays a key role in adapting them to the extreme of environments. Hence, it is pertinent to understand, explore, and exploit the plant-microbe interactions. The biochemical dynamics of these interactions are very intricate and specific to the type of plant root exudates. These are useful nutrient and energy sources for soil microorganisms, with which they establish an explicit communication systems. There are some beneficial bacteria and fungi, which act as plant growth-promoting microorganisms, may reduce phytotoxicity, and stimulate plant growth indirectly through the induction of defense mechanisms against phytopathogens and/or directly through the solubilization of mineral nutrients (nitrogen, phosphate, potassium, iron, etc.), production of plant growth-promoting substances, and secretion of specific enzymes (e.g., 1-aminocyclopropane-1-carboxylate deaminase). This chapter focuses on the biochemical dynamics of beneficial plant-microbe interaction, which is important in increasing the crop productivity.

14.1 Introduction

Soil which can be defined as a blend of minerals, organic matter, liquids, and gases is a platform of a living space of soil microflora comprising of microorganisms such as earthworm, insects, nematodes, protozoa, mites, etc. Hence the soil microflora ranges from prokaryotes to eukaryotes in which prokaryotes form the major contribution than other constituents present in the soil microflora (Hinsinger et al. 2009; Curtis et al. 2002; Crawford et al. 2005).

Rhizosphere, a highly dynamic region present in and around the roots, was portrayed by Lorenz Hiltner, who characterized rhizosphere as a microbial hotspot in which numerous biological, chemical, and physical processes occur, and these processes differ from bulk soil (Fig. 14.1) (Darrah 1993; Hinsinger 1998; Hartmann

P. Lonakadi · R. Jobby · N. Desai · P. Jha (✉)
Amity Institute of Biotechnology, Amity University Mumbai, Mumbai, Maharashtra, India
e-mail: pjha@mum.amity.edu

A. Varma et al. (eds.), *Plant Microbiome Paradigm*,
https://doi.org/10.1007/978-3-030-50395-6_14

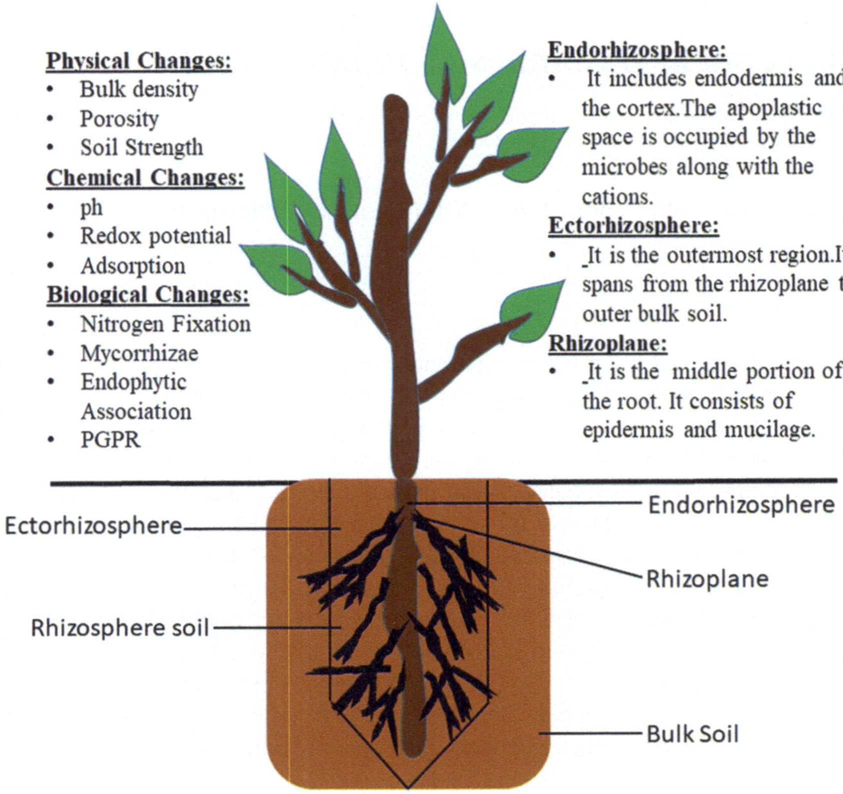

Fig. 14.1 Schematic representation of different processes in rhizosphere

et al. 2008; Berendsen et al. 2012). Hiltner also stated that rhizosphere is essentially a root-encompassing soil which is affected by root exudates (Hartmann et al. 2008). Hence, it is of most extreme significance not just because of microbial root inhabitants for plant development and well-being yet additionally as a defense mechanism against soil-borne pathogens (Weller et al. 2002).

The microbial activity that occurs in and around the underlying roots of the plants, i.e., the rhizosphere, is critical since it contributes towards plant defense responses, nutrient uptake, and so on. The microbial networks that are available inside the rhizosphere include symbionts, mutualists, antagonists, etc. (Singh et al. 2019; Prasad et al. 2020). The region specific around the roots cannot be just characterized as a rhizosphere; rather it reflects a zone of gradient spreading over the root with the microorganisms and physical and chemical factors. The biochemical elements in rhizosphere influence us to comprehend the diverse association of living systems, which has major effect on yield profitability. Numerous interactions like among the microbes themselves and also with plants are examined inside and out for better profitability with long-term sustainability.

This chapter centers around the different sorts of plant-microbe interactions since the underlying roots of the plants release diverse kinds of substances which help in attracting microorganisms. These microorganisms can be either advantageous, neutral, or pathogenic to plants (Badri et al. 2009b).

14.1.1 Plant-Microbe Interaction

The interaction between the microorganisms present and the plants is vital not just in impacting the soil's biological, chemical, and physical processes but on the other hand is directly associated with growth promotion, stress alleviation, biocontrol, nutrition, etc (Varma et al. 2019, 2020).

There are various advantages related with plants because of the microorganisms. Some of them include disease suppression, growth development, increment in crop yield, increase in stress resistance, etc. (Lugtenberg and Kamilova 2009; Yang et al. 2009). Therefore, the microbiome of the plant forms one of the most essential factors in determining the well-being and profitability of the plant.

14.1.2 Factors Influencing Plant-Microbe Interaction

The microbial groups including algae, nematodes, protozoa, bacteria, fungi, etc. which establish the rhizosphere help in influencing the overall development of the plant and also its health. There are various factors which influence the plant-microbe interaction (Fig. 14.2). The microorganisms in charge of influencing the plant well-being and development unfavorably are bacteria, oomycetes, pathogenic parasites, and so forth. The microorganism which has a positive connection incorporates the nitrogen-fixing bacteria, endomycorrhizal and ectomycorrhizal organisms, PGPR, etc. (Prasad et al. 2015). Due to the processes that decide the structure, dynamics, and development of the rhizosphere, microflora has pulled a genuine concern for scientists from various disciplines, and hence it can be exploited for the progression of new methods to propel the development of the plant and also its well-being.

14.1.3 Root Exudation

The roots emit certain chemicals which are called as root exudates. Thus with the assistance of root exudation, such as ions, water, free oxygen, mucilage, enzymes, metabolites, etc., the development of a variety of microorganisms can be supported

Fig. 14.2 Factors influencing plant-microbe interaction

(Nardi et al. 2000; Walker et al. 2003). The root exudates can be divided into low and high molecular weight compounds (Fig. 14.3). This further aides in the accompanying:

- Nutrient uptake
- Promotion of the growth of the plants
- Alleviation of stress and suppression of the disease (Jung et al. 2012)

For the import and export of various compounds into the rhizosphere, plants use a range of diverse types of transport mechanisms. The mechanism through which the plant roots release root exudates can be either an active or a passive process.

The passive process which depends on polarity of exuded compounds, membrane polarity, cytosolic proteins, etc. basically includes the low weight molecular organic compounds. These transporter proteins are linked with the transport of different types of compounds into the rhizosphere (Weston et al. 2012).

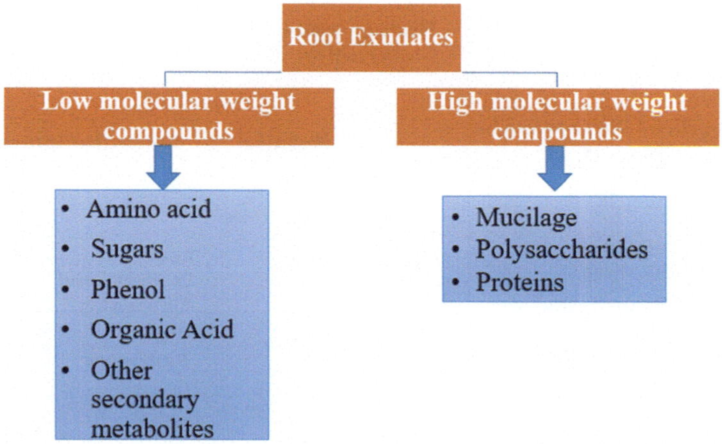

Fig. 14.3 Classification of root exudates

With the help of many transporter proteins such as ATP-binding cassette (ABC) transporter, the multidrug and toxic compound extrusion (MATE) family, the major facilitator superfamily, and the aluminum-activated malate transporter family, the plant root cells secrete other compounds such as polysaccharides, secondary metabolites, proteins, etc. (Weston et al. 2012).

From the above transporter protein, MATE is responsible for exporting different types of substrates across the membrane with the help of electrochemical gradient of ions (Badri et al. 2008, 2009a; Loyola-Vargas et al. 2007; Sugiyama et al. 2008; Yazaki 2005; Reddy et al. 2012; Weston et al. 2012).

14.1.4 Classification of Plant-Microbe Interaction

Over the most recent couple of years, different sorts of rhizospheric interactions have been considered which predominantly incorporates the interactions among the plants, interactions between the microbe and the plant, and the plant-faunal interactions (Badri et al. 2013; Broeckling et al. 2008; Chaparro et al. 2013; Doornbos et al. 2012). These interactions can be either positive, negative, or pathogenic based on the conditions of the environment (Fig. 14.4). Some of the positive plant-microbe interactions are mentioned below:

- Nitrogen fixation
- Mycorrhizal interaction
- Endophytic interaction
- PGPR

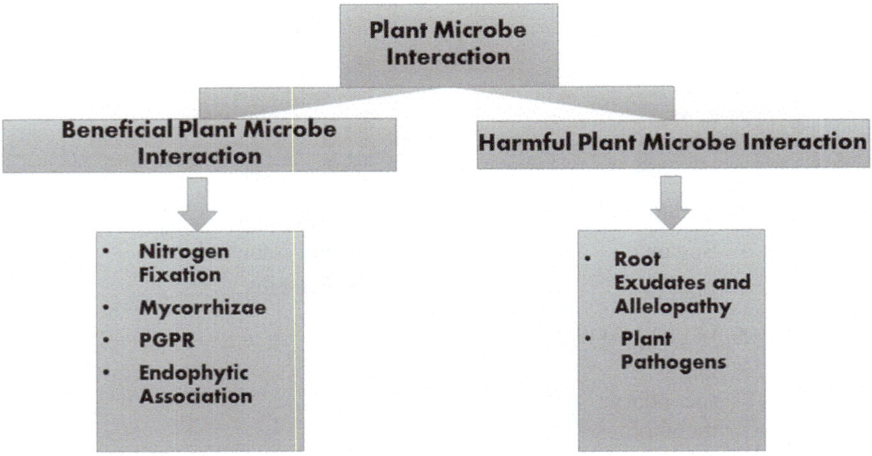

Fig. 14.4 Classification of plant-microbe interaction

14.2 Beneficial Plant-Microbe Interaction

14.2.1 Nitrogen Fixation

The most essential element of all forms of life, nitrogen, which is also the most important nutrient for the productivity and growth of plants constitutes about 78% of the atmosphere, but it is still not directly available for plants.

Nitrogen fixation is required for biosynthesis of DNA and RNA nucleotides and also for biosynthesis of amino acids for proteins. The atmospheric nitrogen does not react readily with other compounds to form new compounds, and therefore this nitrogen is relatively inert due to the strength of its triple bond. Hence, nitrogen fixation is performed by various microorganisms which functions as the process of freeing up the N atoms from their diatomic form.

It is defined as a process wherein the molecular nitrogen present in the Earth's atmosphere is converted into ammonia so that the utilization of organic nitrogen is possible in biological process. Hence, to make the nitrogen available to the plants, biological nitrogen fixation (BNF) is performed by nitrogen-fixing bacteria in which the nitrogen is converted into ammonia by an enzyme called as nitrogenase (Arora et al. 2012).

Biological nitrogen fixation occurs at mild temperatures by nitrogen-fixing bacteria, which are comprehensively distributed in nature. The nitrogenase complex is a compound which aids in nitrogen fixation. Structure of nitrogenase is delegated as a two-section metalloenzyme containing (1) dinitrogenase reductase (iron protein) and (2) dinitrogenase which involves a metal cofactor.

Dinitrogenase reductase helps in donating electrons which have a high reducing power, while dinitrogenase utilizes these electrons to reduce nitrogen to ammonia. This procedure devours huge amount of energy as ATP. The nitrogen fixation thus

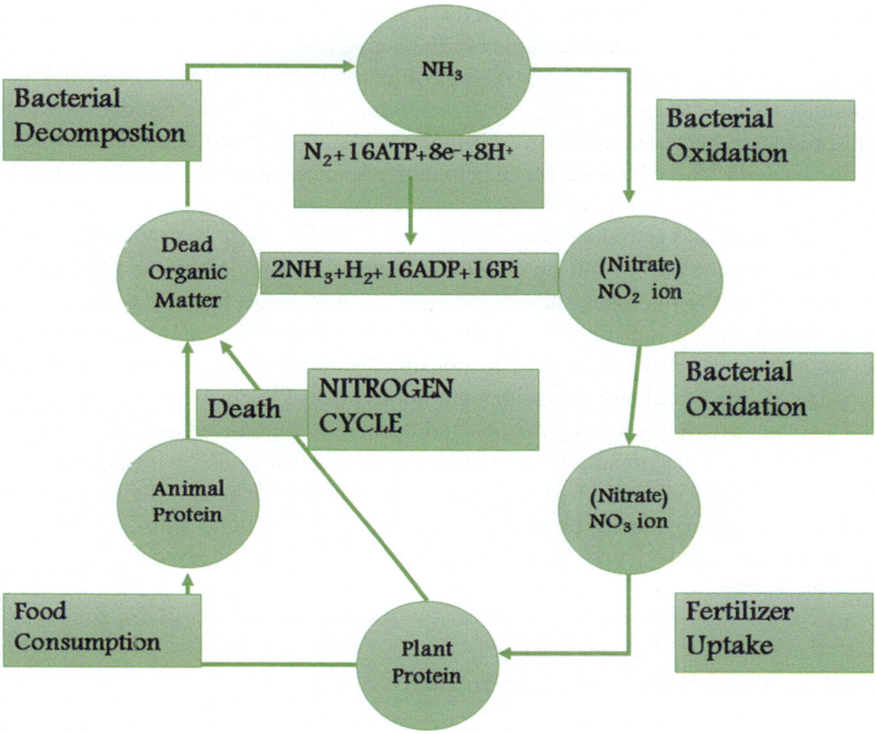

Fig. 14.5 Nitrogen cycle

requires nitrogenase (nif) which is sensitive to oxygen; and also to keep oxygen from hindering nitrogen fixation and hence simultaneously providing adequate oxygen to the bacteroides inside the nodule to breathe (Prasad et al. 2020).

The nif genes incorporate structural genes that activate molybdenum Fe protein and other regulatory genes that are connected with the synthesis and function of the catalyst and are being accessible in both symbiotic and free-living systems. Since nitrogen fixation is a high energy-using process, requiring 16 moles of ATP (Fig. 14.5) for each mole of nitrogen that is reduced, it would be advantageous if bacterial carbon sources are composed towards oxidative phosphorylation, which lead to ATP synthesis, rather than the synthesis of glycogen and consequently using glycogen as the store of energy.

Nitrogen fixation takes place by two mechanisms: symbiotic and non-symbiotic.

14.2.1.1 Symbiotic Nitrogen Fixation

The biological nitrogen fixation discovered by German agronomist Hermann Hellriegel and Dutch microbiologist Martinus Beijerinck takes place under the influence of enzyme nitrogenase which has a sulfur- and an iron-containing cofactor which includes a heterometal complex in the active site.

This fixation is carried out by two types of microorganisms, which are symbiotic in nature, such as *Rhizobium* and *Bradyrhizobium*. *Rhizobium* is a rapidly growing acid-producing bacteria, and *Bradyrhizobium* is a slow-growing and non-acid-producing bacteria. These bacteria hence help in the development of root nodules which in turn serve as nitrogen fixation (Marschner 1995).

In this type of symbiotic relationship, the fixed nitrogen is provided by the bacteria, while the energy for the growth of bacteria is provided by the plant itself. The plants and the microbes share a mutualistic relationship because of symbiotic nitrogen fixation in which there is an entry of the microbes followed by the formation of root nodules. The rhizobacteria is responsible for the symbiotic interaction which is made possible by colonization with leguminous plants (Ahemad and Kibret 2014).

The plant growth-promoting rhizobacteria generally introduced as symbionts are *Rhizobium*, *Bradyrhizobium*, *Sinorhizobium*, and *Mesorhizobium* with leguminous plants and *Frankia* with non-leguminous trees and shrubs (Gaby and Buckley 2012). The other microorganisms performing this type of symbiotic nitrogen fixation include those under the genus of *Alnus*, *Ceanothus*, *Myrica*, *Coriaria*, *Elaeagnus*, etc.

14.2.1.2 Non-symbiotic Nitrogen Fixation

Certain examples of nitrogen fixation by free-living bacteria are *Azolla*, *Anabaena*, *Spirillum*, *Azotobacter*, and *Beijerinckia.* Some bacteria receive energy from plant residues, are heterotrophic, such as *Clostridium*, and are also able to fix nitrogen. Other examples of non-symbiotic nitrogen fixation include diazotrophs which help in stimulating non-legume plant growth such as rice and radish.

Genera such as *Azoarcus*, *Acetobacter*, *Burkholderia*, *Azospirillum*, *Enterobacter*, *Gluconacetobacter*, and *Pseudomonas* are involved in non-symbiotic nitrogen fixation (Ahemad and Kibret 2014).

14.2.2 Mycorrhizae

The term mycorrhiza is derived from a Greek word (myos—fungus, rhiza—roots). Mycorrhizal associations are found in more than 80% of angiosperms and almost all gymnosperms (Varma et al. 2017a, b, c). A mutualistic relationship between the higher plants and organisms is mycorrhizae, which is the most fundamental groups of soil microorganisms that change broadly in capacity and structure (Morgan et al. 2005; Prasad et al. 2017). Symbiotic association as ancient as land plants, i.e., mycorrhizae, is assumed to be solely beneficial (Pirozynski and Malloch 1975; Wagner and Taylor 1981). The mycorrhizal association can be basically characterized into endomycorrhizae or arbuscular mycorrhizae and ectomycorrhizae. These associations help in expanding the surface of roots and proficiency of mineral uptake. VAM (vesicular arbuscular mycorrhizae) fungi are available inside the underlying roots of the plants, and this sort of fungi is named as obligate plant symbionts. The impact of VAM fungi can extend from parasitic to mutualistic (O'Bannon et al. 1980; Modjo and Hendrix 1986). To diminish the reliance on chemical fertilizers and furthermore to improve the yield of harvest, broad investigations are being performed in the zone of inoculation of crops with VAM. Plants which are capable of growing in mildly nutrient stress condition make a better VAM host since more soluble carbohydrates are released into the root exudates.

There are hypothetical reasons to estimate that fertilizing soil may choose for VAM fungal strains that are subpar mutualists or even parasites. It has been seen that nutrient-stressed plants release soluble carbohydrates in root exudates and improve VAM hosts than unstressed plants. A strong pressure will be applied on VAM fungal population when fertilization causes host plants to circulate less carbohydrates to root exudates. As a result of fertilization, the abundance of VAM fungal strains that most compellingly procure host carbohydrate will augment to the cost of less aggressive strains. Comparable properties that make a VAM parasite productive in a low-carbohydrate condition moreover decline its mutualistic impacts. To be explicit, an effective fungus could acquire carbohydrates that the host plant has not assigned to it and, therefore, parasitically organize its own growth and development without adding to the well-being of the plant. Therefore, less valuable VAM developments could be chosen in fertilized soils where plants obtain little from VAM associations. The harmfulness of various parasites diminishes with time, and over an adequately long time period, various parasitic associations form into progressively stable mutualistic associations (Fig. 14.6).

A mycorrhiza incorporates both a plant and a parasite, so an "adapted mycorrhiza-soil complex" can be portrayed as a dynamic system in which both plant and fungal systems have adjusted, and continue modifying, to the soil conditions and to one another so that the mycorrhizal relationship inside the complex ends up being logically more mutualistic after some time. There are various advantages of mycorrhizal relationship. Most common association is AMF (arbuscular mycorrhizal fungi), particularly in terrestrial crops (Harrier and Watson 2003). AMF growth when related with a host plant is very intricate since it comprises of numerous developmental stages which are as per the following:

- Spore germination
- Differentiation of the hyphae
- Penetration of the roots
- Intercellular growth
- Intracellular arbuscular formation
- Exchange of nutrients (Harrier and Watson 2003)

Certain examples of AMF come under the genera of *Gigaspora*, *Glomus*, *Scutellospora*, etc. (Bagyaraj 2011). The principle function of mycorrhizae is separating more measure of nutrients and water from the soil and consequently plays an important job in incitement of microbial activity and aggregation process.

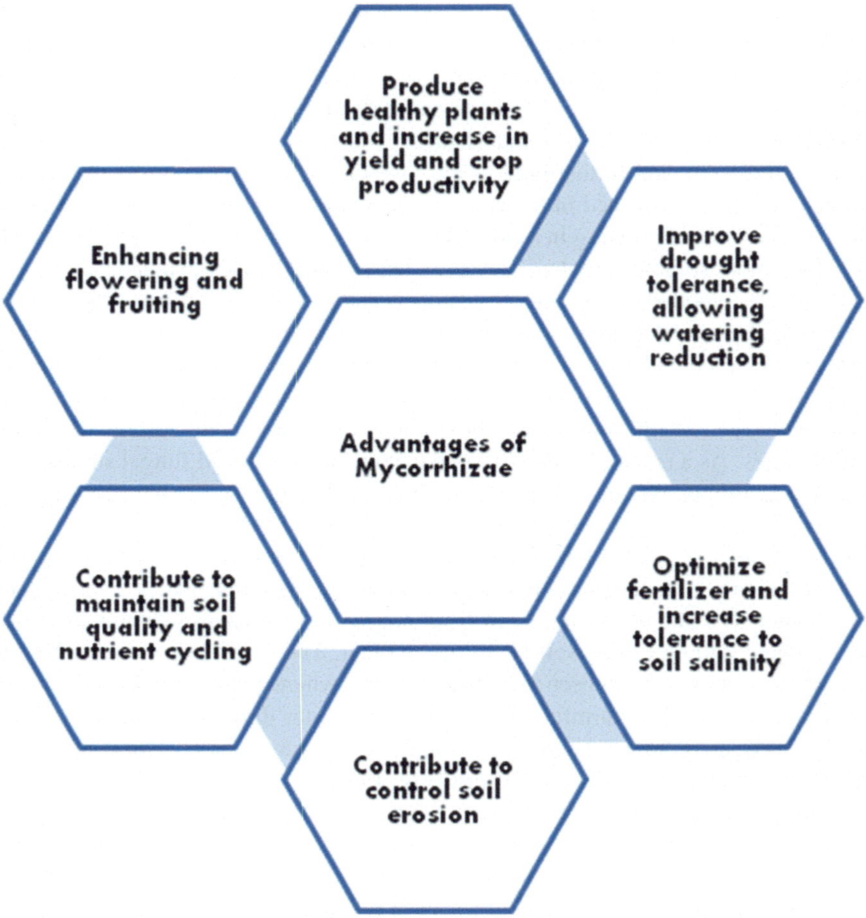

Fig. 14.6 Advantages of mycorrhizae

14.2.2.1 Mechanism and Role of Mycorrhizae in Growth of Plants

AM fungi are viewed as obligate biotrophs which are subject to be dependent on plants for survival. The symbiotic association occurs in various steps as shown below:

14.2.2.1.1 Search for Roots of Host Plants by Fungi

Some chemicals released from tissues or specialized cells which help in the induction of function in cells or tissues which are present nearby of the same or different organism are called bioactive molecules. Examples of such molecules include strigolactones which are secreted by the roots. These strigolactones are responsible for initial recognition of host plants by fungi. They also help in the stimulation of growth of AM along with its branching. Mycorrhizal factors (Myc) are secreted by the fungi which help in communication between nitrogen-fixing bacteria and AM fungi. Seven genes called as SYM genes are further induced which help in establishing the AM interactions. Cytosolic Ca secretion can be induced in root cells when MFR (Myc factor receptor) comes in contact with Myc signals. Due to the above interaction, a protein which codes for a receptor-like kinase responsible for recognition of AM fungal signal called as SYMPK (second membrane protein kinase) is activated. The main function of SYMPK is transduction of these signals into the nucleus from cytoplasm by phosphorylation of an unknown substrate by kinase. A rapid signal transduction into the nucleus can be activated by localizing all the downstream elements present in the cytoplasm. Calcium channels and transporters present for the entry into nucleus help in oscillations of the calcium which are decoded by CaMK (calmodulin-dependent protein kinase). It helps in the phosphorylation of CYCLOPS (SYM gene product). This finally leads to other gene regulation followed by root colonization.

14.2.2.1.2 Penetration of Fungi and Mycorrhizal Symbiosis Establishment

An interaction is seen between the root of the host and the hyphae of the fungi in which fungal hyphae help in the formation of hypopodium by propagation of the hyphae into the host root. The above process is known as the primary step of colonization which is generally followed by a formation of PPA (pre-penetration apparatus) which helps in the development of fungi inside the plant. Development of arbuscules which help in the accommodation of fungi into the host cell cytoplasm is the last step of this symbiotic process. The principle function of these arbuscules is transport and obtaining of nutrients. Numerous sorts of proteins and genes are included for the procedure of take-up of nutrients and consequently it helps in the achievement of symbiotic interaction.

14.2.3 Endophytic Association

It is hard to fuse bacterial components which are non-resident into adjusted and established microbial networks, and subsequently numerous endeavors have been made in the introduction of bacteria which are beneficial into the rhizosphere (Brockwell et al. 1988; Thies et al. 1991). One of the methodologies in the establishment of the preselected beneficial organisms in rhizosphere is by the introduction of early establishment of selected communities of endophytic microorganism underlying the root frameworks. The term endophyte predominantly refers to fungi (Carroll 1988). Endophyte basically includes bacteria or fungi which invades the plant tissues and causes no symptoms of any disease (Wilson 1995). The idea of recovery of bacterial population from root cortex and endodermis of plants has caused the penetration and colonization of root tissues.

Darbyshire and Greaves with the help of Nicolson and old proposed the incorporation of endophytic bacteria into the bacterial rhizosphere. To acquire a nonstop apoplastic pathway from the root epidermis to the shoot, the root cortex must be fused into the soil root microbial condition (Peterson et al. 1981). Hence, for fruitful manipulation of endophytic bacteria, many variables are responsible such as ability to choose, incorporate, and maintain beneficial microbial populaces.

Endophyte derived from a Greek word meaning "in the plant" can be defined as a microbe that lives within the plant cells or in the tissue without having any adverse effect on them and hence falls under the spectrum of microbes such as bacteria, fungi, etc. (Kobayashi and Palumbo 2000; Stone et al. 2000; Marler et al. 1999; Peters 1991). Endophyte in the early developmental stages can be used for mutualistic endophytic bacteria, pathogenic endophytic algae, parasitic endophytic plants, and pathogenic bacteria (Chanway 1996; Adhikari et al. 2001; Bai et al. 2002). Therefore dependence on host plant for its nutrition and protection is seen in many ways. Endophytes perform various types of functions which are as follows:

- Acceleration of seedling emergence
- Promotion of plant growth
- Yield enhancement
- Bioremediation
- Improvement in nutrient cycling
- Reduction of proliferation of pathogen (Arnold 2007)

Certain examples of bacterial endophytes come under the genera of *Azospirillum*, *Gluconacetobacter*, *Herbaspirillum*, *Achromobacter*, *Acinetobacter*, etc. (Coombs and Franco 2003).

Apart from bacterial endophytes, fungal endophytes are also present which come under the taxa of *Exophyla*, *Cladophialophora*, *Periconia*, etc. Among the above, *Basidiomycetes* were proved to be the most dominant endophyte (Jumpponen 2001).

Colonization of endophytes can be seen in various parts of the plants such as roots, stem, leaves, bark, seed, floral parts, etc. (Ganley et al. 2004). Therefore, the whole microbiome is associated with diverse types of endophytes which are solely responsible in the nutrition of plants and also function in eliciting the defense mechanism through modulation of gene expression (Ganley et al. 2004).

The most generally considered endophytic bacteria incorporates nitrogen fixers for example *Rhizobium* which grows successfully in nitrogen restricted agricultural soils due to its significant contribution in zones of induction of root nodules on legumes plants of agricultural significance and furthermore giving them fixed nitrogen. *Rhizobium* is additionally in charge of colonization of underlying roots of certain cereal crop plants and therefore advancing the yield and development.

14.2.3.1 Mechanism of Endophytes in Growth of the Plants

The mechanism of endophytes in the stimulation of plant growth can be divided into direct and indirect mechanism (Fig. 14.7).

The direct mechanism includes the following:

Biological Nitrogen Fixation
The second most imperative biological process after photosynthesis is viewed as biological nitrogen fixation (BNF) which is confined just to prokaryotic life forms. The endophytic bacteria help in fixing nitrogen without the nodule-like structure formation. The interior of the plants, low in oxygen, and rich in carbon help in maintaining a favorable environment for nitrogen fixation, while the bacterial and fungal endophytes which live in the interior of the plants are protected from competition, and hence direct nutrition is supplied from the host plants (Prasad et al. 2020).

Phosphate Solubilization
One of the crucial macronutrients required for plant advancement in higher measure is phosphorus (P). It isn't found in a form that is readily accessible for plant uptake. The deficiency of phosphorus can be cured by the utilization of chemical phosphate fertilizers or biofertilizers. Insoluble inorganic phosphorus compounds can be made dissolvable by a few endophytic bacteria & fungi; consequently make them open for plant uptake (Malla et al. 2004). Such microorganisms are called as PSM (phosphate-solubilizing microorganism).

Siderophores Synthesis
There are some iron-binding compounds which are of low molecular weight that are made by many microorganisms present in the rhizosphere under iron deficiency

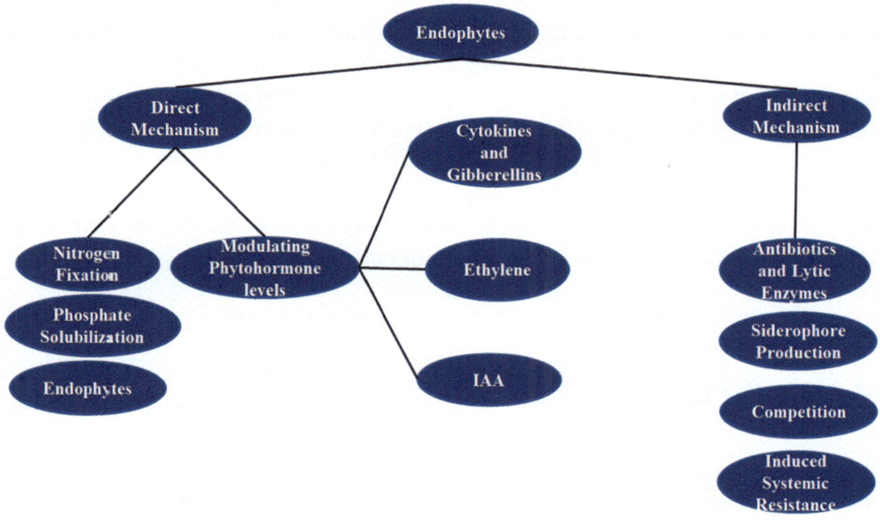

Fig. 14.7 Characterization of endophytes (on the basis of mechanisms)

conditions called as siderophores which are integrated by microorganisms that assist in chelation of Fe^{3+} and transport it back to their cells where it ends up open for advancement of the microbes (Das et al. 2007). Endophytic microbes have been accounted to make siderophores, a mechanism which is significant for their advancement. Siderophore delivering bacteria can help in the improvement of their host plant either as biofertilizers (i.e., increase in iron availability in the encompassing region of their host plant roots) or by their biocontrol activities.

Production of Phytohormones (IAA)

Different fungal and bacterial species have the ability to make diverse plant development controllers or phytohormones, for instance, auxins, gibberellins, cytokinins, ethylene, and abscisic acids. Different genera of microorganisms are in charge of the generation of indole acetic acid. Endophytic bacteria and fungi are moreover in charge of synthesis of indole acetic acid. Indole-3-acetic acid related with cell division and differentiation in the long run expands root length and root hair abundance giving more sites to ailment and nodulation. This difference in root structures upgrade root surface area and thus increase plant ability to ingest more nutrients, which hence help in incitement of plant growth and development. The indirect growth development of the host plants by their endophytes occurs through concealment of phytopathogenic microorganisms in a methodology named as biological control, in which the endophytes make biocontrol attributes lethal to the pathogenic microorganisms or fight with them for root colonization sites and supply of nutrients.

The indirect mechanism incorporates the following:

Biological Control of Plant Pathogens

Soil-borne plant pathogens are a significant hazard to rural improvement and profitability around the globe. A couple of plant infection control procedures have been put into action to guarantee crops against a wide extent of phytopathogens, but they come with a variety of side effects. An engaging strategy to control plant infection is the usage of plant rhizosphere-related beneficial microorganisms, which are called biological control agents (BCAs). Various biological controls are known to reduce the rate of plant disease. *Bacillus* and *Pseudomonas* spp. are the common bacterial control agents; however, *Trichoderma* spp. is the most essential fungal biological control agent. Bacterial endophytes in a likewise manner show antagonistic activities against a wide scope of parasitic pathogens.

14.2.4 PGPR

Plant growth-promoting rhizobacteria (PGPR), a heterogeneous group of beneficial root-related microorganisms, are fundamentally in charge of improving the well-being and the development of plants with the guide of different mechanisms. PGPR can be categorized on the basis of its location (Table 14.1). Numerous soluble and

Table 14.1 Characterization of PGPR on the basis of its location with examples

	Extracellular PGPR	Intracellular PGPR	References
Location	Rhizosphere/rhizoplane	Inside the specialized nodular structure of root cells	Martínez-Viveros et al. (2010)
Examples	*Bacillus*, *Burkholderia*, *Agrobacterium*, *Erwinia*, *Arthrobacter*, *Azotobacter*, etc.	*Allorhizobium*, *Bradyrhizobium*, *Mesorhizobium*, *Rhizobia*, *Frankia*	Ahemad and Kibret (2014) Bhattacharyya and Jha (2012)

volatile compounds are discharged by PGPR, which are embroiled in cell signaling, antibiosis, induction of tolerance, and opposition in plants against biotic and abiotic stresses. PGPR upgrade plant development and phytoremediation effectiveness as follows:

Secretion of plant growth-promoting substances such as indole-3-acetic acid (IAA), cytokinin, and gibberellins

1. Excretion of stress-alleviating metabolites such as 1-aminocyclopropane-1-carboxylic acid deaminase (ACC deaminase)
2. Alteration of the metal bioavailability by secretion of certain chelators such as siderophores and organic acids, altering soil pH
3. Solubilization of nutrients such as phosphorus and nitrogen fixation

The reduction of plant toxicity is done by plant growth-promoting bacteria and fungi and hence promote

1. Indirectly through the induction of defense mechanisms against phytopathogens
2. Directly by
    - Mineral solubilization nutrients (nitrogen, phosphate, potassium, iron, etc.)
    - Secretion of certain enzymes (e.g.,1-aminocyclopropane-1-carboxylate deaminase) and production of plant growth-promoting substances

Plant growth-promoting rhizobacteria as the name proposes helps in plant development as well as plays a vital job in establishing of plants and their development in nutrient-insufficient conditions (Bloemberg and Lugtenberg 2001; Prasad et al. 2005).

The main functions of PGPR are as follows:

- Reduction of agrochemical usage
- Root hair proliferation
- Seedling emergence increase
- Early nodulation
- Enhancement of the surface area of leaf
- Increasing indigenous plant hormone levels (Cooke et al. 2002; Singh et al. 2007)

Because of the above advantages of PGPR, the demand for its biofertilizer has been increasing at alarming rate day by day since it involves minimum use of chemicals.

Table 14.2 Different plant growth-stimulating phytohormones produced by PGPR

Phytohormones	PGPR	References
Gibberellin	*Acetobacter diazotrophicus*	Bastián et al. (1998)
	Herbaspirillum seropedicae	
	Bacillus licheniformis	Gutiérrez-Mañero et al. (2001)
	B. pumilus	
	B. cereus MJ-1	Joo et al. (2004)
	B. macroides CJ-29	
	B. pumilus CJ-69	
IAA	*Agrobacterium* sp.	
	Alcaligenes piechaudi	
	Comamonas acidovorans	Kaushik et al. (2000)
	Azospirillum brasilense	
	Aeromonas veronii	Mehnaz et al. (2001)
	Enterobacter cloacae	
	Enterobacter sp.	Mirza et al. (2001)
	Comamonas acidovorans RC41	Erturk et al. (2008)
	Paenibacillus polymyxa RC05	
	Bacillus RC23	
	Bacillus simplex RC19	
	Bacillus RC03	
	Bacillus megaterium RC01	
Cytokinin	*Paenibacillus polymyxa*	Timmusk et al. (1999)
	Pseudomonas fluorescens	García de Salamone et al. (2001)
ACC deaminase	*Pseudomonas putida*	Mayak et al. (1999)
	P. cepacia	Cattelan et al. (1999)
	Enterobacter cloacae	Saleh and Glick (2001)
	Pseudomonas brassicacearum Am3	Belimov et al. (2007)
	Variovorax paradoxus 5C-2	Belimov et al. (2009)
	Pseudomonas putida biovar B	Rodriguez et al. (2008)
	P. putida N21	
	P. aeruginosa N39	Zahir et al. (2009)

Therefore, these rhizospheric microbes increase the nutrient uptake of plants and also help in the secretion of plant growth-promoting hormones such as IAA, cytokines, GA, ethylene, etc. (Kloepper 1992). There are various plant growth-stimulating phytohormones produced by PGPR (Table 14.2).

The development of the plants can likewise be improved by inhibition of explicit plant pathogen through its biocontrol action.

The mechanisms involved in PGPR (plant growth-promoting bacteria) which has increased overall significance have both direct and indirect impacts.

Phytohormone generation is seen by PGPR during direct effects. These microorganisms can be used for sustainable agriculture and also for the future use (Siddiqui et al. 2007; Dubeikovsky et al. 1993). Recent investigations affirm that when the seeds are treated, bacteria which are non-pathogenic in nature, for example, *Agrobacterium*, *Bacillus*, *Streptomyces*, *Pseudomonas*, *Alcaligenes*, etc., instigated root development in a few plants as a result of regular auxin production of bacteria (Srinivasan et al. 1996). Despite the fact that the mechanism is not totally understood, root induction by PGPR is the accepted result of phytohormones, for example, auxin production, ethylene synthesis inhibition, and mineralization of nutrients by PGPR. Considering the different interactions between the various hormonal signaling pathways in plants, it is difficult to assess which of these pathways is the basic focus of PGPR. This could represent the particular morphological changes seen, for example, horizontal root prolongation and root hair advancement.

One of the trademark effects of PGPR is an expanded elongation rate, and furthermore the initiation rate, of lateral roots bringing about more extended root framework architecture.

There are different uses of PGPR which can be comprehensively grouped into two classifications:

1. Improvement of the plant nutrient acquisition
2. Induction of plant resistance

Moreover, the PGPR provides cross-protective properties such as resistance towards pathogen and abiotic resistance by the various hormonal pathways present in the plants.

Since the last few years, PGPR finds its various applications in terms of research as a result of increment in the yield of crops, plant growth, being less unsafe to the earth, and furthermore lessening the expense of chemical fertilizers. There are numerous mechanisms through which PGPR helps in the growth of plants. These mechanisms can be basically characterized by direct and indirect mechanism. The direct mechanism can be as follows:

1. Production of plant hormones
2. Nitrogen fixation
3. Phosphorus solubilization

The indirect mechanism can be as follows:

1. Lytic and antibiotic enzymes
2. Induced resistance
3. HCN production

The direct mechanisms such as plant hormone production, nitrogen fixation, and phosphorus solubilization have already been discussed in the above sections. A detailed study about the types of phytohormone responsible for the stimulation of plant growth using PGPR is given below.

14.2.4.1 Indole Acetic Acid

Also called IAA, it is extensively studied and the most common auxin whose main function is cell extension, elongation, differentiation, and division. However, the developmental process of plants can be interfered by IAA which is released by rhizobacteria since the endogenous pool can be changed by the IAA acquisition which is secreted by the bacteria present in the soil. The function of IAA can be stated as follows: axillary bud promotion, bud formation, apical dominance, and lateral and adventitious root development.

The level of IAA synthesis can also be regulated with the help of tryptophan, the precursor of IAA. This is done with the help of inhibition of anthranilate.

14.2.4.2 Ethylene

This is a hormone which is endogenously produced and is responsible for induction of various physiological changes at the molecular level. It has many types of biological activities and is also one of the most simplest gaseous hormones. Ethylene is formed when methionine that is present in the cells is broken down to form ADOMET (*S*-adenosylmethionine) ACC synthase followed by the formation of ACC (1-aminocyclopropane-1-carboxylate), eventually leading to the formation of ethylene. Amid unfavorable conditions, the synthesis of ethylene increments and hence adversely affects the development of underlying roots of plants. PGPR comprises of an enzyme called as ACC deaminase which helps in ethylene synthesis by development of alpha ketoglutarate and ammonia from ACC. Hence, by decreasing the amount of ACC, the root ethylene production can be decreased thereby alleviating the repressing effect of ethylene on growth of roots. For support of ideal development and improvement of plants under unfavorable condition, it is fundamental to introduce genes of ACC deaminase for the regulation of level of ethylene present in the plants.

14.2.4.3 Gibberellin and Cytokinin

Gibberellin and cytokinin are both phytohormones having varied functions in plants. The main functions of gibberellin are as follows:

- Stimulation of alpha amylase followed by starch hydrolysis that is present in seeds into glucose
- Stem elongation, dormancy, germination
- Act as chemical messengers by breaking dormancy

The main functions of cytokinin are as follows: cell differentiation and delay of senescence.

PGPR helps in production of cytokinin and consequently helps in the upgradation of plant development to improve the quality and yield of crop.

The limit of PGPR to lessen the adverse effects of plant pathogens on the development can be named as indirect mechanisms which are as follows:

14.2.4.3.1 Production of Lytic Enzymes and Antibiotics

This incorporates the lytic proteins synthesis including cellulases, chitinases, proteases 1,3-glucanases, and lipases that can lyse a fragment of the cell wall of various pathogenic life forms. Because of the expansion in plant pathogens, antibiotic production is observed. Regardless, overreliance on antimicrobial-producing microorganisms as biocontrol agents may be an impediment because of the resistance against antibodies. The production of at least one antibiotic is the mechanism most associated with the limit of PGPR to act as antagonist agents against phytopathogens. This arrangement of antibiosis is to make low molecular weight compounds that are poisonous and essential to major enzymes and metabolism of various microorganisms in this manner hinder the development.

14.2.4.3.2 Induced Systemic Response

Induced systemic resistance (ISR) is another type of indirect mechanism in which an increased resistance at specific sites of plants is involved. When an assault of pathogenic agent occurs, the defense mechanism of ISR is activated, and hence ISR is not explicit against specific pathogen and, however, encourages the plant to control diseases. ISR includes ethylene and jasmonate signaling inside the plant, and these hormones induce the host plant's defense reactions to a range of pathogen.

14.2.4.3.3 Production of HCN

The development of the plant can be stifled by the plant root surfaces colonization by the pernicious rhizobacteria which act as biocontrol agents of weeds. Cyanide being toxic is conveyed by most microorganisms including algae, bacteria, plants, and fungi as a strategy for survival by competing with the partners. No negative effect on the host plants is seen by cyanide-producing bacterial strain inoculation and host-explicit rhizobacteria which can additionally act as biological weed control agents. Additionally, the secondary metabolite that is conveyed, which is a powerful agent for the biocontrol of weeds, is HCN whose synthesis is completed by *Bacillus* and *Pseudomonas* species. The restraint of the HCN and vitality supply to cell is performed by electron transport chain. The inhibition of suitable working of enzymes and characteristic receptors is finished by PGPR.

14.2.5 ACC Deaminase

Different types of mechanisms are used by the PGPR for the plant growth promotion. The key characteristic in encouraging plant development is the presence of the catalyst 1-aminocyclopropane-1-carboxylate (ACC) deaminase which is exclusively in charge of cleavage of ACC (ethylene forerunner) into ammonia and alpha-ketobutyrate (Fig. 14.8). If the levels of ACC are decreased, then subsequently the levels of ethylene are also decreased by the ACC deaminase-producing organisms.

On increasing the levels of ethylene, the growth of the plant is inhibited and can also cause the death of the plant. By checking the production of either ammonia or a-ketobutyrate, which are the results of ACC, enzymatic action of ACC deaminase can be measured.

Exactly when plants are presented to conditions that compromise their ability to endure, a comparative mechanism that produces ethylene for development produces "stress ethylene," which may be described as an increment in ethylene biosynthesis related with natural and ecological stresses and pathogenic attack. Ethylene is synthesized from *S*-adenosyl-L-methionine (AdoMet) by the intermediate ACC. While working at the ethylene biosynthesis pathway, it was found that when ACC

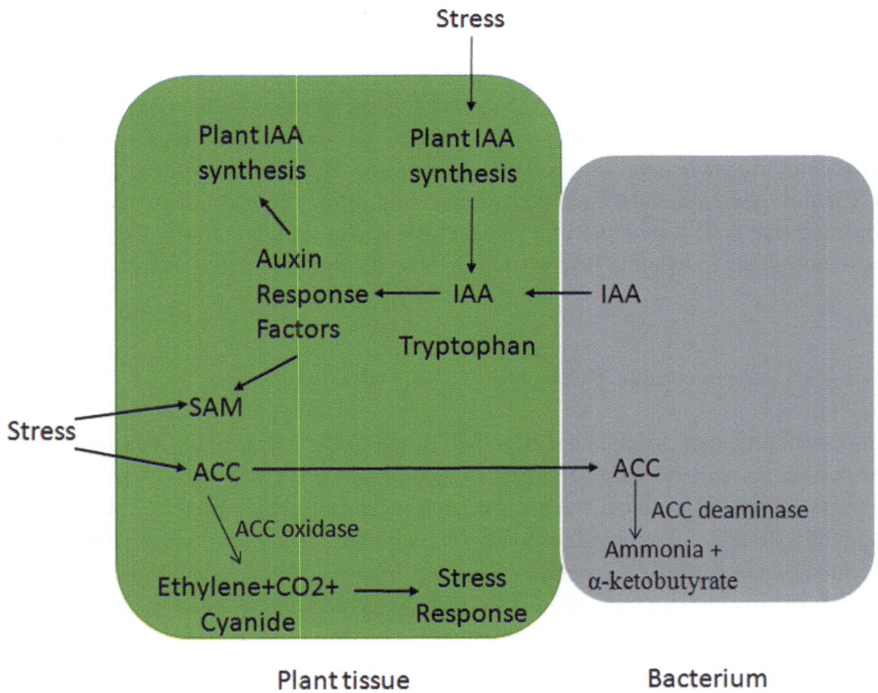

Fig. 14.8 A schematic model of how plant growth-promoting bacteria that both produce ACC deaminase and synthesize IAA may facilitate plant growth (*SAM S*-adenosylmethionine, *ACC* 1-aminocyclopropane-1-carboxylic acid)

was connected to various plant organs, an increment in ethylene generation was gained. Hence, ACC, as a key intermediate that associated the methionine cycle and ethylene biosynthesis, was considered to be the immediate precursor of ethylene biosynthesis with its level specifically controlling ethylene synthesis in plants.

ACC deaminase is found solely in microorganisms, and hence no microorganisms can be found wherein the synthesis of ethylene can take place via ACC (Fukuda et al. 1993). ACC deaminase is an enzyme which is multimeric in nature, which means it can be either homodimeric or homotrimeric with a subunit of molecular mass of approximately 35–42 kDa. It is a sulfhydryl catalyst in which one PLP is firmly bound to every subunit. ACC deaminase is localized cytoplasmically, and thus the substrate ACC can be exuded by plant tissues and in this way taken up by an ACC deaminase-containing microorganism before it is cleaved (Glick et al. 1998). Consequently this enzyme and substrate relationship help in the Km estimations of ACC deaminase for ACC assessed at pH 8.5, in all instances inspected to be roughly 1.5–17.4 mM which demonstrates that the enzyme does not have any affinity for ACC (Honma and Shimomura 1978). It is seen that ACC levels in plants are generally in mM; along these lines in most plant tissues, the ACC concentration is less the Km of ACC deaminase for this substrate. This can be demonstrated by the Michaelis–Menten rate condition for compound catalyzed response; a small increment in the ACC focus will result in a parallel increase in the rate of ACC cleavage.

14.3 Conclusion

Amid the increment of existing plant species, a variety of frameworks has been adopted by plant systems to acclimate to unfavorable ecological conditions. In the adaptability procedure, the soil region enveloping the roots is at risk to various physical, chemical, and biological changes. The most detectable physical changes are root temperature, water-holding point of the underlying roots, and soil structure.

The chemical changes associated with adaptability of plants are pH, redox potential, root exudates, nutrient concentration, organic matter content, and allelopathy. Microbial association with root expects a basic role in nutrient availability to plants. The microorganisms that are useful to plants are *Rhizobium* or *Frankia* genera and mycorrhiza developments, which are prepared to develop a cooperative association with their host plant.

The plausibility of nutrients is extended by the association with these rhizobacteria or fungi. These progressions are accountable for nutrient solubility and plant accessibility. The size of physical, chemical, and organic changes varies with plant species, soil type, and biological factors and their affiliations.

In this way, rhizosphere changes are particularly confusing in nature and components, and complete data about them is still not available completely. More research is relied upon to appreciate or clear up these progressions and their relationship with plant advancement.

Thus, having a whole understanding of the plant microorganism communication is basic for the improvement of assurance procedures and reasonable yield generation. Their examination is important since they help in extending the yield of items and continuing soil ripeness besides improvement in harvests.

References

Adhikari TB, Joseph CM, Yang G, Phillips DA, Nelson LM (2001) Evaluation of bacteria isolated from rice for plant growth promotion and biological control of seedling disease of rice. Can J Microbiol 47(10):916–924

Ahemad M, Kibret M (2014) Mechanisms and applications of plant growth promoting rhizobacteria: current perspective. J King Saud Univ Sci 26(1):1–20

Arnold AE (2007) Understanding the diversity of foliar endophytic fungi: progress, challenges, and frontiers. Fungal Biol Rev 21(2–3):51–66

Arora NK, Tewari S, Singh S, Lal N, Maheshwari DK (2012) PGPR for protection of plant health under saline conditions. In: Maheshwari DK (ed) Bacteria in agrobiology: stress management. Springer, Berlin, pp 239–258

Badri DV, Loyola-Vargas VM, Broeckling CD, De-la-Peña C, Jasinski M, Santelia, Vivanco JM (2008) Altered profile of secondary metabolites in the root exudates of Arabidopsis ATP-binding cassette transporter mutants. Plant Physiol 146(2):762–771

Badri DV, Quintana N, El Kassis EG, Kim HK, Choi YH, Sugiyama A, Verpoorte R, Vivanco JM (2009a) An ABC transporter mutation alters root exudation of phytochemicals that provoke an overhaul of natural soil microbiota. Plant Physiol 151(4):2006–2017

Badri DV, Weir TL, Van der Lelie D, Vivanco JM (2009b) Rhizosphere chemical dialogues: plant–microbe interactions. Curr Opin Biotechnol 20(6):642–650

Badri DV, Chaparro JM, Zhang R, Shen Q, Vivanco JM (2013) Application of natural blends of phytochemicals derived from the root exudates of Arabidopsis to the soil reveal that phenolic-related compounds predominantly modulate the soil microbiome. J Biol Chem 288 (7):4502–4512

Bagyaraj DJ (2011) Microbial biotechnology for sustainable agriculture, horticulture & forestry. Nipa

Bai Y, D'Aoust F, Smith DL, Driscoll BT (2002) Isolation of plant-growth-promoting Bacillus strains from soybean root nodules. Can J Microbiol 48(3):230–238

Bastián F, Cohen A, Piccoli P, Luna V, Bottini R, Baraldi R (1998) Production of indole-3-acetic acid and gibberellins A 1 and A 3 by *Acetobacter diazotrophicus* and *Herbaspirillum seropedicae* in chemically-defined culture media. Plant Growth Regul 24(1):7–11

Belimov AA, Dodd IC, Safronova VI, Hontzeas N, Davies WJ (2007) Pseudomonas brassicacearum strain Am3 containing 1-aminocyclopropane-1-carboxylate deaminase can show both pathogenic and growth-promoting properties in its interaction with tomato. J Exp Bot 58(6):1485–1495

Belimov AA, Dodd IC, Hontzeas N, Theobald JC, Safronova VI, Davies WJ (2009) Rhizosphere bacteria containing 1-aminocyclopropane-1-carboxylate deaminase increase yield of plants grown in drying soil via both local and systemic hormone signalling. New Phytol 181 (2):413–423

Berendsen RL, Pieterse CM, Bakker PA (2012) The rhizosphere microbiome and plant health. Trends Plant Sci 17(8):478–486

Bhattacharyya PN, Jha DK (2012) Plant growth-promoting rhizobacteria (PGPR): emergence in agriculture. World J Microbiol Biotechnol 28(4):1327–1350

Bloemberg GV, Lugtenberg BJ (2001) Molecular basis of plant growth promotion and biocontrol by rhizobacteria. Curr Opin Plant Biol 4(4):343–350

Brockwell J, Herridge DF, Morthorpe LJ, Roughley RJ (1988) Numerical effects of rhizobium population on legume symbiosis. In: Beck DP, Materon LA (eds) Nitrogen fixation by legumes in Mediterranean agriculture. Springer, Dordrecht, pp 179–193

Broeckling CD, Broz AK, Bergelson J, Manter DK, Vivanco JM (2008) Root exudates regulate soil fungal community composition and diversity. Appl Environ Microbiol 74(3):738–744

Carroll G (1988) Fungal endophytes in stems and leaves: from latent pathogen to mutualistic symbiont. Ecology 69(1):2–9

Cattelan AJ, Hartel PG, Fuhrmann JJ (1999) Screening for plant growth–promoting rhizobacteria to promote early soybean growth. Soil Sci Soc Am J 63(6):1670–1680

Chanway CP (1996) I Endophytes: they're not just fungi! Can J Bot 74:321–322

Chaparro JM, Badri DV, Bakker MG, Sugiyama A, Manter DK, Vivanco JM (2013) Root exudation of phytochemicals in Arabidopsis follows specific patterns that are developmentally programmed and correlate with soil microbial functions. PLoS One 8(2):e55731

Cooke TJ, Poli D, Sztein AE, Cohen JD (2002) Evolutionary patterns in auxin action. Plant Mol Biol 49:319–338

Coombs JT, Franco CM (2003) Isolation and identification of actinobacteria from surface-sterilized wheat roots. Appl Environ Microbiol 69(9):5603–5608

Crawford JW, Harris JA, Ritz K, Young IM (2005) Towards an evolutionary ecology of life in soil. Trends Ecol Evol 20(2):81–87

Curtis TP, Sloan WT, Scannell JW (2002) Estimating prokaryotic diversity and its limits. Proc Natl Acad Sci 99(16):10494–10499

Darrah PR (1993) The rhizosphere and plant nutrition: a quantitative approach. Plant Soil 155 (1):1–20

Das A, Prasad R, Srivastava A, Giang PH, Bhatnagar K, Varma A (2007) Fungal siderophores: structure, functions and regulations. In: Varma A, Chincholkar SB (eds) Microbial siderophores, vol 12. Springer-Verlag, Berlin-Heidelberg, pp 1–42

Doornbos RF, van Loon LC, Bakker PA (2012) Impact of root exudates and plant defense signaling on bacterial communities in the rhizosphere. A review. Agron Sustain Dev 32(1):227–243

Dubeikovsky AN, Mordukhova EA, Kochetkov VT, Polikarpova FY, Boronin AM (1993) Growth promotion of blackcurrant softwood cuttings by recombinant strain *Pseudomonas fluorescens* BSP53a synthesizing an increased amount of indole-3-acetic acid. Soil Biol Biochem 25 (9):1277–1281

Erturk Y, Ercisli SEZAI, Sekban R, Haznedar A, Donmez MF (2008) The effect of plant growth promoting rhizobacteria (PGPR) on rooting and root growth of tea (*Camellia sinensis* var. Sinensis) cuttings. Roum Biotech Lett 13:3747–3756

Fukuda H, Ogawa T, Tanase S (1993) Ethylene production by micro-organisms. In: Rose AH (ed) Advances in microbial physiology, vol 35. Academic, New York, pp 275–306

Gaby JC, Buckley DH (2012) A comprehensive evaluation of PCR primers to amplify the nifH gene of nitrogenase. PLoS One 7(7):e42149

Ganley RJ, Brunsfeld SJ, Newcombe G (2004) A community of unknown, endophytic fungi in western white pine. Proc Natl Acad Sci 101(27):10107–10112

García de Salamone IE, Hynes RK, Nelson LM (2001) Cytokinin production by plant growth promoting rhizobacteria and selected mutants. Can J Microbiol 47(5):404–411

Glick BR, Penrose DM, Li J (1998) A model for the lowering of plant ethylene concentrations by plant growth-promoting bacteria. J Theor Biol 190(1):63–68

Gutiérrez-Mañero FJ, Ramos-Solano B, Probanza AN, Mehouachi J, Tadeo R, Talon M (2001) The plant-growth-promoting rhizobacteria *Bacillus pumilus* and *Bacillus licheniformis* produce high amounts of physiologically active gibberellins. Physiol Plant 111(2):206–211

Harrier LA, Watson CA (2003) The role of arbuscular mycorrhizal fungi in sustainable cropping systems. Adv Agron 79(79):185–225

Hartmann A, Rothballer M, Schmid M (2008) Lorenz Hiltner, a pioneer in rhizosphere microbial ecology and soil bacteriology research. Plant Soil 312(1–2):7–14

Hinsinger P (1998) How do plant roots acquire mineral nutrients? Chemical processes involved in the rhizosphere. Adv Agron 64:225–265
Hinsinger P, Bengough AG, Vetterlein D, Young IM (2009) Rhizosphere: biophysics, biogeochemistry and ecological relevance. Plant Soil 321(1–2):117–152
Honma M, Shimomura T (1978) Metabolism of 1-aminocyclopropane-1-carboxylic acid. Agric Biol Chem 42(10):1825–1831
Joo GJ, Kim YM, Lee IJ, Song KS, Rhee IK (2004) Growth promotion of red pepper plug seedlings and the production of gibberellins by *Bacillus cereus*, *Bacillus macroides* and *Bacillus pumilus*. Biotechnol Lett 26(6):487–491
Jumpponen A (2001) Dark septate endophytes–are they mycorrhizal? Mycorrhiza 11(4):207–211
Jung SC, Martinez-Medina A, Lopez-Raez JA, Pozo MJ (2012) Mycorrhiza-induced resistance and priming of plant defenses. J Chem Ecol 38(6):651–664
Kaushik R, Saxena AK, Tilak KVBR (2000) Selection of Tn5:: lacZ mutants isogenic to wild type *Azospirillum brasilense* strains capable of growing at sub-optimal temperature. World J Microbiol Biotechnol 16(6):567–570
Kloepper JW (1992) Plant growth-promoting rhizobacteria as biological control agents. In: Metting FB (ed) Soil microbial ecology. Mercel Dekker, New York
Kobayashi DY, Palumbo JD (2000) Bacterial endophytes and their effects on plants and uses in agriculture. In: Bacon CW, White JF (eds) Microbial endophytes. CRC, Boca Raton, pp 213–250
Loyola-Vargas VM, Broeckling CD, Badri D, Vivanco JM (2007) Effect of transporters on the secretion of phytochemicals by the roots of *Arabidopsis thaliana*. Planta 225(2):301–310
Lugtenberg B, Kamilova F (2009) Plant-growth-promoting rhizobacteria. Annu Rev Microbiol 63:541–556
Malla R, Prasad R, Kumari R, Giang PH, Pokharel U, Oelmueller R, Varma A (2004) Phosphorus solubilizing symbiotic fungus *Piriformospora indica*. Endocytobiosis Cell Res 15(2):579–600
Marler M, Pedersen D, Mitchell-Olds T, Callaway RM (1999) A polymerase chain reaction method for detecting dwarf mistletoe infection in Douglas-fir and western larch. Can J For Res 29 (9):1317–1321
Marschner H (1995) Mineral nutrition of higher plants, 2nd edn. Academic, London
Martínez-Viveros O, Jorquera MA, Crowley DE, Gajardo GMLM, Mora ML (2010) Mechanisms and practical considerations involved in plant growth promotion by rhizobacteria. J Soil Sci Plant Nutr 10(3):293–319
Mayak S, Tirosh T, Glick BR (1999) Effect of wild-type and mutant plant growth-promoting rhizobacteria on the rooting of mung bean cuttings. J Plant Growth Regul 18(2):49–53
Mehnaz S, Mirza MS, Haurat J, Bally R, Normand P, Bano A, Malik KA (2001) Isolation and 16S rRNA sequence analysis of the beneficial bacteria from the rhizosphere of rice. Can J Microbiol 47(2):110–117
Mirza MS, Ahmad W, Latif F, Haurat J, Bally R, Normand P, Malik KA (2001) Isolation, partial characterization, and the effect of plant growth-promoting bacteria (PGPB) on micropropagated sugarcane in vitro. Plant Soil 237(1):47–54
Modjo HS, Hendrix JW (1986) The mycorrhizal fungus *Glomus macrocarpum* as a cause of tobacco stunt disease. Phytopathology 76(7):688–691
Morgan JAW, Bending GD, White PJ (2005) Biological costs and benefits to plant–microbe interactions in the rhizosphere. J Exp Bot 56(417):1729–1739
Nardi S, Concheri G, Pizzeghello D, Sturaro A, Rella R, Parvoli G (2000) Soil organic matter mobilization by root exudates. Chemosphere 41(5):653–658
O'bannon JH, Evans DW, Peaden RN (1980) Alfalfa varietal response to seven isolates of vesicular-arbuscular mycorrhizal fungi. Can J Plant Sci 60(3):859–863
Peters AF (1991) Field and culture studies of *Streblonema macrocystis* sp. nov. (Ectocapales, Phaeophyceae) from Chile, a sexual endophyte of giant kelp. Phycologia 30(4):365–377

Peterson CA, Emanuel ME, Humphreys GB (1981) Pathway of movement of apoplastic fluorescent dye tracers through the endodermis at the site of secondary root formation in corn (*Zea mays*) and broad bean (*Vicia faba*). Can J Bot 59(5):618–625

Pirozynski KA, Malloch DW (1975) The origin of land plants: a matter of mycotrophism. Biosystems 6(3):153–164

Prasad R, Garg AP, Varma A (2005) Interaction of medicinal plants with plant growth promoting rhizobacteria and symbiotic fungi. In: Podila GK, Varma A (eds) Basic research and applications of Mycorrhizae, vol 1. IK International, Delhi, pp 363–407

Prasad R, Kumar M, Varma A (2015) Role of PGPR in soil fertility and plant health. In: Egamberdieva D, Shrivastava S, Varma A (eds) Plant-growth-promoting rhizobacteria (PGPR) and medicinal plants. Springer, Cham, pp 247–260

Prasad R, Bhola D, Akdi K, Cruz C, Sairam KVSS, Tuteja N, Varma A (2017) Introduction to mycorrhiza: historical development. In: Varma A, Prasad R, Tuteja N (eds) Mycorrhiza-function, diversity, state of the art. Springer, Cham, pp 1–7

Prasad R, Chhabra S, Gill SS, Singh PK, Tuteja N (2020) The microbial symbionts: potential for the crop improvement in changing environments. In: Tuteja N, Tuteja R, Passricha N, Saifi SK (eds) Advancement in crop improvement techniques. Elsevier, Amsterdam, Netherlands, pp 233–240

Reddy VS, Shlykov MA, Castillo R, Sun EI, Saier MH Jr (2012) The major facilitator superfamily (MFS) revisited. FEBS J 279(11):2022–2035

Rodriguez H, Vessely S, Shah S, Glick BR (2008) Effect of a nickel-tolerant ACC deaminase-producing Pseudomonas strain on growth of nontransformed and transgenic canola plants. Curr Microbiol 57(2):170–174

Saleh SS, Glick BR (2001) Involvement of gacS and rpoS in enhancement of the plant growth-promoting capabilities of *Enterobacter cloacae* CAL2 and UW4. Can J Microbiol 47 (8):698–705

Siddiqui ZA, Baghel G, Akhtar MS (2007) Biocontrol of *Meloidogyne javanica* by rhizobium and plant growth-promoting rhizobacteria on lentil. World J Microbiol Biotechnol 23(3):435–441

Singh R, Behl R, Jain P, Narula N, Singh K (2007) Performance and gene effects for root characters and micronutrient uptake in wheat inoculated with arbuscular mycorrhizal fungi and *Azotobacter chroococcum*. Acta Agronomica Hungarica 55(3):325–330

Singh D, Raina TK, Kumar A, Singh J, Prasad R (2019) Plant microbiome: a reservoir of novel genes and metabolites. Plant Gene 18:100177

Srinivasan M, Holl FB, Petersen DJ (1996) Influence of indoleacetic-acid-producing Bacillus isolates on the nodulation of *Phaseolus vulgaris* by *Rhizobium etli* under gnotobiotic conditions. Can J Microbiol 42(10):1006–1014

Stone JK, Bacon CW, White JF Jr (2000) An overview of endophytic microbes: endophytism defined. In: Bacon CW, White JF (eds) Microbial endophytes. CRC, Boca Raton, pp 17–44

Sugiyama A, Shitan N, Yazaki K (2008) Signaling from soybean roots to rhizobium: an ATP-binding cassette-type transporter mediates genistein secretion. Plant Signal Behav 3 (1):38–40

Thies JE, Singleton PW, Bohlool BB (1991) Influence of the size of indigenous rhizobial populations on establishment and symbiotic performance of introduced rhizobia on field-grown legumes. Appl Environ Microbiol 57(1):19–28

Timmusk S, Nicander B, Granhall U, Tillberg E (1999) Cytokinin production by *Paenibacillus polymyxa*. Soil Biol Biochem 31(13):1847–1852

Varma A, Prasad R, Tuteja N (2017a) Mycorrhiza: function, diversity and state-of-art. Springer International Publishing, Switzerland. ISBN 978-3-319-53064-2. http://www.springer.com/us/book/9783319530635

Varma A, Prasad R, Tuteja N (2017b) Mycorrhiza: eco-physiology, secondary metabolites, nanomaterials. Springer International Publishing, Switzerland, ISBN 978-3-319-57849-1. http://www.springer.com/us/book/9783319578484

Varma A, Prasad R, Tuteja N (2017c) Mycorrhiza: nutrient uptake, biocontrol, ecorestoration. Springer International Publishing, Switzerland. ISBN 978-3-319-68867-1. http://www.springer.com/us/book/9783319688664

Varma A, Swati T, Prasad R (2019) Plant microbe interface. Springer International Publishing, Switzerland. ISBN 978-3-030-19831-2. https://www.springer.com/gp/book/9783030198305

Varma A, Swati T, Prasad R (2020) Plant microbe symbiosis. Springer International Publishing, Switzerland. ISBN 978-3-030-36247-8. https://www.springer.com/gp/book/9783030362478

Wagner CA, Taylor TN (1981) Evidence for endomycorrhizae in Pennsylvanian age plant fossils. Science 212(4494):562–563

Walker TS, Bais HP, Grotewold E, Vivanco JM (2003) Root exudation and rhizosphere biology. Plant Physiol 132(1):44–51

Weller DM, Raaijmakers JM, Gardener BBM, Thomashow LS (2002) Microbial populations responsible for specific soil suppressiveness to plant pathogens. Annu Rev Phytopathol 40 (1):309–348

Weston LA, Ryan PR, Watt M (2012) Mechanisms for cellular transport and release of allelochemicals from plant roots into the rhizosphere. J Exp Bot 63(9):3445–3454

Wilson D (1995) Endophyte: the evolution of a term, and clarification of its use and definition. Oikos 73:274–276

Yang J, Kloepper JW, Ryu CM (2009) Rhizosphere bacteria help plants tolerate abiotic stress. Trends Plant Sci 14(1):1–4

Yazaki K (2005) Transporters of secondary metabolites. Curr Opin Plant Biol 8(3):301–307

Zahir ZA, Ghani U, Naveed M, Nadeem SM, Asghar HN (2009) Comparative effectiveness of Pseudomonas and Serratia sp. containing ACC-deaminase for improving growth and yield of wheat (*Triticum aestivum* L.) under salt-stressed conditions. Arch Microbiol 191(5):415–424

Chapter 15
Endophytic Secondary Metabolites for Biological Control: A Latest Perspective

G. V. S. Bhanu Prakash and T. Srinivasan

Abstract Microorganisms are being used as biological control agents (BCA) for years to control diseases and pest infestation on the economically important crop. This has opened new avenues for sustainable agriculture by using eco-friendly methods. It can be achieved by further understanding the approaches, life cycle, and mechanism of action, etc. of natural BCA species. Most microorganisms in the environment which infest the target pest and cause deterioration of its physiology or its death can be of potent biological control of the target pest. There are numerous microorganisms antagonistic to different types of pests that have been studied extensively for their contribution to plant health and for their unique modes of action in the areas of plant protection and management system. However, so far there is very little information about the intimate interrelationships between the endophytic fungi and pest and their host plants. The mode of action of these endophytic fungi against the target pest is unique, and the mechanism of action may be by the production of toxic secondary metabolites, by production of repellent compounds, or by a combination of toxic and repellent compounds. In recent years, endophytes and their bioactive products have garnered significant attentiveness from the various research groups. In this article, a short review of the impact of endophytic fungi on plant parasite infection will be discussed.

15.1 Introduction

The natural and biological control methods of pests and pathogens have gained much attention in the past few decades as a way of reducing the use of chemicals in agriculture, thereby protecting the environment. The capability of endophytes to

G. V. S. B. Prakash
Neelakanta Vidyapeeth, Hyderabad, Telangana, India

T. Srinivasan (✉)
Department of Botany, Indira Gandhi National Tribal University, Amarkantak, Madhya Pradesh, India
e-mail: srinivasan@igntu.ac.in

A. Varma et al. (eds.), *Plant Microbiome Paradigm*,
https://doi.org/10.1007/978-3-030-50395-6_15

colonize internally in the host tissues has gained the attention and outmost interest of many researchers as a tool for sustainable auricular practices. In this regard, entomopathogenic microorganisms are those that antagonize other pathogenic microorganisms of host plants which can be an alternative way to reduce or eliminate the use of chemicals in agriculture.

In the early 1980s, the published literature indicates that endophytic microorganisms could play an important role in plants to protect them from natural enemies like pests, pathogens, etc. It was also demonstrated that the presence of these microorganisms in their respective hosts could result in the reduction of insect attacks. The mechanisms by which endophytic fungi control insect infestation on plants are by producing toxic secondary metabolites. The nature of the protection and the variables involved in the process were also addressed.

Microbial endophytes associated with the majority of plant species are considered as extremely important plant partners with low-stress induction on the host (Hallmann et al. 1997; Arnold et al. 2000). However, these endophytes, having effective bioactive products, have received attention from the researchers as knowledge on increased neighboring and beneficial qualities thus conferred were immense (Azevedo et al. 2000; Schulz et al. 2002; Aly et al. 2010). Endophytes are a class of microorganisms that are extremely diverse, restricted to small area infections in host tissues.

This group of microbes shows a great difference in their biological actions based on the symbiotic relationship of their hosts, lifestyle associated to their life cycles. This particular biological behavioral character exhibited by endophytes or groups of endophytes with specific biocontrol capability ultimately makes these organisms capable as plant protective agents. Apparently, endophytes must exhibit one or more bioactive properties which ultimately benefit to its host plant; as a result, they can be used as a potent BCA. The wide range of biological control nature of these endophytes is based on one or more mechanisms of action towards the target pathogen or pest of a plant. Based on this, they are classified into four main categories: (1) antibiosis; (2) competition; (3) direct parasitism; and (4) host-induced resistance. Sometimes, they may simultaneously exhibit multiple mechanisms of action; it is an added advantage so that this may increase the efficacy of the BCA endophyte, the response of the target pest/pathogen may delay, or they may develop resistance in the population that assist the organism in controlling multiple unrelated pests/pathogens on different host plants (Punja 1997). Consequently, indirect mechanisms of antagonism like antibiosis and host-induced resistance are more effective when compared to the direct mechanisms like parasitism and competition of an endophyte. Antibiosis is the most predominant form of antagonism expressed in endophytes. Several bacterial as well as fungal endophytes produce numerous bioactive secondary metabolites and show antagonistic and inhibitory deterrent properties. The mechanism of action is extremely associated because these secondary metabolites show dual roles and also function as elicitors of plant-induced resistance (Danielsson et al. 2007), by acting as signaling pathways between the endophyte and its host (Granér et al. 2003).

Ownley et al. (2008), Vega (2008), Lohse et al. (2015) reported that *Beauveria bassiana* shows endophytic lifestyle in tissues of host plant, it also colonizes in a range of plant tissues, across various hosts, including leaves, shoots, roots and seed of wide range of monocots and dicots (Bing and Lewis 1992; Posada and Vega 2005, 2006; Akello et al. 2008; Gurulingappa et al. 2011). Clark et al. (1989) isolated 900 samples of fungal isolates from *Abies balsamea* and red spruce *Picea rubens.* Out of which five produced toxic metabolites and three of them are highly effective. They proved to be very fatal to the insects and also curtailed the growth and development of *C. fumiferana.*

15.2 Source of Endophytes

There is diverse microbial population in plants consisting of archaeal, bacterial, fungal, and protist groups (Hardoim et al. 2015). More than one million endophytic fungal species were isolated, identified, and studied (Strobel and Daisy 2003; Ganley et al. 2004). Endophytic fungi are universal in distribution, as they are isolated from plants that are adapted to wide range of ecosystems (Arnold 2007, 2008; Arnold and Lutzoni 2007). These are present in all the major groups of plants, viz. bryophyte, pteridophyta, and spermatophyta (Arnold 2007). The reports on diversity of fungal endophytes across geographical areas are increasing on the other hand; the endophyte diversity in plants has been decreasing from the tropics to northern boreal forests (Arnold 2007; Arnold and Lutzoni 2007). Further, the spectrum of fungi within a plant varies by area, age, season, and its part of localization.

Webber (1981) was the first researcher who isolated the endophytic fungus, *Phomopsis oblonga*, to protect elm trees against the beetle (*Physocnemum brevilineum*) infestation. Similarly, Johnson et al. (1985) reported that endophytic fungus, *Acremonium coenophialum*, exhibited insecticidal activity against aphids (*Rhopalosiphum padi, Schizaphis graminum*) and milkweed bug (*Oncopeltus fasciatus*). Schardl (2001) stated that endophytes can reduce herbivory by producing alkaloids which are toxic to insects and vertebrates. Larran et al. (2002) and dos Santos et al. (2003) reported that *Cladosporium herbarum, A. alternata, Rhodotorula rubra, Epicoccum nigrum, Cryptococcus* sp., *Penicillium* sp., and *Fusarium graminearum* act as plant protectants from herbivores. Posada and Vega (2006) found *Beauveria bassiana* as an endophytic fungus in coffee seedlings to control the borer in the coffee plantation. Vega et al. (2008) isolated different genera of entomopathogenic fungi *Acremonium, Beauveria, Cladosporium, Clonostachys,* and *Paecilomyces* from the coffee plants; among them, *Beauveria bassiana* and *Clonostachys rosea* have shown pathogenicity towards coffee berry borer. Similar results were also obtained by Baskar et al. (2012). *B. bassiana* isolated from Puleny exhibited larvicidal and growth inhibitory activities against *Spodoptera litura.* Amatuzzi et al. (2018) isolated 517 fungal colonies belonging to 13 genera from the leaves of strawberry. Eight isolates belonging to the genera *Aspergillus, Cladosporium, Diaporthe,* and *Paecilomyces* were tested for pathogenicity against third instar larvae of *Duponchelia fovealis*, where *Paecilomyces* exhibited the highest mortality rate. The endosymbionts produce novel compounds from toxicants like formilonine, paxilline analogous, heptelidic acid, and rugulosin and repellants like naphthalene, beta-pinene, gamma-terpinene, and limonene (Moloinyane and Nchu 2019) that could help in controlling the pest. Similarly it also helps in the protection of the plant from various biotic and abiotic stress conditions (Aly et al. 2013), so that they can be used as biocontrol agents and also enhancers of crop productivity and soil fertility.

15.3 Classification of Fungal Endophytes

Endophytes are a diverse group of microbes; only a few species have been isolated and completely characterized till date. Endophytic fungi are classified based on their diversity or functional roles. According to Rodriguez et al. (2009), they have been grouped into two major groups as clavicipitaceous (common in grasses) and non-clavicipitaceous (vascular and non-vascular plant species). On the other hand, Rodriguez et al. (2009), Purahong and Hyde (2011), Brem and Leuchtmann (2001), Saikkonen et al. (2002), and Varma et al. (1999) classified endophytes taking several criteria into consideration. They are the host range, source of nutrition, part of plant colonized, mode of reproduction, mode of transmission, and also on the symptoms of infection. The major criteria are:

15.3.1 Source of Nutrition

Endophytes are heterotrophs in nature; the organic compounds produced by the host plant act as their carbon source. Biotrophic endophytic fungi obtain required nutrients from the tissue of the living host and thus establish a long-term feeding relationship with the host rather than killing them as part of its infection process. Necrotrophic endophytic fungi mortify and grow on the dead tissues of the host (Kemen and Jones 2012; Delaye et al. 2013). Based on the periodic evolution as well as ecological changes, some of these endophytic fungi switches between the two lifestyles, i.e., from biotrophic to necrotrophic lifestyle (Promputtha et al. 2007; Purahong and Hyde 2011; Delaye et al. 2013; Junker et al. 2012).

15.3.2 Part of Plant Colonized

The colonization of these endophytes has been widely reported based on the tissue of localization within their host plant. Popularly they are classified as root and foliar endophytes. Meyling et al. (2011) and Behie et al. (2015) reported that the colonization of *B. bassiana* and *P. chlamydosporia* was apparently confined to the stems and leaves (foliar part of the plant) and they are known as foliar endophytes, whereas *Metarhizium* spp. are confined to roots and are known as root endophytes (Varma et al. 1999; Wilberforce et al. 2003; Wyrebek et al. 2011).

15.3.3 Mode of Transmission

Transmission of endophytic fungi is of two types: (1) vertical transmission (transfer directly from the parental host plants to their progenies) and (2) horizontal transmission (transfer among different individuals of a given population). In general, the transmission of endophytic fungus takes place through airborne spores (Hartley and Gange 2009). *Neotyphodium* mostly shows vertical transmission through seeds from one plant to another and is known as seed-transmitted endophytes; this can also be observed in *Epichloë* endophytes (Dongyi and Kelemu 2004; Bennett et al. 2008; Hartley and Gange 2009; Schardl et al. 2013). Vertical transmission of *B. bassiana* through seeds in *opium poppy* plants has also been done artificially via seed soaking (Quesada-Moraga et al. 2014). On the other hand, these fungi usually propagate through vegetative propagules or transmitted by spores (Faeth and Fagan 2002). Endophytes present in most woody and herbaceous plants are transmitted horizontally, and these plants are reservoirs for different species of unspecialized endophytes, and they generally exhibit weak pathogenicity against insect herbivores (Higgins et al. 2007; Sieber 2007). Horizontal transmission in endophytic fungi occurs through sexual reproduction, and they transmit via spores, soil, wind, or insect vectors (Sánchez Márquez et al. 2012); they are closely related to pathogenic fungi, although they are not pathogenic themselves.

15.3.4 Based on the Expression of Infection

Endophytic organisms are classified as asymptomatic (symptomless) or symptomatic (expressing symptoms) based on the symptoms of the infection in the host plant (Pinto et al. 2000). Generally, the maximum number of endophytes infects the aerial parts of plant asymptomatically. They live as symbiotic organisms, and they are on focus mainly due to ubiquitous nature, their vast diversity, and multiple roles (Saikkonen et al. 2006; Arnold and Lutzoni 2007). Symptomatic endophytes can be considered as asymptomatic in some cases when the host plant is resistant to them. This phenomenon may be affected by change of its micro- and macroenvironments. Delaye et al. (2013) reported that some symptomless endophytes yet became pathogenic under changed environmental conditions. Apart from this, age of the host plant plays an important role in the portrayal of symptoms (Saikkonen et al. 1998; Schulz and Boyle 2005; Hyde and Soytong 2008; Porras-Alfaro and Bayman 2011).

15.4 Types of Secondary Metabolites from Endophytes

Biocontrol agents are occupying a small part of the insecticide field, but their usage is on upsurge (Demain 2000). Endophytic fungi can protect their host plants from pathogens and pests by secreting potent secondary metabolites (Arnold et al. 2013). Evidence of endophytes dates back to more than 400 million years, implicating that these microorganisms played a significant role in host plant adaptation to habitat transitions, but in the recent year, only their bioactive products gained significant attention from the scientific community. Endophytes are a diverse group of organisms and can exhibit different biological behavior; they produce a vast variety of novel secondary metabolites. In addition to protecting plants against pathogens and pests, some endophytes can be used to make plants tolerant to a range of biotic as well as abiotic stresses and also for improved management of post-harvest control. The production of bioactive substances by endophytes is directly related to the evolution of these organisms, which might have integrated genetic information from higher plants. They participate in a variety of host-pathogen interactions to carry out functions such as protection of plants from pathogens, insects, and grazing animals (Strobel and Daisy 2003). Some of these endophytic secondary metabolites showing anti-feeding habit and insecticidal activity are listed in (Fig. 15.1).

Several endophytes show anti-insecticidal properties in which novel compounds like indole diterpenes and nodulisporic acids exhibit potential insecticidal activity against the larvae of blow fly (Calliphoridae). Generally, these compounds act on activating glutamate-gated chloride channel of insects especially that control locomotion, feeding, and mediating sensory inputs into behavior. Demain (2000) for the first time isolated nodulisporic compounds from an endophyte, *Nodulisporium* sp., from the plant *Bontia daphnoides*. This has ensued in an intensive search for more *Nodulisporium* spp. and other producers of more potent nodulisporic acid analogs. The endophytic fungi *Claviceps purpurea* whose secretions contain ergotoxine and related alkaloids that stimulate smooth muscles also shows significant insecticidal activity against *A. gossypii* Glover (Hemiptera: Aphididae) (Shi et al. 2013). Senthilkumar et al. (2014) isolated different types of phytochemicals (ethyl ester, phthalic acid, octyl 2-pentyl ester, and dodecanoic acid) from *Phomopsis* sp. isolated from *Tectona grandis* which show insecticidal activity. Similarly, Bensaci et al. (2015) reported that *Cladosporium oxysporum* also showed insecticidal activity against *A. fabae.* The topical application of the extracts of *Emericella nidulans*, *A. oryzae*, *A. tamarii*, and *A. versicolor* on *Spodoptera litura* larvae showed insecticidal activity (Abraham et al. 2015). Li et al. (2012) reported that *Aspergillus fumigatus* isolated from the bark of *Melia azedarach* produced 39 secondary metabolites. Nine of them steered antifeedant activity against armyworm (*Mythimna separata*) larvae. Among these nine, fumitremorgin B (50.0%) and verruculogen (55.0%) exhibited the best activity.

Findlay (1997) isolated two new benzofuran compounds from an endophytic fungus of wintergreen (*Gaultheria procumbens*), which showed notable insecticidal activity against larvae of spruce budworm (*Choristoneura fumiferana* Clem).

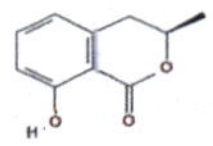

(3R,4aS,8S, 8aR)-8-hydroxy-3-methyl-3,4,4a,5,6,7,8,8a-octahydro-1H-2-benzopyran-1-one)

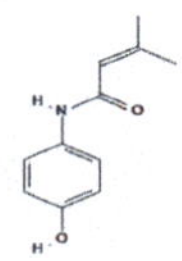

(lS,3R,4S,5S)-4-hydroxy-3-methyl-2-oxobicyclo-[3.3.1]non-6-one

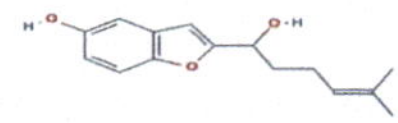

5-hydroxy-2-(1-hydroxy-5-methyl- 4-hexenyl) benzofuran

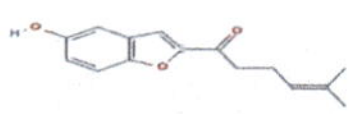

5-Hydroxy-2-(5-Methyl-1-Oxo-4-Hexenyl)Benzofuran

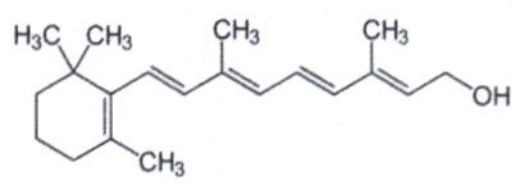

Diterpenoid

Dodecanoic acid

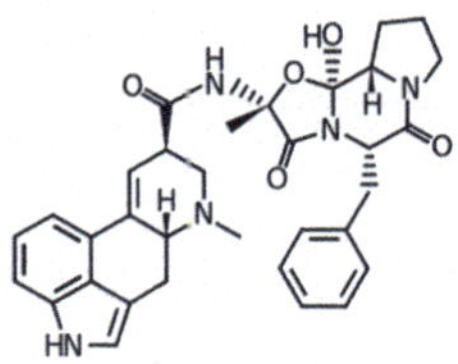

Ergotamine

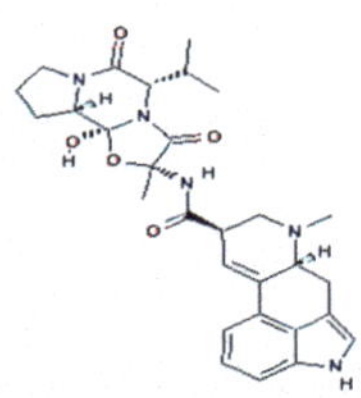

Ergovaline

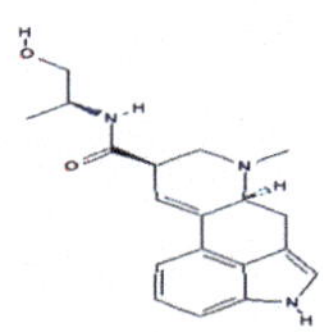

Ergovine

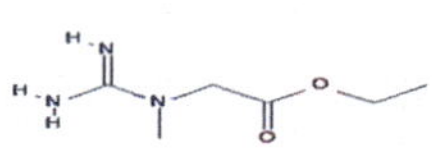

Ethylester

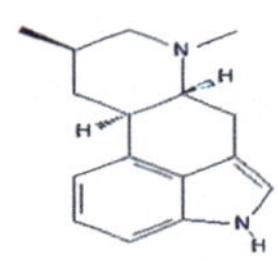

Festuclavine

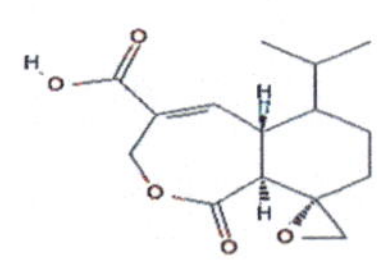

heptelidic acid

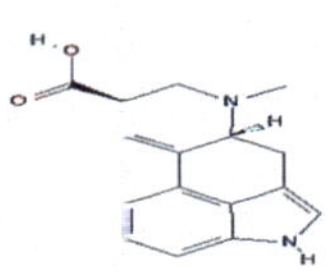

lisergic acid

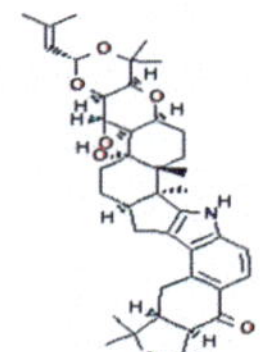

lolitrem B

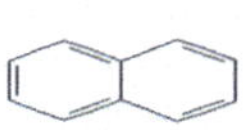

Naphthalene

Fig. 15.1 Secondary metabolites showing anti-feeding and insecticidal activity from endophytic fungi

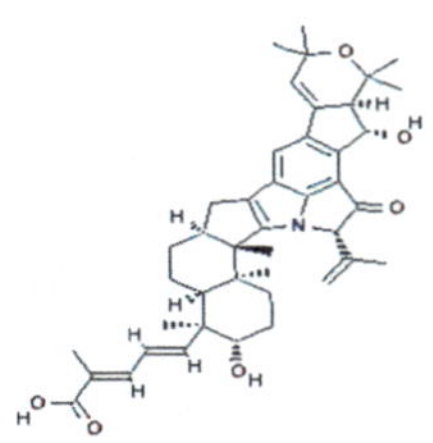

Nodulisporic acids

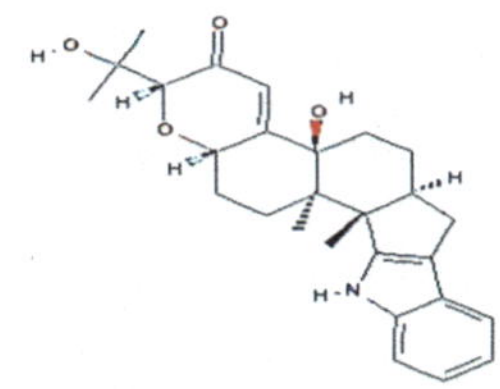

paxiline analogous

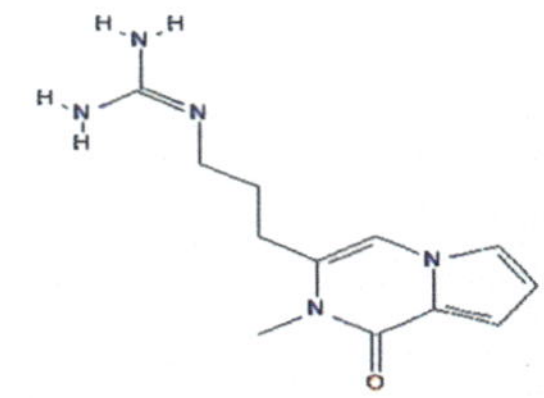

Peramine

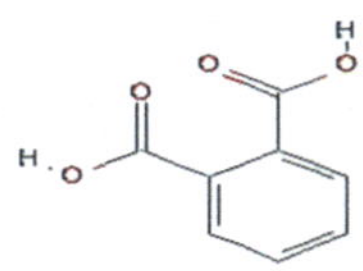

Phthalic acid

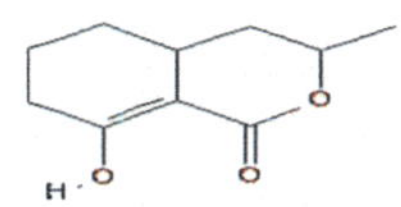

Ramulosin

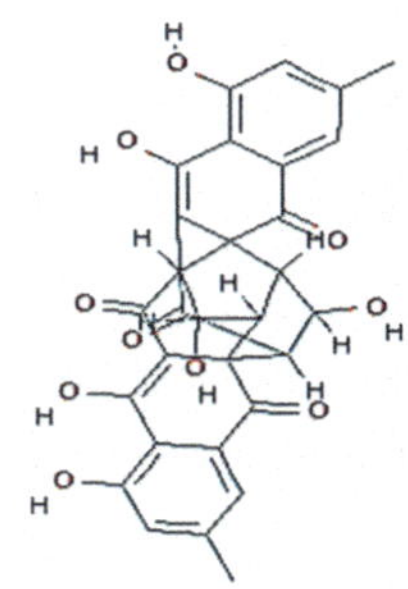

Rugulosinex

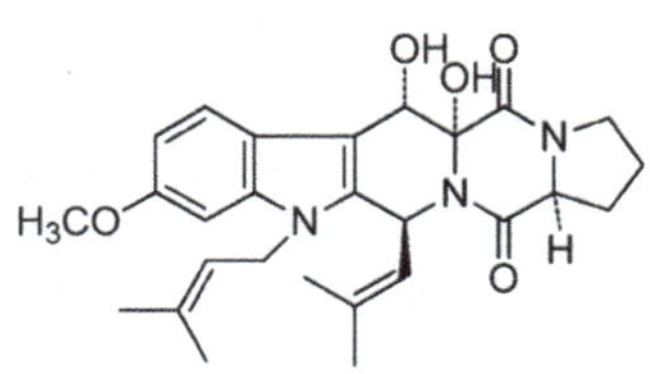

Fumitremorgin B

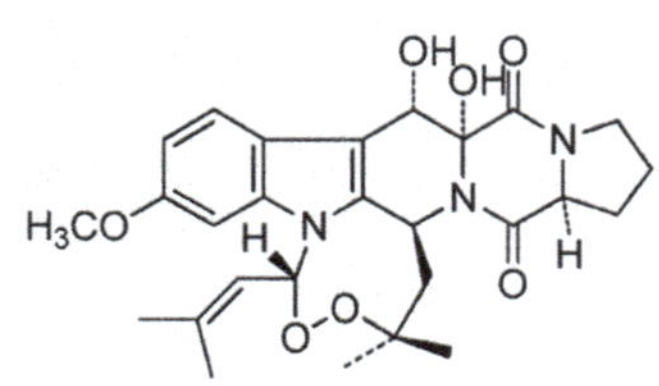

Verruculogen

Fig. 15.1 (continued)

Findlay et al. (1995b) isolated two novel diterpenoid toxins such as 9α-hydroxy-1,8-(14),15-isopimaratriene-3,7,11-trione and 9α-hydroxy-1,8(14),15-isopimaratrien-3,11-dione from an endophyte derived from a needle of a balsam fir. These compounds also showed similar toxicity towards larvae of *Choristoneura fumiferana*. Ramulosin and mellein analogs which were isolated from conifer endophytes also showed toxicity to spruce budworm larvae (*C. fumiferana*) (Findlay et al. 1995a).

Prestidge and Gallagher (1988) stated that a strong toxin, lolitrem B, from fungus *A. lolii* in *Lolium perenne* once added to *Listronotus bonariensis* diets and reduced insect growth, feeding behavior, and survival of larvae. The absorption of the toxin occurs only by ingestion but not by absorption through the insect integument. Siegel et al. (1990) reported the release of the alkaloids like peramine, lolitrem B, *N*-formyl, *N*-acetylcholine, and ergovaline by the fungal endophytes of grasses during plant attack by aphids. Correspondingly, when several types of grasses infected with

Acremonium spp. and *E. typhina* were analyzed, various types of alkaloids like peramine and ergovaline were produced by these symbiotic fungi. Peramine, lolitrem B, and ergovaline were also produced by *Lolium* infected with *A. coenophialum* and *A. lolii* and *F. longiflora* infection with *E. typhina*. The pests like *Rhopalosiphum padi* and *Schizaphis graminum* did not survive when they attack grasses containing the alkaloid loline. On the other hand, ergovaline did not show any adverse effect towards these insect species. Where as *F. arundinacea* treated with methanolic extracts of *A. coenophialum* containing lolines, is able to alter feed behavior and reduction in the weight of the insect. When diet is supplemented with extracts of loline derivatives, notable changes were observed in a reduction in weight of insect and altered behaviors in *S. frugiperda* and *O. nubilalis* (Riedell et al. 1991).

Behavioral changes were observed in insects of Coleoptera, when they were fed with the diet supplement with alkaloids produced by *Neotyphodium lolii* and *L. perenne*. Ergonovine showed moderate effects on *Heteronychus arator* larvae, whereas ergotamine, ergovaline from the ergot-type alkaloid family produced from perennial ryegrass, is responsible for the plant resistance towards adult *Heteronychus arator* (Ball et al. 1997). The alkaloids like lolitrem B, lysergol, peramine, and festuclavine and lysergic acid did not show any effect on the insect. Various types of ergot alkaloids are produced by some endophytic fungi, and the biosynthetic pathway of these ergot alkaloids has been studied extensively by Lorenz et al. (2009), Panaccione (2010), and Wallwey and Li (2011). The diverse metabolites in the ergot that belong to the alkaloid family have been extensively studied by many researchers, and they were grouped as clavines, lysergic acid (simple amides), or ergopeptines based on their complexity and relative location in the biosynthetic pathway (Lorenz et al. 2009; Ortel and Keller 2009; Coyle et al. 2010).

According to Lorenz et al. (2009), Panaccione (2010), and Wallwey and Li (2011), ergot alkaloids interact in many pathways as a protagonist or as an antagonist which actively oppose the receptors for the monoamine neurotransmitters. Hence, we can observe various consequences in the organisms like uncontrolled muscle contraction, vasoconstriction, improper functioning of the central nervous, and reproductive systems. These alkaloids also affect the feeding and development of insects and nematodes, thus leading to the increased mortality (Clay and Cheplick 1989; Ball et al. 1997; Potter et al. 2008). This ergot alkaloid pathway is eminent for the accumulation of intermediate metabolites and enhances the production of intermediates beyond the concentrations of the end product/s in the pathway (Panaccione et al. 2003; Panaccione and Coyle 2005).

The isolates of *Neotyphodium* sp. produce two types of toxic secondary metabolites like *N*-formilonine and a paxilline analogous in the host plant *Echinopogon ovatus* which showed insecticidal activity against *L. bonariensis* (Miles et al. 1998). This endophyte is also reported to produce aminopyrrolizidine alkaloids loline, in the plants *Adenocarpus decorticans* (Fabaceae) and *Argyreia mollis* (Convolvulaceae) (Schardl et al. 2007; Tofern et al. 1999). Calhoun et al. (1992) isolated endophytic fungus of woody plants (*Phyllosticta* and *Hormonema dematioides*) that produce toxic products and were able to alter the growth and lethality rates in larvae of the spruce budworm *C. fumiferana* when they feed on balsam fir. The toxic compounds produced by *Phyllosticta* and *Hormonema dematioides* are heptelidic acid and rugulosin. Bills et al. (1992) also isolated a toxic tremorgenic compound from a woody plant infected with an endophytic fungus from the genus *Phomopsis*, which showed toxicity against some pests and nematodes.

Indole diterpenes are representing as another important class of diverse alkaloids produced by some filamentous fungi and endophytic spp. (Saikia et al. 2008). Similar to the ergot alkaloids, these indole diterpenes are diverse in nature, based on the oxidation and prenylation of intermediate products like terpendole I, and its subsequent metabolites independently result in formation of different end products like janthitrems, lolitrems, and terpendoles. The less tremorgenic indole diterpenes are beneficial to host plant by showing protection against insects, and it was demonstrated for its structural dissimilarity with nodulisporic acid (Young et al. 2006, 2009). *Nodulisporium* sp. produces nodulisporic acid and has shown good insecticidal activity against a range of insects (Byrne et al. 2002); janthitrems also show insecticidal activity towards *Wiseana cervinata* (porina). The activity and efficacy of lolines against insects were confirmed by both the purified lolines and endophyte-infested plants (Yates et al. 1989; Siegel et al. 1990; Riedell et al. 1991; Jensen et al. 2009). However, the confirmation of host plant or endophyte as the source of lolines is yet to be determined in case of *Adenocarpus* and *Argyreia* species.

Peramine is unique and widely distributed among the four major classes of alkaloids produced by Epichloë (ascomycete endophyte) (Schardl et al. 2011). Peramine shows strong anti-feeding habit for Argentine stem weevil and several other insects (Clay et al. 1985; Johnson et al. 1985; Rowan et al. 1986, 1990; Rowan 1993); the feeding deterrent effects of these peramine are not universal (Johnson et al. 1985; Gaynor and Rowan 1986). Naphthalene is another major type of secondary metabolite produced by *Muscodor vitigenus*, from a liana (*Paullinia paullinioides*); this is an active ingredient in common mothballs and widely exploited as an insect repellant. *M. vitigenus* showed an insect deterrent and insect repellency activity against the wheat stem sawfly (*Cephus cinctus*) (Daisy et al. 2002a, b).

Hu et al. (2005) isolated a strain of *Penicillium* sp. from the fresh roots of *Derris elliptica*; a total of 12 fractions were isolated from the chloroform extracts. The biological activity of these extracts was tested against the adult turnip aphid, *Lipaphis erysimi*, by dipping the insect in 1 mg/ml solution. Among these fractions, D, E, and J showed corresponding mortalities 57.68%, 63.28%, and 69.74%, respectively, after 48 h of treatment. They also showed strong anti-feeding activity against third instar larvae of *Plutella xylostella*. Hu et al. (2005) accomplished that the bioactive compounds in the fraction D could be rotenone or its analogous.

As the world ecological balance is getting damaged by synthetic pesticides, this endophytic research is an alternative source for the discovery of novel powerful, selective, and safe methods for an integrated pest management system.

15.5 Conclusions and Perspective

The overexploitation of nature for human existence/welfare has caused ecological imbalance. The need to fulfill the basic requirements like food, clothing, etc. has proliferated the use of chemicals in growth and cultivation of crops, trees, etc. The majority of chemical usage is for the control of pests and pathogens. As an alternative method for the usage of the synthetic chemicals, the endophyte research has shown lot of promise. Thus many endophytic fungi were isolated, identified, and characterized. These studies have shown that the secondary metabolites produced by the endophytes are main reason for their activity. These compounds have not only acted in control of the biological agents but also promoted other biotic and abiotic stress tolerance. These reports have kept the trust on the endophyte research, but still the knowledge of effective usage of specific endophytes and their combinations on different plant systems is still eluding the scientific community and farmers. These kind of articles and future studies may help in achieving the required in near future.

References

Abraham S, Basukriadi A, Pawiroharsono S, Sjamsuridzal W (2015) Insecticidal activity of ethyl acetate extracts from culture filtrates of mangrove fungal endophytes. Mycobiol 43:137–149

Akello J, Dubois T, Coyne D, Kyamanywa S (2008) Endophytic *Beauveria bassiana* in banana (*Musa* spp.) reduces banana weevil (*Cosmopolites sordidus*) fitness and damage. Crop Prot 27:1437–1441

Aly A, Debbab A, Kjer J, Proksch P (2010) Fungal endophytes from higher plants: a prolific source of phytochemicals and other bioactive natural products. Fungal Divers 41:1–16

Aly AH, Debbab A, Proksch A (2013) Fungal endophytes – secret producers of bioactive plant metabolites. Pharmazie 68:499–505

Amatuzzi RF, Cardoso N, Poltronieri AS, Poitevin CG, Dalzoto P, Zawadeneak MA, Pimentell C (2018) Potential of endophytic fungi as biocontrol agents of *Duponchelia fovealis* (Zeller) (Lepidoptera:Crambidae) Braz. J Biol 78:429–435

Arnold AE (2007) Understanding the diversity of foliar endophytic fungi: progress, challenges, and frontiers. Fungal Biol Rev 21:51–66

Arnold A (2008) Hidden within our botanical richness, a treasure trove of fungal endophytes. Plant Press, pp 13–15

Arnold AE, Lutzoni F (2007) Diversity and host range of foliar fungal endophytes: are tropical leaves biodiversity hotspots? Ecology 88:541–549

Arnold AE, Maynard Z, Gilbert GS, Coley PD, Kursar TA (2000) Are tropical fungal endophytes hyperdiverse? Ecol Lett 3:267–274

Arnold AE, Mejia LC, Kyllo D, Rojas EI, Maynard Z, Robbins N, Herre EA (2013) Fungal endophytes limit pathogen damage in a tropical tree. Proc Natl Acad Sci 100:15649–15654

Azevedo JL, Maccheroni W Jr, Pereira JO, de Araújo WL (2000) Endophytic microorganisms: a review on insect control and recent advances on tropical plants. Electron J Biotechnol 3:15–16

Ball OJP, Miles CO, Prestidge RA (1997) Ergopeptine alkaloids and Neotyphodiumlolii-mediated resistance in perennial ryegrass against adult *Heteronychus arator* (Coleoptera, Scarabaeidae). J Econ Entomol 90:1382–1391

Baskar K, Raj GA, Mohan PM, Lingathurai S, Ambrose T, Muthu C (2012) Larvicidal and growth inhibitory activities of entomopathogenic fungus, *Beauveria bassiana* against Asian army worm, Spodoptera litura fab. (Lepidoptera: Noctuidae). J Entomol 9:155–162

Behie SW, Jones SJ, Bidochka MJ (2015) Plant tissue localization of the endophytic insect pathogenic fungi Metarhizium and Beauveria. Fungal Ecol 13:112–119

Bennett R, Hutmacher R, Davis R, Bennett R (2008) Seed transmission of *Fusarium oxysporum* f. sp. vasinfectum race 4 in California. J Cotton Sci 12:160–164

Bensaci OA, Daoud H, Lombarkia N, Rouabah K (2015) Formulation of the endophytic fungus Cladosporium oxysporum Berk. & M.A. Curtis, isolated from Euphorbia bupleuroides subsp. luteola, as a new biocontrol tool against the black bean aphid (Aphis fabae Scop.). J Plant Prot Res 55:80–87

Bills GF, Giacobbe RA, Lee SH, Pelaez F, Tkacz JS (1992) Tremorgenic mycotoxins paspalitrem A and C from a tropical *Phomopsis*. Mycol Res 96:977–983

Bing LA, Lewis LC (1992) Endophytic *Beauveria bassiana* (Balsamo) *Vuillemin* in corn: the influence of the plant growth stage and *Ostrinia nubilalis* (Hübner). Biocontrol Sci Tech 2:39–47

Brem D, Leuchtmann A (2001) Epichloë grass endophytes increase herbivore resistance in the woodland grass *Brachypodium sylvaticum*. Oecologia 126:522–530

Byrne KM, Smith SK, Ondeyka JG (2002) Biosynthesis of nodulisporic acid a: precursor studies. J Am Chem Soc 124:7055–7060

Calhoun LA, Findrlay JA, Miller JD, Whitney NJ (1992) Metabolites toxic to spruce budworm from balsam fir needle endophytes. Mycol Res 96:281–286

Clark CL, Miller JD, Whitney NL (1989) Toxicity of conifer needle endophytes to spruce budworm. Mycol Res 93:508–512

Clay K, Cheplick GP (1989) Effect of ergot alkaloids from fungal endophyte-infected grasses on fall armyworm (*Spodoptera frugiperda*). J Chem Ecol 15:169–182

Clay K, Hardy TN, Hammond AM (1985) Fungal endophytes of grasses and their effects on an insect herbivore. Oecologia 66:1–5

Coyle CM, Cheng JZ, O'Connor SE, Panaccione DG (2010) An old yellow enzyme gene controls the branch point between *Aspergillus fumigatus* and *Claviceps purpurea* ergot alkaloid pathways. Appl Environ Microbiol 76:3898–3903

Daisy BH, Strobel GA, Castillo U, Ezra D, Sears J, Weaver D, Runyon JB (2002a) Naphthalene, an insect repellent, is produced by *Muscodor vitigenus*, a novel endophytic fungus. Microbiology 148:3737–3741

Daisy B, Strobel G, Ezra D, Castillo U, Baird G, Hess WM (2002b) Muscodor vitigenus anam. sp. nov., an endophyte from *Paullinia paullinioides*. Mycotaxon 84:39–50

Danielsson J, Reva O, Meijer J (2007) Protection of oilseed rape (*Brassica napus*) toward fungal pathogens by strains of plant-associated *Bacillus amyloliquefaciens*. Microb Ecol 54:134–140

Delaye L, García-Guzmán G, Heil M (2013) Endophytes versus biotrophic and necrotrophic pathogens—are fungal lifestyles evolutionarily stable traits? Fungal Divers 60:125–135

Demain AL (2000) Microbial biotechnology. Trends Biotechnol 18:26–31

Dongyi H, Kelemu S (2004) *Acremonium implicatum*, a seed-transmitted endophytic fungus in Brachiaria grasses. Plant Dis 88:1252–1254

dos Santos RMG, Rodrigues-Fo E, Rocha WC, Teixeira MFS (2003) Endophytic fungi from Melia azedarach. World J Microbiol Biotechnol 19:767–770

Faeth SH, Fagan WF (2002) Fungal endophytes: common host plant symbionts but uncommon mutualists. Integr Comp Biol 42:360–368

Findlay JA, Buthelezi S, Lavoie R, Rodriguez L (1995a) Bioactive isocoumarins and related metabolites from conifer endophytes. J Nat Prod 58:1759–1766

Findlay JA, Li G, Penner PE (1995b) Novel diterpenoid insect toxins from a conifer endophyte. J Nat Prod 58:197–200

Findlay JA, Bethelezi S, Li G, Sevek S (1997) Insect toxins from an endophyte fungus from wintergreen. J Nat Prod 60:1214–1215

Ganley RJ, Brunsfeld SJ, Newcombe G (2004) A community of unknown, endophytic fungi in western white pine. Proc Natl Acad Sci U S A 101:10107–10112

Gaynor DL, Rowan DD (1986) Insect resistance, animal toxicity and endophyte-infected grass. Proc NZ Grassland Assoc 47:115–120

Granér G, Persson P, Meijer J, Alström S (2003) A study on microbial diversity in different cultivars of *Brassica napus* in relation to its wilt pathogen, *Verticillium longisporum*. FEMS Microbiol Lett 224:269–276

Gurulingappa P, McGee PA, Sword G (2011) Endophytic *Lecanicillium lecanii* and *Beauveria bassiana* reduce the survival and fecundity of *Aphis gossypii* following contact with conidia and secondary metabolites. Crop Protect 30:349–353

Hallmann J, Quadt-Hallmann A, Mahaffee WF, Kloepper JW (1997) Bacterial endophytes in agricultural crops. Can J Microbiol 43:895–914

Hardoim PR, van Overbeek LS, Berg G et al (2015) The hidden world within plants: ecological and evolutionary considerations for defining functioning of microbial endophytes. Microbiol Mol Biol R 79:293–320

Hartley SE, Gange AC (2009) Impacts of plant symbiotic fungi on insect herbivores: mutualism in a multitrophic context. Annu Rev Entomol 54:323–342

Higgins KL, Arnold AE, Miadlikowska J, Sarvate SD, Lutzoni F (2007) Phylogenetic relationships, host affinity, and geographic structure of boreal and arctic endophytes from three major plant lineages. Mol Phylogenet Evol 42:543–555

Hu MY, Zhong GH, Sun ZhT SG, Liu HM, Liu XQ (2005) Insecticidal activities of secondary metabolites of endophytic *Pencillium* sp. in *Derris elliptica* Benth. J Appl Entomol 129 (8):413–417

Hyde K, Soytong K (2008) The fungal endophyte dilemma. Fungal Divers 33:163–173

Jensen JG, Popay AJ, Tapper BA (2009) Argentine stem weevil adults are affected by meadow fescue endophyte and its loline alkaloids. NZ Plant Prot 62:12–18

Johnson MC, Dahlman DL, Siegel MR, Bush LP, Latch GCM, Potter DA, Varney DR (1985) Insect feeding deterrents in endophyte-infected tall fescue. Appl Environ Microbiol 49:568–571

Junker C, Draeger S, Schulz B (2012) A fine line – endophytes or pathogens in *Arabidopsis thaliana*. Fungal Ecol 5:657–662

Kemen E, Jones JD (2012) Obligate biotroph parasitism: can we link genomes to lifestyles? Trends Plant Sci 17:448–457

Larran S, Perelló A, Simón M, Moreno V (2002) Isolation and analysis of endophytic microorganisms in wheat (*Triticum aestivum* L.) leaves. World J Microbiol Biotechnol 18:683–686

Li XJ, Zhang Q, Zhang AL, Gao JM (2012) Metabolites from Aspergillus fumigatus, an endophytic fungus associated with *Melia azedarach*, and their antifungal, antifeedant, and toxic activities. J Agric Food Chem 60(13):3424–3431

Lohse R, Jakobs-Schönwandt D, Vidal S, Patel AV (2015) Evaluation of new fermentation and formulation strategies for a high endophytic establishment of *Beauveria bassiana* in oilseed rape plants. Biol Control 88:26–36

Lorenz N, Haarmann T, Pazoutov S, Jung M, Tudzynski P (2009) The ergot alkaloid gene cluster: functional analyses and evolutionary aspects. Phytochemistry 70:1822–1832

Meyling NV, Thorup-Kristensen K, Eilenberg J (2011) Below-and aboveground abundance and distribution of fungal entomopathogens in experimental conventional and organic cropping systems. Biol Control 59:180–186

Miles CO, diMena ME, Jacobs SWL, Garthwaite I, Lane GA, Prestidge RA, Marshal SL, Wilkinson HH, Schardl CL, Ball OJP, Latch CM (1998) Endophytic fungi in indigenous Australasian grasses associated with toxicity to livestock. Appl Environ Microbiol 64:601–606

Moloinyane S, Nchu F (2019) The effects of endophytic *Beauveria bassiana* inoculation on infestation level of *Planococcus ficus*, growth and volatile constituents of potted greenhouse grapevine (*Vitis vinifera* L.). Toxins 11:E72

Ortel I, Keller U (2009) Combinatorial assembly of simple and complex D-lysergic acid alkaloid peptide classes in the ergot fungus *Claviceps purpurea*. J Biol Chem 284:6650–6660

Ownley BH, Griffin MR, Klingeman WE, Gwinn KD, Moulton JK, Pereira RM (2008) *Beauveria bassiana*: endophytic colonization and plant disease control. J Invertebr Pathol 98:267–270
Panaccione DG (2010) Ergot alkaloids. In: Hofrichter M (ed) The Mycota, Vol. X, Industrial applications, 2nd edn. Springer-Verlag, Berlin-Heidelburg, pp 195–214
Panaccione DG, Coyle CM (2005) Abundant respirable ergot alkaloids from the common airborne fungus *Aspergillus fumigatus*. Appl Environ Microbiol 71:3106–3111
Panaccione DG, Tapper BA, Lane GA, Davies E, Fraser K (2003) Biochemical outcome of blocking the ergot alkaloid pathway of a grass endophyte. J Agric Food Chem 51:6429–6437
Pinto LSRC, Azevedo JL, Pereira JO, Vieira MLC, Labate CA (2000) Symptomless infection of banana and maize by endophytic fungi impairs photosynthetic efficiency. New Phytol 147:609–615
Porras-Alfaro A, Bayman P (2011) Hidden fungi, emergent properties: endophytes and microbiomes. Annu Rev Phytopathol 49:291–315
Posada F, Vega FE (2005) Establishment of the fungal entomopathogen *Beauveria bassiana* (Ascomycota: Hypocreales) as an endophyte in cocoa seedlings (*Theobroma cacao*). Mycologia 97:1195–1200
Posada F, Vega FE (2006) Inoculation and colonization of coffee seedlings (*Coffea arabica* L.) with the fungal entomopathogen *Beauveria bassiana* (Ascomycota: Hypocreales). Mycoscience 47:284–289
Potter DA, Stokes JT, Redmond CT, Schardl CL, Panaccione DG (2008) Contribution of ergot alkaloids to suppression of a grassfeeding caterpillar assessed with gene-knockout endophytes in perennial ryegrass. Entomol Exp Appl 126:138–147
Prestidge RA, Gallagher RT (1988) Endophyte conifers resistance to ryegrass: argentine stem weevil larval studies. Ecol Entomol 13:429–435
Promputtha I, Lumyong S, Dhanasekaran V, Mckenzie EHC, Hyde KD, Jeewon R (2007) A phylogenetic evaluation of whether endophytes become saprotrophs at host senescence. Microb Ecol 53:579–590
Punja ZK (1997) Comparative efficacy of bacteria, fungi, and yeasts as biological control agents for diseases of vegetable crops. Can J Plant Pathol 19:315–323
Purahong W, Hyde KD (2011) Effects of fungal endophytes on grass and non-grass litter decomposition rates. Fungal Divers 47:1–7
Quesada-Moraga E, López-Díaz C, Landa BB (2014) The hidden habit of the entomopathogenic fungus *Beauveria bassiana*: first demonstration of vertical plant transmission. PLoS One 9: e89278. https://doi.org/10.1371/journal.pone
Riedell WE, Kieckhefer RE, Petroski RJ, Powell RG (1991) Naturally occurring and synthetic loline alkaloid derivatives: insect feeding behavior modification and toxicity. J Entomol Sci 26:122–129
Rodriguez R, White J Jr, Arnold A, Redman R (2009) Fungal endophytes: diversity and functional roles. New Phytol 182:314–330
Rowan DD (1993) Lolitrems, peramine and paxilline: mycotoxins of the ryegrass/endophyte interaction. Agric Ecosyst Environ 44:103–122
Rowan D, Hunt M, Gaynor D (1986) Peramine, a novel insect feeding deterrent from ryegrass infected with the endophyte *Acremonium loliae*. J Chem Soc Chem Commun 142:935–936. https://doi.org/10.1039/c39860000935
Rowan DD, Dymock JJ, Brimble MA (1990) Effect of fungal metabolite peramine and analogs on feeding and development of Argentine stem weevil (Listronotusbonariensis). J Chem Ecol 16:1683–1695
Saikia S, Nicholson MJ, Young C, Parker EJ, Scott B (2008) The genetic basis for indole-diterpene chemical diversity in filamentous fungi. Mycol Res 112:184–199
Saikkonen K, Faeth SH, Helander M, Sullivan TJ (1998) Fungal endophytes: a continuum of interactions with host plants. Annu Rev Ecol Syst 29:319–343
Saikkonen K, Ion D, Gyllenberg M (2002) The persistence of vertically transmitted fungi in grass metapopulations. Proc Biol Sci 269:1397–1403

Saikkonen K, Lehtonen P, Helander M, Koricheva J, Faeth SH (2006) Model systems in ecology: dissecting the endophyte–grass literature. Trends Plant Sci 11:428–433
Sánchez Márquez S, Bills GF, Herrero N, Zabalgogeazcoa I (2012) Non-systemic fungal endophytes of grasses. Fungal Ecol 5:289–297
Schardl CL (2001) Epichloe festucae and related mutualistic symbionts of grasses. Fungal Genet Biol 33:69–82
Schardl CL, Grossman RB, Nagabhyru P, Faulkner JR, Mallik UP (2007) Loline alkaloids: currencies of mutualism. Phytochemistry 68:980–996
Schardl CL, Young CA, Faulkner JR, Florea S, Pan J (2011) Chemotypic diversity of epichloae, fungal symbionts of grasses. Fungal Ecol 5:331–344
Schardl CL, Young CA, Pan J, Florea S, Takach JE, Panaccione DG et al (2013) Currencies of mutualisms: sources of alkaloid genes in vertically transmitted epichloae. Toxins 5:1064–1088
Schulz B, Boyle C (2005) The endophytic continuum. Mycol Res 109:661–686
Schulz B, Boyle C, Draeger S, Römmert A-K, Krohn K (2002) Endophytic fungi: a source of novel biologically active secondary metabolites. Mycol Res 106:996–1004
Senthilkumar N, Murugesan S, Babu DS, Rajeshkannan C (2014) GC-MS analysis of the extract of endophytic fungus, Phomopsis sp. isolated from tropical tree species of India, Tectona grandis L. Int J Innov Res Sci Eng Tech 3:10176–10179
Shi YW, Zhang X, Lou K (2013) Isolation, characterization, and insecticidal activity of an endophyte of drunken horse grass, Achnatherum inebrians. J Insect Sci 13(1):151
Sieber TN (2007) Endophytic fungi in forest trees: are they mutualists? Fungal Biol Rev 21:75–89
Siegel MR, Latch GCM, Bush LP, Fannin FF, Rowan DD, Tapper BA, Bacon CW, Johnson MC (1990) Fungal endophyte-infected grasses: alkaloid accumulation and aphid response. J Chem Ecol 16:3301–3316
Strobel G, Daisy B (2003) Bioprospecting for microbial endophytes and their natural products bioprospecting for microbial endophytes and their natural products. Microbiol Mol Biol Rev 67:491–502
Tofern B, Kaloga M, Witte L, Hartmann T, Eich E (1999) Phytochemistry and chemotaxonomy of the Convolvulaceae part 8 – occurrence of loline alkaloids in Argyreiamollis (Convolvulaceae). Phytochemistry 51:1177–1180
Varma A, Verma S, Sahay N, Bütehorn B, Franken P (1999) *Piriformospora indica*, a cultivable plant-growth-promoting root endophyte. Appl Environ Microbiol 65:2741–2744
Vega FE (2008) Insect pathology and fungal endophytes. J Invertebr Pathol 98:277–279
Vega FE, Posada F, Aime MC, Pava-Ripoll M, Infante F, Rehner SA (2008) Entomopathogenic fungal endophytes. Biol Control 46:72–82
Wallwey C, Li SM (2011) Ergot alkaloids: structure diversity, biosynthetic gene clusters and functional proof of biosynthetic genes. Nat Prod Rep 28:496–510
Webber J (1981) A natural control of Dutch elm disease. Nature 292:449–451
Wilberforce E, Boddy L, Griffiths R, Griffith G (2003) Agricultural management affects communities of culturable root-endophytic fungi in temperate grasslands. Soil Biol Biochem 35:1143–1154
Wyrebek M, Huber C, Sasan RK, Bidochka MJ (2011) Three sympatrically occurring species of Metarhizium show plant rhizosphere pecificity. Microbiology 157:2904–2911
Yates SG, Fenster JC, Bartelt RJ (1989) Assay of tall fescue seed extracts, fractions, and alkaloids using the large milkweed bug. J Agric Food Chem 37:354–357
Young CA, Felitti S, Shields K, Spangenberg G, Johnson RD, Bryan GT, Saikia S, Scott B (2006) A complex gene cluster for indolediterpene biosynthesis in the grass endophyte Neotyphodiumlolii. Fungal Genet Biol 43:679–693
Young CA, Tapper BA, May K, Moon CD, Schardl CL, Scott B (2009) Indole-diterpene biosynthetic capability of Epichloë endophytes as predicted by ltm gene analysis. Appl Environ Microbiol 75:2200–2211

The manufacturer's authorised representative in the EU is Springer Nature Customer Service Centre GmbH, Europaplatz 3, 69115 Heidelberg, Germany. If you have any concerns regarding our products, please contact ProductSafety@springernature.com

Printed and bound by CPI Group (UK) Ltd, Croydon, CR0 4YY
30/09/2026
02225504-0010